PLURALISTIC CASUISTRY

Philosophy and Medicine

VOLUME 94

PLURALISTIC CASUISTRY

MORAL ARGUMENTS, ECONOMIC REALITIES, AND POLITICAL THEORY

Edited by

MARK J. CHERRY

Department of Philosophy, St. Edward's University, Austin, Texas

and

ANA SMITH ILTIS

Center for Health Care Ethics, Saint Louis University, St. Louis, Missouri

A C.I.P. Catalogue record for this book is available from the Library of Congress.

ISBN 978-1-4020-6259-9 (HB)
ISBN 978-1-4020-6260-5 (e-book)

Published by Springer,
P.O. Box 17, 3300 AA Dordrecht, The Netherlands.

www.springer.com

Printed on acid-free paper

Baruch Alter Brody, Ph.D.
Mesivta Theological Seminary, 1959-1963
B.A., Brooklyn College, 1962
M.A., Princeton University, 1965
Fulbright Fellow, Oxford University, 1965-1966
Ph.D., Princeton University, 1967

Leon Jaworski Professor of Biomedical Ethics
Director, Center for Medical Ethics and Health Policy
and
Distinguished Service Professor
Baylor College of Medicine
Houston, Texas

Andrew Mellow Professor of Humanities
Department of Philosophy
Rice University
Houston, Texas

TABLE OF CONTENTS

1. MORAL CASUISTRY, MEDICAL RESEARCH AND INNOVATION, AND RABBINICAL DECISION-MAKING 1

MARK J. CHERRY AND ANA SMITH ILTIS

SECTION I PLURALISTIC MORAL CASUISTRY 21

2. NOTES TOWARD A PLURALISTIC PROFESSIONAL MEDICAL ETHICS 23

LAURENCE B. MCCULLOUGH

3. ETHICS AND DEEP MORAL AMBIGUITY 37

KEVIN WM. WILDES, S.J.

4. MORAL JUDGMENT AND THE IDEAL INTUITOR: DEALING WITH MORAL CONFUSION AND MORAL DISAGREEMENT 49

JANET MALEK

5. CASUISTRY NATURALIZED 65

MAUREEN KELLEY

SECTION II JEWISH MEDICAL ETHICS 83

6. INTUITIONISM, DIVINE COMMANDS, AND NATURAL LAW 85

B. ANDREW LUSTIG

7. IN CASE: CONTINGENCY AND PARTICULARITY IN BIOETHICS: DISCURSIVE METHOD AND CLINICAL MIDRASH BRIEF NOTES ON A LESSON FROM MY TEACHER BARUCH BRODY 99

LAURIE ZOLOTH

8. THE EUTHYPHRO'S DILEMMA RECONSIDERED: A VARIATION ON A THEME FROM BRODY ON HALAKHIC METHOD 109

H. TRISTRAM ENGELHARDT, JR.

SECTION III BIOMEDICAL PUBLIC POLICY 131

9. THE GOOD (PHILOSOPHY), THE BAD (PUBLIC POLICY) AND THE UGLINESS OF BLAMING FAMILIES FOR INEFFECTUAL TREATMENTS 133

ROBERT M. ARNOLD

10. A MATTER OF OBLIGATION: PHYSICIANS VERSUS CLINICAL INVESTIGATORS 149

E. HAAVI MORREIM

11. IS WITHHOLDING ARTIFICIAL NUTRITION AND HYDRATION FROM PVS PATIENTS ACTIVE EUTHANASIA? 167

LORETTA M. KOPELMAN

SECTION IV CRITICAL APPLICATION AND ANALYSIS 179

12. THE VIRTUE OF INTEGRITY IN BARUCH BRODY'S MORAL FRAMEWORK 181

J. CLINT PARKER

13. PARADIGMS, PRACTICES AND POLITICS: ETHICS AND THE LANGUAGE OF HUMAN EMBRYO TRANSFER/DONATION/RESCUE/ADOPTION 191

SARAH-VAUGHAN BRAKMAN

14. BRODY ON PASSIVE AND ACTIVE EUTHANASIA 211

F.M. KAMM

SECTION V RESPONSE TO FRIENDS AND CRITICISMS 219

15. COMMENTS ON THE ESSAYS 221

BARUCH A. BRODY

APPENDIX: SELECTED BIBLIOGRAPHY OF THE WORKS OF BARUCH A. BRODY 233

INDEX 243

NOTES ON CONTRIBUTORS 253

ACKNOWLEDGEMENTS

The development of this volume benefited from the kind efforts of many. We are deeply thankful to the contributors, many of whom recast their essays several times over the course of the project, to craft the final versions contained herein. A particular debt is owed to H. Tristram Engelhardt Jr., Lisa Rasmussen, and Jeremy Garrett, who are excellent colleagues and the best of friends, as well as to the anonymous reviewers, who helped us to see the challenges raised by the contributions anew.

Mark J. Cherry wishes to recognize the on-going generosity of St. Edward's University, the School of Humanities, and the Department of Philosophy, especially Donna Jurick, SND, Louis T. Brusatti, William J. Zanardi, Peter Wake, and Jack Green-Musselman. Each has been instrumental, though in diverse capacities, to the success of this project. Thanks are also due to my research assistant, Adrienne Carpenter. As with all of my projects, this volume would not exist without the support, kindness, and love of Mollie E. Cherry.

Ana Iltis wishes to recognize the generous support of the Saint Louis University Center for Health Care Ethics, especially its director James DuBois, as well as the ongoing support of the Graduate School through Dean Donald Brennan. Finally, it would not be possible to pursue a career without the generosity and love of my husband, Steven Iltis.

The volume is dedicated to the life and work of our teacher: Professor Baruch A. Brody, a true scholar among scholars, whom we thank for granting us permission to prepare this festschrift in his honor.

CHAPTER 1

MORAL CASUISTRY, MEDICAL RESEARCH AND INNOVATION, AND RABBINICAL DECISION-MAKING

MARK J. CHERRY* AND ANA SMITH ILTIS**

Department of Philosophy, St. Edward's University, Austin, Texas
Center for Health Care Ethics, Saint Louis University, St. Louis, Missouri

I. A PHILOSOPHICAL INTRODUCTION

> The scope of bioethics has expanded over the years from its original emphasis on questions arising out of clinical encounters and biomedical research to an increasing emphasis on questions about health policy and health care reform. I have long been convinced, however, that even this expanded scope is still conceived far too narrowly. While few technological advances in medicine take place without public commentary by bioethicists, major structural changes in the delivery of health care or in the conduct of biomedical research often go unnoticed by the bioethics community, despite the fact that these changes often raise profound moral and social issues (Brody, 1996, p. 5).

The impressive range and depth of Baruch Brody's writings in biomedical ethics and the philosophy of medicine illustrate his deep appreciation that thorough and critical scientific research and philosophical analysis are central to reigning in the untutored human desire to ameliorate pain and suffering so that medical treatments and health care policy do more good than harm. On the one hand, medicine must abandon the presupposition that customary or accepted treatments are good simply because they are customary and accepted.[1] Nor, should medicine assume that new pharmaceuticals and medical technologies constitute the best practices simply because they are new. And, on the other hand, well-meaning but heavy handed paternalistic and moralistic public policy, frequently given social political credence through the endorsement of bioethicists, has at times failed to protect the long term interests of persons. For example, if medical practice is to advance, if there is to be development of more efficient and effective pharmaceuticals and medical devices, there must be significant incentives to stimulate innovation: medical advancement requires

---

*Mark J. Cherry, Ph.D. is the Dr. Patricia A. Hayes Professor in Applied Ethics and Associate Professor of Philosophy, St. Edward's University, Austin, Texas.
**Ana Smith Iltis, Ph.D. is Associate Professor, Center for Health Care Ethics, Saint Louis University, St. Louis, Missouri.

M.J. Cherry and A.S. Iltis (eds.), Pluralistic Casuistry, 1–20.

an engine for innovation. Yet, as Brody recognizes, one must both overcome the usual chronic menace to innovation—the interests of the status quo in suppressing change[2]—and correct for the ill consequences resulting from the all too human impulses to increase professional status and income, wield social political authority, and to help those in need. As the history of medicine pays witness, the human suffering caused by ill-founded, but well-meaning interventions—both medical and political—have been significant.

As the contributions to this volume consistently illustrate, few who work in biomedical ethics are as well aware of the moral and political challenges that medical research and bioethical policy represents as Baruch Brody. He has recognized, where others have failed adequately to do so, the intimate connection between bioethics and other areas of philosophy: e.g., political theory, metaphysics, epistemology, axiology, and the philosophy of religion. His appreciation of the importance of such connections has led to his critical exploration of core questions, which others have failed to appreciate. This skillful analysis time and again permitted the fields of bioethics, the philosophy of medicine, and health care policy to see seemingly intractable problems in a new light. Such philosophical insights frequently led to very practical solutions to medical and political challenges. Indeed, Brody has called for bioethical reflections to be deeply philosophical *so that* they can be practical (Brody, 1989).

Contrary to a common stereotype of philosophy professors, philosophical analysis does not inevitably lead to abstract insights and utopian recommendations. Instead, done well, philosophical work can clarify complex issues, facilitate creative problem solving, and lead to real-world solutions to difficult situations. Brody's work exemplifies the ways careful philosophical examination of biomedical issues can help generate practical solutions. He has been one of the most important voices in bioethics over the last several decades, asking new and challenging questions about a range of problems, examining recalcitrant issues in novel ways, always with the goal of offering practical solutions to complex problems. A number of contributors to this volume address Brody's epistemological and methodological contribution to bioethics, namely his normative moral theory—pluralistic casuistry (Kelley, Wildes, Malek, McCullough)—and his discussions of Jewish medical ethics (Engelhardt, Lustig). Others critically reflect on many of the areas he has addressed or employ his model of bioethical analysis to analyze particular issues, including human embryo transfer (Brakman), medical futility (Arnold), life and death decisions in pediatrics (Zoloth), euthanasia and end-of-life decision-making (Kamm, Kopelman), the obligations of clinical researchers toward study participants (Morreim), and professional integrity (Parker). Brody also has made important contributions to discussions of philosophy of law, philosophy of medicine, philosophy of religion, metaphysics, epistemology, abortion, access to health care and justice.

In this brief introduction to Brody's contributions to the field of bioethics, we draw the reader's attention to two areas in which he has made especially important contributions—human research ethics and end-of-life decision-making—to illustrate the relationship among Brody's normative moral theory, his careful analytic work, and the crafting of practical solutions to complex bioethical problems. Thereafter, we encourage readers to address the chapters throughout this volume, as each carefully and critically assess in greater detail Brody's philosophical thought and bioethical research.

II. RESEARCH ETHICS

> These observations are about research in developing countries in general, and not just about research on vertical transmission [of HIV]. Three lessons have emerged. The standard for when a placebo group is justified is a normative standard (what they should have received if they were not in the trial) rather than a descriptive standard (what they would have received if they were not in the trial). Coercion is not a serious concern in trials simply because attractive offers are made to the subjects. Legitimate concerns about exploiting subjects should be addressed by ensuring their future treatment, rather than by asking what will happen in their community at large (Brody, 2002a, pp. 2857–2858).

Throughout much of the 20th century, concerns emerged about the ethical conduct of biomedical, behavioral, and social sciences research on human beings, often because of specific practices and problematic cases that became widely known.[3] Research ethics has become a major sub-field within bioethics, and Brody has made significant contributions to a number of debates, often identifying issues and questions not previously addressed or creating pathways for resolving disagreements over what is and is not permissible. Some of the most important issues he has addressed include ethical issues regarding controlled clinical trials in general (e.g., Brody, 1995 and 1998a) and placebo controlled trials in particular (1997a; 2003), including trials that call for placebo surgery (Moseley et al., 2002); research in developing nations (2002a and b); emergency research (1997b); informed consent (2000; 2001); research on vulnerable subjects (1998b); and the recruitment of human subjects (2002c). Brody also has helped us understand and evaluate the regulation of biomedical research, the drug approval process (1995), and intellectual property and technology transfer law in the United States (2006), as well as research ethics guidelines and regulations from around the world (1998a). Here we consider his contributions to the field regarding consent for emergency research and conflicts of interest, where he has made extraordinarily important contributions to the literature and where the rigor of his philosophical insight combined with his commitment to resolving conflicts and crafting practical solutions are evident.

To provide safe and effective medical care, interventions must be evaluated and tested using research methods, and to use humans in research generally requires the consent of the research subjects, or, in the case of non-competent humans, permission of their legal guardians or representatives. For many years, the need for research coupled with the importance of consent or permission made it nearly impossible to conduct research on emergency treatments. Observing that overall it is not in the public interest that new treatments for life-threatening emergency medical conditions either not be developed or not be tested prior to widespread use, in 1996 the Food and Drug Administration (FDA) issued regulations that would allow institutional review boards (IRBs) to waive or alter informed consent requirements for some emergency research under specific, narrow conditions (FDA, 1996). Controversy surrounded and continues to surround such waivers (see, for example, Kipnis, King, and Nelson, 2006; Carnahan, 1999; Fost, 1998; Biros et al., 1998; Karlawish and Hall, 1996). In the wake of the new regulations, Brody offered one of the most cogent defenses of waivers of consent for emergency research and he predicted accurately some of the areas that would pose the most difficulty in interpreting and applying the regulations (Brody, 1997b). As with so much of his work, Brody focused his analysis of the ethics of emergency research on the non-ideal nature of medicine and the statistical nature of medical knowledge, on the need to make decisions and promote the good of individuals and society in an imperfect world, and on the multiplicity of values that ought to be respected:

> (a) the desperate social need for research in the emergency setting to test promising treatments for patients presenting with acute crises such as strokes and closed head injuries for which there are few treatments of proven value that can limit the damage; (b) the potential benefit to some patient-subjects (those in the treatment group) who receive promising new therapies if those therapies fulfill their promise; (c) the need to protect these individuals from being exploited and harmed by researchers when (as often happens) promising new therapies do not fulfill their promise and turn out to be harmful; (d) the right of all individuals not to be used as research subjects without their consent or the consent of those who speak for them (Brody, 1997b, p. 7).

In ideal circumstances, it would be possible to respect all of these values, but, Brody argues, emergency circumstances sometimes make it impossible to do so. Furthermore, he holds, these values cannot be rank-ordered in an absolute fashion. Instead, their significance in individual cases or circumstances must be assessed and a judgment made. The moral world, Brody has long argued, is not governed by absolute values but by a series of independent, non-absolute values. He argues that the FDA regulations reflect the importance of pluralism as well as of assessing and comparing the significance of competing values. Sound ethical judgment requires that we consider competing values when it is not possible to respect them all fully. The FDA regulations on emergency research, Brody argues, reflect just such a consideration competing values. They acknowledge the importance of research subject consent but recognize that sometimes it is not possible to fulfill the need for certain types of biomedical research and grant individuals access to potentially beneficial interventions, if one must obtain consent. The regulations also acknowledge the possibility of exploitation and the need to protect persons from exploitation by requiring community consultation.

Brody's assessment of the ethical issues regarding research in emergency settings reflects his emphasis on the importance philosophical analysis has for generating practical solutions to serious issues. Brody defends the FDA regulations on emergency research by returning to his philosophical account of the moral world according to which multiple values are relevant and at stake. The consideration of competing values relevant to emergency research justify the FDA regulations permitting waivers of consent in some cases, Brody argues. At the same time, those regulations leave open important questions that also must be addressed, such as the permissibility of conducting placebo-controlled trials with waivers of consent. Resolution of these concerns will require careful deliberation using the same process of identifying the competing underlying values and assessing their relative significance. Ethical concerns emerge because there are multiple values that, in an ideal world, all would be fully respected. In our non-ideal world, we must identify those values and assess their relative significance. Doing so, Brody has demonstrated, allows us to build practical and ethically sound solutions.

Failure adequately to analyze the nature of ethical controversies puts us at risk for trying to solve the wrong problem, leaving undiscovered related ethical concerns, ignoring morally significant dimensions of issues, and missing potential mechanisms for resolving controversies. Understanding the nature of ethical controversies is a necessary first step toward crafting sound and practical solutions. In examining conflicts of interest in research, for example, Brody's analysis demonstrates that the scope of issues related to conflicts of interest are much broader than typically acknowledged. Much of the literature on conflicts of interest in research focuses on financial conflicts of interest, such as stock ownership and consulting/speaking fees, which may affect investigators' decisions to enroll subjects into studies and the integrity of data, potentially compromising subject safety and the validity of results. These issues are, according to Brody, both real and significant. But they hardly represent the full range of concerns raised by conflicts of interest in research. Conflicts of interest, Brody argues, can affect a number of other decisions, including:

> (1) Which treatments will be tested in the proposed trial, and which will not be tested?; (2) Will there be a placebo control group as well, or will the treatments be tested against each other or against some active control group?; (3) What will be taken as the favorable endpoint (the result constituting the evidence of the dangerousness of the treatment)?; (4) What will be the condition for inclusion or exclusion of subjects from the trial?; (5) What provisions will be made for informed consent?; (6) Under what conditions will the trial be stopped or modified because there have been too many adverse endpoints in one or more arms of the trial or because the preliminary data have shown that one of the treatments is clearly the most efficacious treatment?; (7) Under what conditions will the trial be stopped or modified because of the newly available results of other trials?; (8) Which patients who meet the criteria will actually be enrolled, and which ones will not? (Brody, 1996, p. 409).

Until we acknowledge the full range of sources of conflicts of interest and the many ways in which such conflicts can affect the research enterprise, it will remain impossible to manage such conflicts appropriately or to understand the impact they may have on research results (Brody, 1996, p. 407).

Brody's analysis of the scope and impact of conflicts of interest in trials of various thrombolytic agents demonstrates that to focus strictly on stockholding and other financial relationships between investigators and sponsors will lead one to miss important areas of conflict and thus to propose inadequate solutions (Brody, 1996, p. 408). Much of the anxiety over conflicts of interest typically concerns financial conflicts that, some fear, will result in fraud. This, Brody argues, is not where our energies should be focused. Instead, he says "the concern should be with how conflicts of interest may lead investigators, perhaps unconsciously, to make inappropriate decisions about the design and conduct of clinical trials" (Brody, 1996, p. 408). Fraud, he argues, can be managed relatively well through a number of mechanisms, such as double-blinding and independent data analysis. As noted, the focus of concerns to control for potential conflicts of interest should be the many decisions about trial design and study implementation that can affect the outcome of a study. Once we understand the multiple ways in which conflicts of interest can affect research, we might find ourselves convinced that the obvious solution is to abolish them or to require disclosure of conflicts so as to alert readers that they should be cautious in their reading and interpretation of study results. He argues that the various "obvious" solutions that have been proposed are flawed because they focus on commercial relationships, such as stock ownership and consulting or speaking fees, rather than on broader conflicts, such as those generated by the importance of grantsmanship and of grant/sponsorship income for physician practices, departments, and institutions (Brody, 1996, pp. 413–414). They reflect an incomplete analysis of the problem. Thus, Brody argues, we need more data on how much profit grants/sponsored studies generate so that we can develop "detailed guidelines for avoiding conflicts of interest from grant income" (Brody, 1996, p. 416). In evaluating approaches to conflicts of interest in research, we should heed Brody's call to identify and assess the multiple values relevant to ethical analysis of an issue and refrain from treating our interest in avoiding the difficulties associated with conflicts of interest as absolute.

If one applied Brody's framework for analyzing competing values, in developing such guidelines one would want to consider the value of a research enterprise free of conflicts of interest as well as relevant values, such as the importance of an efficient and effective research enterprise. We would need critically to assess the relative significance of each of those values and consider how different proposed responses to conflicts of interest in research would respect such competing values. Brody is careful to point out that to craft a solution requires empirical evidence we lack; to attempt to solve a problem without acknowledging the multiple values at stake and without facts about the current circumstances and thus the ways in which different proposals might affect those values is irresponsible.[4]

III. END-OF-LIFE CARE

> There is, moreover, a practical reason for such an approach. The varying criteria for brain death were developed in response to the emergence of life support systems and transplantation technology. Three basic clinical questions emerged. One question is old: When is a patient ready for the services of the undertaker rather than those of the clinician? Two questions are new: When is it appropriate to *unilaterally* stop supporting patients (as opposed to stopping support at the request of a patient or surrogate)? and When can organs be obtained for transplantation? (Halevy and Brody, 1993, p. 523).

As Brody is aware the determination of "death" is not merely descriptive, as if it straightforwardly described an objective fact of the world; but is in addition both evaluative and performative. The now classic fourth edition of *Black's Law Dictionary* defined death as "the cessation of life; the ceasing to exist; defined by physicians as a total stoppage of the circulation of the blood, and a cessation of the animal and vital functions consequent thereon, such as respiration, pulsation, etc." (4th ed., rev., 1968). This description of whole body death emphasizes the presence of circulatory and respiratory functions. The advent of intensive care units and respirators capable of sustaining brain-dead but biologically alive bodies and the significant costs that such treatment engendered, placed whole body definitions of death under intense conceptual scrutiny. Developments in kidney transplantation in the 1950's and heart transplantation in 1967 further underscored a perceived need conceptually to place the death of the person over against the death of the human body—i.e., a need for a brain oriented definition of death. As the transplantation community well understood, brain dead, but otherwise biologically alive, human bodies could be excellent sources of transplantable organs. In 1981, the *President's Commission for the Study of Ethical Problems in Medicine and Biomedical and Behavioral Research* endorsed a change that would incorporate the importance of brain stem function for the life of persons into its definition of death: "An individual who has sustained either (1) irreversible cessation of circulatory and respiratory functions, or (2) irreversible cessation of all functions of the entire brain, including the brain stem is dead. A determination of death must be made in accordance with accepted medical standards" (1981, p. 3). With brain death whatever is distinctly personal in the organism is dead. Both *Black's Law Dictionary* and the *President's Commission* point to physician judgment and medical standards for the declaration of death.

The determination of death is a performative judgment: it changes the social significance and place of the human body. Dead bodies are not treated as living persons. Thus, as Brody notes, controversies regarding defining the moment that "death" occurs turn on how little life activity, and of what kind, may remain before the person may appropriately be declared dead. To be alive is to be able to gear into the world of experience and action, where death is characterized by the irreversible loss of those characteristics that are essential and necessary to be so engaged. Yet, the debate among whole-body-oriented, whole-brain-oriented, and neocortical-oriented definitions of death turns on this very question of how much life activity, and of what type, is essential to sustain the life of a person. As Brody lays out these three competing definitions: Death can be understood as the 1) "permanent cessation of the flow of vital bodily fluids" (whole-body death); 2) "permanent cessation of the integrative functioning of the organism as a whole" (whole-brain death); or 3) "permanent loss of what is essential to the nature of humans" (neocortical-oriented or higher-order brain death) (Halevy & Brody, 1993, p. 520). The President's Commission, Brody notes, chose to focus on whole-brain death, in part because of the brain's primacy in integrative functioning and partly because the brain is the sponsor and location of human consciousness.

While many continue to urge the adoption of a higher-order brain death criterion, Brody has argued that the challenge of a neocortical-oriented brain criterion is that there is no wide-spread consensus on what portions of the brain are required for cognition and consciousness; moreover, even when such sites can be found, their cessation often cannot be assessed with the certainty that a legislative and statutory definition requires (1993, p. 519). Here, the challenge even to whole-brain death, Brody concludes, is a set of adequate medical tests:

> A review of published reports about brain death shows that many patients who meet the standard clinical tests for brain death still maintain some brain functioning and therefore do not satisfy the whole-brain criterion of death. Three areas of persistent functioning are neurohormonal regulation, cortical functioning as shown by significant nonisoelectirc electroencephalograms, and brain stem functioning as shown by evoked responses (1993, p. 520).

A difficulty with the standard definition of whole-brain death is that it treats the line between life and death as if it were a sharply defined physical moment, and then attempts to identify that moment with the lose of brain function. However appealing, Brody argues, such an attempt fails to recognize the complexity of the dying process:

> The data we have presented challenge this consensus by showing that different aspects of brain functioning cease at many different times. Thus, any sharp dichotomy between life and death based on brain functioning, although convenient and appealing, is biologically artificial (1993, p. 523).

What is needed is more careful philosophical appreciation of the types of questions that are being asked and the appropriate physical and medical context for answering them. If death is appreciated as a process, such as irreversible cessation of conscious functioning, satisfaction of clinical tests for brain death, and asystole, then particular concerns, such as unilaterally withholding treatment, harvesting organs, and burying the body, can be addressed separately with more careful nuance. The implications of such an approach to understanding death and the dying process are significant.

Consider, for example, its implications for the allocation of scarce health care resources to patients in a permanently vegetative state (PVS). While some have argued that care can be withheld from permanently vegetative patients because it is futile, as Brody has noted, if the goal of the patient's family or surrogate decision-maker is the extension of mere biological life, then care of a permanently vegetative patient that accomplishes this goal is not futile.[5] Others have argued that patients in PVS may, at least sometimes, be permissibly euthanasized.[6] Others argue that PVS patients should simply be understood as dead and thereby be so declared. Life-sustaining therapy can be unilaterally withheld from patients who have died, without the need for any elaborate decision-making process. The challenge is that PVS patients remain living human beings, and do not even have a terminal illness in the usual meaning of the term, since they can go on living for many years if they are given certain types of medical care, such as artificial nutrition and hydration. Unless consensus is reached regarding higher-order brain death criteria, which is unlikely, PVS patients may not permissibly be declared "dead" without straightforwardly begging the question.

Here Brody argues that it is unnecessary fully to resolve the controversies regarding medical futility, permissible euthanasia, or higher-order brain death, adequately to address the issues which the care of PVS patients raises: "Because PVS patients are not persons, in any plausible account of personhood, society should give their life-prolonging care a very low priority as it develops priorities for the allocation of health care resources, and individual providers should give them a very low priority in clinical triaging decisions" (1992, p. 104). If care of PVS patients is seen as regarding the allocation of resources, rather than a definition of death issue, then given scarce medical resources:

> appropriate use of social resources should serve as the justification for the unilateral withholding or withdrawing of care. For example, irreversible cessation of conscious functioning is a point on the continuum where the need to rationally use societal resources outweighs the desires of some persons for unlimited care. In such cases, the question of the unilateral withholding or withdrawing of care can be answered without any appeal to a criterion for death (1993, p. 524).

While this position may give rise to further questions regarding the appropriate use of scarce health care resources and whether individuals may permissibly purchase additional care for PVS patients utilizing private resources, the discussion is more appropriately focused on the permissible use of public resources verses private resources as well as on maintaining the integrity of the medical profession.[7] Whereas some hold that PVS patients constitute among the most vulnerable members of society and that their basic dignity requires that they be kept alive more or less indefinitely, even on artificial nutrition and hydration, it may still be appropriate to place limits on the use of public funds to care for PVS patients, leaving such extended and potentially very expensive care to private resources, if available.[8]

Consider also organ harvesting. Some commentators have begun arguing in favor of utilizing higher-order brain criterion of death because it would enable the harvesting of a greater number and healthier vital organs for transplantation.[9] Here Brody argues that the question of when it is appropriate to harvest human organs should not be confused or conflated with the very different question of how to conceptualize death. What really is at stake, he argues, is a balancing of the public acceptance of organ harvesting with encouraging the availability of organs for transplant to save lives:

> The shortage of available organs has led to the consideration of using organs from vegetative patients and to the proposal that we use organs from anencephalic infants. It might be suggested that organs can be obtained from such patients if we adopt a new criterion for death. We rejected that argument above. *But we also feel that the criterion for death is not where the discussion should be centered.* For us, it should center around the attempt to balance the advantage of lives saved through increased organ availability (which argues for harvesting in such cases) against the need for public acceptance of organ donation (which may require forgoing harvesting organs in such cases). We feel, in view of these considerations, that the combination of irreversible cessation of conscious functioning with apnea is the appropriate point on the continuum for organ harvesting. This is, in fact, close to the point at which we currently harvest organs, using the whole-brain criterion and the standard clinical tests (Halevy and Brody, 1993, p. 524, emphasis added).

The attempt to redefine death so as to be able to harvest organs from living humans is deeply to misunderstand—or even culpably to misrepresent—this core metaphysical and moral issue. If we begin taking internal organs from living humans, then the proximate cause of death is not the underlying disease or infirmity, but the act of taking the vital organs. It is, in effect, an attempt to side-step the dead donor rule, by broadening the definition of who is considered to be dead. Still, it may make sense to rethink the dead donor rule, and to permit the earlier harvesting of vital organs, at least with the consent of the individual or surrogate decision maker, and the transplant team.

Similar, very practical questions can be raised about when to call the undertaker. We might be comfortable with declaring death once the criteria for whole-brain death are met, and even unilaterally to withdraw and to withhold care. Some might even be comfortable with the higher-order brain death criterion. But, are we comfortable calling for the undertaker? "Little is to be gained in terms of conserving social resources by using the services of the undertaker before the classic criterion [whole-body death] is met because the social costs of minimal care are relatively low and do not outweigh respecting the intuitive social feeling that breathing bodies should not be cremated or buried" (Halevy and Brody, 1993, p. 524). Since in each case we are asking quite different questions (when unilaterally to cease care, when to harvest organs, when to call the undertaker) it is not unreasonable to believe that we will reach quite different moments on the continuum from life to death as plausible answers.

IV. PHILOSOPHICAL AND BIOMEDICAL IMPACT

A. *Pluralist Moral Casuistry*

> The tremendous scientific and technological advances of the second half of the twentieth century have given physicians and other health-care providers the ability to keep patients alive who would have died relatively quickly in the past. Successful techniques for resuscitating patients whose cardiopulmonary functioning has ceased, for monitoring and responding to major organ failures in the intensive care unit, and for transplanting kidneys, livers, and hearts all exemplify the advances in medical knowledge and technology that have enabled patients to survive for a considerable period of time. ...Physicians and other health-care providers, patients and their families, and the general public eventually came to accept the idea that the technological imperative which urges the use of these techniques on all occasions should be resisted. Out of that recognition arose one of the fundamental problems of modern biomedical ethics; when to strive to preserve the life of the patient and when to simply allow the patient to die (Brody, 1988, p. 3).

The first brace of chapters articulate, explore, and critically assess Brody's foundational moral theory: pluralistic moral casuistry. As Laurence McCullough argues, Brody's articulation of a pluralistic moral theory is not premised on the assumption that single-component moral theories, such as those that appeal uniquely to consequences, or to rights, or to virtues, and so forth, are necessarily false. Rather, each failed adequately to account for the complexity of the moral world because it recognized only one of the many legitimate and core moral appeals; i.e., each appealed to only a portion of the truth. Brody's theory underscores the complexity of the fundamental philosophical questions regarding the moral world, including his willingness to acknowledge, at times, the existence of deep moral ambiguity, Kevin Wm. Wildes, S.J. notes; it thereby "...provides a sharp challenge to the desire for simplicity and sound bite answers" (p. 39). It is, as McCullough makes the point, by combining each of the various moral appeals that Brody has been able critically to articulate and defend the use of his moral theory. It is the theory's ability to capture important elements of our moral experience and to utilize those various elements, through the careful consideration of different appeals, in the resolution of moral controversies and moral decision-making, that the defense of Brody's pluralistic moral casuistry is found.

McCullough argues, for example, that the core professional virtues of physicians include integrity, compassion, self-effacement, and self-sacrifice. Such virtues, however, are not sufficient in themselves to provide comprehensive guides for care in the medical clinic or while engaged in medical research—additional moral guidance is necessary. He argues:

> Here the key insight of Brody's pluralistic moral theory comes to bear powerfully. Rather than adding what could be, at worst, endless or, at best, unmanageable specification to virtues to make them do all of the work needed to fulfill the commitments to competence and protection of the patient's health-related interests, a pluralistic professional medical ethics recognizes the need for complementary action guides based on specified ethical principles such as beneficence, respect for persons, and respect for autonomy. These ethical principles, when specified to the demands of clinical practice or research, provide concrete guides to and the basis for critical assessment of clinical judgment, decision making, and behavior (McCulllough, p. 31).

For example, while the professional virtues direct physicians to address the patient's health related concerns, the ethical principle of respect for patient autonomy reminds physicians that patients bring a much broader array of interests and values to the clinical encounter than just health-related concerns. Such attention deepens our appreciation of the clinical encounter as well as of the physician/patient relationship. As Wildes makes the point, unlike those who lament the empirical reality of moral pluralism, Brody offers bioethics a positive account of moral pluralism—a richer understanding of the phenomenological and ontological world of biomedical ethics.

Maureen Kelley and Janet Malek each raise important objections to Brody's theory and method of pluralistic casuistry. Malek notes, for example, that Brody's moral theory relies on an account of moral intuitionism. In *Life and Death Decision Making* (1988), Brody posits the existence of a fundamental cognitive capacity "...which enables us to recognize the moral value of individuals, actions, and social arrangements" (p. 12). Here a central question regards whether there is reason to believe that such intuitions ever reveal objective moral truths. Kelley's suggestion is that a detailed account of the training for moral judgment and expertise is needed further to develop the theory. We ought to consider, she argues, issues in social moral psychology and seek to develop the virtues of an expert casuist: courage, mental flexibility, integrity, and emotional sensitivity. Yet, as Malek argues, at the heart of the theory of conflicting appeals is human judgment on which there are constraints imposed by our moral cognitive faculties. Her questions: Do natural cognitive constraints exist? And, if so, is there reason to believe that those constraints lead us to moral truth?

B. Jewish Medical Ethics

> I must confess that I have little sympathy for the parents' decision. Coming from a Jewish theological background, I find their position objectionable on theological grounds. The objection was best put by Rabbis Akiva and Ishmael when they said: 'just as one does not weed, fertilize, and plow, the trees will not produce, and if fruit is produced but is not watered or fertilized it will not live but die, so with regard to the body. Drugs and medicines are the fertilizer and the physician is the tiller of the soil.' It is strange to find people actively intervening through natural means to produce desired results in all areas but matters of life and death, and insisting that in those areas alone man should merely pray and leave himself in the hands of God (Brody, 1981, p. 10).

The second section, with chapters authored by Laurie Zoloth, H. Tristram Engelhardt, Jr., and B. Andrew Lustig, turns to an exploration of Brody's robustly Orthodox Jewish bioethics, especially his forthright use of halakhic material and traditional Jewish casuistry to offer insight into modern biomedical moral controversies, such as life and death decision making, suicide and euthanasia, abortion, and even gender reassignment surgery (see Brody, 2003, chapters 16, 17, and 18). For example, Lustig explores Brody's comments on the status of the Noahide commandments, the Bnai-Noah, the commandments of God for Gentiles. Lustig considers whether such commandments should be characterized as a Jewish version of natural law conclusions. His exploration includes, in turn, Brody's arguments regarding: 1) Maimonides' statement that the appropriate motives for the observance of these commandments is because God has commanded them—not because they are inherently reasonable; 2) that central details concerning interpretations and application of these commandments are derived from biblically or Talmudically based arguments, rather than from "pure reason"; and 3) that not all of the given commandments comport with natural law accounts based on considerations of practical reason.

Zoloth highlights Brody's ability to draw on Jewish casuistry for understanding and appreciating the moral complexities of cases, including the often conflicting commitments of the moral decision makers. In pediatric cases, for example, medical teams often forget that parents of critical ill children may have deep and important duties to other members of their families:

> This is perhaps an idea more theological than philosophical, but here too Brody is a teacher, for his moral philosophy is never far from the Jewish tradition itself, in which the actor is never completely alone...It was Brody who stressed work in justice, impressing a graduate student (this author) with his argument that families in the clinical realm were more than adjuncts to the ill child, but persons with obligations to other children outside our view, and with the capacity to organize and order their lives for those children as well as the ill one we knew (p. 105).

Within the hospital setting we often forget that parents exist in a deeper more complex family context. Zoloth notes that Brody's scholarly familiarity with the narrative stories that illuminate and expand the discourse of the Talmud adds depth to his moral analysis, for he appreciates the case-based examples of halakhic legal decisions as well as the Midrash, the narratives that surround the legal decisions, which in all of their complexity give insight into complex modern cases, including euthanasia and the treatment of PVS patients.[10]

Engelhardt develops Brody's insights into the often stark differences between the assumptions and content of general secular morality and the halakhic requirements for right behavior. As Engelhardt explores Brody's arguments, he appreciates Brody as advancing at least the following three claims. First, that "secular morality does not exhaust the halakhic requirements for right behavior"; second, that "there are conflicts between secular morality and halakhic requirements"; and third, "that the first two points have force in part because Orthodox Jews, and for that matter Orthodox Christians, do not have a morality, a moral philosophy, or a theology, as these practices have come to be understood in Western European culture, especially after the first millennium" (p. 110). Or, to state the point in another fashion, "though Orthodox Judaism and Christianity have a morality in the sense of norms of behavior, and a theology in the sense of a recorded reflection on the experience of God, neither has a morality or theology as a practice independent of the religious life" (p. 110). Moreover, as Engelhardt argues, Orthodox Jewish bioethics affirms one bioethics for Jews and another for Bnai-Noah (non-Jews), and neither of which is compatible with the dominant accounts of secular bioethics. As a consequence, while there are real tensions between the requirements of God, on the one hand, and the requirements of secular morality, on the other, such tensions are not problematic because these two separate sets of requirements are framed by incompatible moral life-worlds. The moral philosophical world of secular bioethics is unable to claim governance over Talmudic argument—the disparate paradigms are incommensurable.

C. *Biomedical Public Policy and Clinical Medical Ethics*

> An infant born with multiple congenital abnormalities that rendered survival unprecedented required high-dose vasopressors to maintain blood pressure. After several days, gangrene developed in the extremities, and the parents sequentially demanded amputations of several limbs in an attempt to "do everything." The surrogate decision maker for a comatose woman dying of multisystem organ failure in an intensive care unit (ICU) was her estranged husband; they separated because of repeated spousal abuse. Despite many conferences with the husband recommending comfort measures and a do-not-resuscitate order, the husband demanded that the medical staff "do everything to my wife." A public hospital serving an indigent community of several hundred thousand had a full ICU, and 3 patients were being kept in the emergency department on ventilators. One of the patients in the ICU was a gentleman who had been ventilator dependent and unresponsive for 4 ½ months after a cardiac arrest; his daughter insisted on full support because she was hoping for a miracle.

> Common to all 3 cases was a health care team that believed that continued aggressive support was inappropriate or futile and a surrogate decision maker who insisted on "everything" being done for the patient (Halevy and Brody, 1996, p. 571).

The next two sections—*Biomedical Public Policy* and *Critical Application and Analysis*—compass a brace of essays concerned with the deployment of Brody's argument on a variety of fronts: from the fashioning of public health care policy to clinical medical ethics decision-making. While Robert Arnold proves critical of the Texas Advance Directive Act of 1999, a state law modeled after the Houston policy crafted by a interdisciplinary committee of which Brody was a chair and central mover, E. Haavi Morreim's analysis of research ethics echoes Brody's rejection of simple mandates and restrictions on research, in favor of working towards reasonable judgments and the consideration of various public interests and moral appeals to fashion non-mechanical conclusions. Loretta Kopelman critically assesses the claim that the removal of artificial nutrition and hydration from PVS patients constitutes active euthanasia. She argues that the burden of proof is on those who wish to forbid guardians of PVS patients from removing such care with the advice and consent of the medical team. Kopelman argues against Brody's conclusion that the delivery of artificial nutrition and hydration to PVS patients is basic care akin to keeping someone warm and dry, and that it is obligatory to provide it.

J. Clint Parker, Sarah-Vaughan Brakman, and Frances Kamm each consider the implications of Brody's arguments for the care and treatment of patients. Parker, for example, applies Brody's pluralistic moral casuistry to an oncology case involving a female elderly patient, to illustrate the explanatory power of Brody's theory to resolve real world cases. And, Brakman's analysis of human embryo adoption echoes Brody's work in clinical ethics as she seeks to couple clinical reality and medical knowledge, with clear analysis and philosophical rigor. Kamm, in turn, explores Brody's views on passive euthanasia and active euthanasia as a way of deepening our understanding of the significance of the moral difference (if any) between killing and letting die.

V. CONCLUSION: A PERSONAL POSTSCRIPT

Some time ago, a first year philosophy graduate student enrolled in a seminar on ethics taught by Professor Brody. The syllabus was daunting: several major articles each week, often an entire book, drawn from the classic works in the history of philosophy and their modern philosophical colleagues. In addition, extensive reading questions were to be completed prior to class and reworked with corrections and revisions after class. Classroom discussion began each week in much the same fashion: What is the author's position? Where is his argument for the position? Can we put it in numbered steps? What criticisms can legitimately be raised to the position? ("I just disagree" never being acceptable, absent extensive further argument, which would also likely need to be put into numbered steps.) What are the alternative positions? Answering each question led to significant, detailed, and philosophically robust discussions, as well as extensive training in sustained philosophical rigor. Brody insisted that we learn the development of philosophically rigorous arguments, drawn from both the history of philosophy and contemporary philosophical analysis, as well as how carefully and critically to assess such positions. Later seminars with Professor Brody would include Philosophy of Law, Ancient Philosophy, and Political Philosophy—each course similarly constructed. Indeed, each semester would inevitably include a couple of extra classes—with extra readings and questions—in the evening at Professor Brody's home, over food and drink, seminars *par excellence*.

Beyond being an inspired and engaging teacher, Brody took a real interest in his students. At a cross-road in graduate school, needing to choose a dissertation specialty, he encouraged me (Mark Cherry) to sit in on his seminar in bioethics for medical students at Baylor College of Medicine. His expressed view: at worst, I would read a series of very interesting and important books in bioethics; at best, I would choose a philosophical career path. Many years later, having nearly finished my dissertation in bioethics and entered the job market, Brody insisted that we meet routinely to hone interview skills and to develop pedagogy for teaching undergraduates. Philosophy has the reputation of being esoteric and isolated from the real world, he pointed out, in large measure because of the behavior of philosophy professors. The challenge is making philosophical insights and challenges interesting and important for the daily lives of our students—a lesson he himself drew on daily as he consulted with physicians, nurses, families, and colleagues in biomedical ethics. Careful and critical philosophical analysis was important for gaining insight into the deep challenges of human life.

What was my philosophy of teaching? He asked one day. A question I had never considered. After much consideration, and gentle but significant prodding, I responded: great books and great ideas. Students should read the great works in the history of philosophy and contemporary philosophy, always guided along with reading questions: What is the author's position? Where is the author's argument....

With the publication of his book *Taking Issue: Pluralism and Casuistry*, we were struck with the extent to which his scholarship had shaped the national and international bioethical debate. This was particularly clear at many of the major professional meetings, where the influence of the method of pluralistic casuistry, as well as his particular insights into complex biomedical cases is always evident in a significant proportion of the papers and presentations. None of the chapters in this volume are merely laudatory; instead, each addresses, explores, and critically assesses the last decade or so of Brody's scholarship from both senior scholars as well as from his former students, individuals whom he has mentored into scholars in their own rights. As is clear from his response to these essays, sometimes he agrees, other times he disagrees but adds depth to his own analysis in his reply. This result is as it should be for a philosopher who has had such an influence. Our hope in presenting this volume to the international world of bioethics is to celebrate a distinguished career marked by a singular depth and breath of scholarship, as well as by exceptional teaching, and generosity to his students and colleagues.

Mentioned here are just a few examples among many gifts of generosity and kindness. We, the editors of this volume, as well as many of the contributors (including Maureen Kelley, J. Clint Parker, Sarah-Vaughan Brakman, Frances Kamm, Kevin Wm. Wildes, S.J., and Janet Malek), have had the distinct honor of being Baruch Brody's students, philosophically engaging with him through classroom lectures and discussions, medical clinical encounters with patients and professional mentoring; it is with sincere honor and, indeed, personal affection that we refer to him as "Professor."[11]

NOTES

[1] Consider his work within the multi-disciplinary community of the Texas Medical Center in patient based research to demonstrate that particular treatments are likely ineffective. For example, patients often report symptomatic relief after undergoing arthroscopy of the knee for osteoarthritis; yet, it is reportedly unclear how the procedure produces this desired result. Indeed, in a randomized, placebo-controlled trial, Brody and colleagues demonstrated that for patients with osteoarthritis of the knee, the outcomes after arthroscopic lavage or arthroscopic debridement failed to be any better than those after a placebo procedure for reported relief of pain and return of level of function (Moseley et al., 2002). Brody's efforts in the testing of thrombolytics presents another excellent example of such critical analysis, see section II.

[2] As Rosenberg and Birdzell make the point: "The diffusion of authority to initiate innovations served also as the West's way of guarding against a chronic menace to innovative change—the interests of the status quo in suppressing innovation. An innovation will seldom be authorized or financed by government or corporate officials whose careers would be adversely affected by the success of the proposal" (2002, p. 22).

[3] Here, in addition to the World War II cases that led to the crafting of the Nuremberg Code, and the Tuskegee syphilis experiments, consider also the cases of Thomas Parham and Jesse Gelsinger. In the case of Thomas Parham, his physician allegedly encouraged him to participate in a clinical trail of a pharmaceutical designed to shrink enlarged prostates. While Parham had not had any previous prostate difficulty, his physician allegedly argued that the drug might prevent future problems. A year earlier, Parham had been hospitalized for a chronic slow heart rate, which should have likely disqualified him for participating in the study. However, his physician sought an exemption from the drug company. After he began taking the study drugs, Parham evidently experienced fatigue, a symptom of a slow heart rate. Eventually, Parham was hospitalized and a pacemaker was implanted. What became clear only after the fact was that Parham's physician was receiving $1,610 for each patient he enrolled in the study. While the pharmaceutical company designed the fee to cover study expenses, it was allegedly sufficient to provide the recruiting physician with a profit per enrolled patient (Goldner, 2000).

In a similar case, Jesse Gelsinger, an 18-year-old, died following his participation in a gene transfer study at the University of Pennsylvania's Institute for Human Gene Therapy. Gelsinger suffered from a rare liver disorder that he managed with drugs and a special diet. Food and Drug Administration (FDA) investigators concluded that researchers placed him on the scientific protocol and gave him an infusion of genetic material, even though his liver was not functioning adequately to meet the minimal level required under study criteria. FDA investigators criticized the study for failure properly to notify the FDA of severe side effects experienced by prior subjects that may have been sufficient to halt the study, failure to notify the FDA of the deaths of four monkeys that were tested using similar treatments, failures in Gelsinger's consent form to notify him of such potential harms, as well as the inability to document that all research subjects had been informed of the risks and benefits of the protocol. Eventually, the FDA suspended all gene therapy studies at this Institute. It emerged that the director of the Institute, James M. Wilson, owned stock in Genovo, the company financing the research. Moreover, Wilson and the former dean of the medical school owned patents on aspects of the procedure. Wilson admitted that he would gain $13.5 million in stock from a biotechnology company in exchange for his shares of

Genovo and that the University has some $1.4 million in equity interest in Genovo. In their agreement, Genovo would receive any rights to gene research discoveries at the Institute in exchange for financial support (Goldner, 2000).

[4] Brody argues, for example, that "...one major role of empirical studies is to help identify the ethical issues that actually arise in the practice of medicine and to find out how they are currently treated. Such findings present ethicists with the opportunity to confront actual questions and to propose defenses of, or alternatives to, current procedures for dealing with these actual questions. This is not, however, the only role of such empirical studies. They can also discover the consequences of alternative ethical policies, and that discovery can provide at least part of the basis for a moral evaluation of the alternatives. Even if one is not a consequentialist, one can at least agree that consequences are morally relevant and are part of the basis for evaluating policies. In this way, then, discoveries about what is the case are relevant for deciding what ought to be the case" (1993, pp. 211–212).

[5] The judgment that a treatment is "futile" is a performative judgment designed to defeat a putative duty to provide the requested healthcare (see Halevy, Neal and Brody, 1996). Part of the difficulty of such judgments, though, as Brody notes is adequately defining "futility". Consider four possibilities: "physiologic futility (the intervention does not have its intended physiologic effect), imminent demise futility (the patient will die before discharge regardless of the intervention), lethal condition futility (the patient has an underlying disease that is not compatible with long-term survival, regardless of the intervention, even if the patient could survive to discharge from this hospitalization), and qualitative futility (the resultant quality of life is too poor)" (Halevy and Brody, 1996, p. 571; see also Halevy and Brody, 1995).

[6] John Harris has argued, for example, that since patients in a PVS state can no longer experience and benefit from their existence that we do them no harm when we cease to support their lives, or indeed to take it from them: "Thus John's critical interest in a further thirty years of life in PVS would give way to the significant critical interests or preferences of any actual persons, persons to whom the satisfaction, or not, of their desires can continue to matter. ... We would not, I imagine, think that someone who could no longer benefit from, or appreciate, the life he was leading should have that life sustained when to do so would cost the lives of others who could appreciate, and benefit from, their existences" (Harris, 1995, p. 19).

[7] It is worth noting that the Society of Critical Care Medicine has argued that PVS patients should not be admitted to the intensive care unit, even if beds are available. "Examples of patients who *should* be excluded from the ICU, whether beds are available or not, include those who competently decline intensive care or request that invasive therapy be withheld; those declared brain dead who are not organ donors; and those in a persistent vegetative or permanently unconscious state" (1994, p. 1202). Their statement appreciated such care as an inappropriate use of scarce health care resources, both physical resources as well as professional expertise.

[8] This viewpoint, for example, was the focus of Pope John Paul II's statement "To the Participants in the International Congress on 'Life-Sustaining Treatments and Vegetative State: Scientific Advances and Ethical Dilemmas'" of March 20, 2004. He urged: "I should like particularly to underline how the administration of water and food, even when provided by artificial means, always represents a *natural means* of preserving life, not a *medical act*. Its use, furthermore, should be considered, in principle, *ordinary* and *proportionate*, and as such morally obligatory, insofar as and until it is seen to have attained its proper finality, which in the present case consists in providing nourishment to the patient and alleviation of his suffering" (para. 4). Here, John Paul II brought together his longstanding claims regarding personal dignity and the value of human life to bear on a very particular judgment regarding the moral obligation to continue artificial nutrition and hydration for patients—even those patients in a permanently vegetative state. There has been sustained and significant debate about the meaning of John Paul II's statement. See, for example, *Christian Bioethics* 12(1), 2006: "Judgments at the Edge of Life and Death: Artificial Nutrition and Hydration" and Cherry (2006).

[9] Consider Robert Veatch: "The relationship of the dead donor rule to the definition of death is complex and not always well understood even by experts in the field. Consider the position of those who believe that organs should be procured from irreversibly comatose persons who still have some residual brain function or from those in a persistent vegetative state. Such patients currently are classified as

alive according to the law in Ohio and all other jurisdictions of the world. Those who conclude that life-prolonging organs can be procured from these people could formulate their position in two quite different ways.

One strategy has been proposed by some defenders of organ procurement from anencephalic infants (American Medical Association 1995; *In re T.A.C.P.* (Baby Theresa) 609 So. 2d 588 (Fla. 1992); cf. Committee Reports: Ethics 1995), irreversibly comatose patients, and those in a persistent vegetative state. These advocates propose 'exceptions' to the DDR to cover such cases. They would procure from special groups of persons classified as living. It seems that this would be the position of at least some of those who responded to Scenarios 2 and 3 by saying that the patient was alive but nevertheless that organs could be procured. Call this 'Option 1.'

A second strategy, however, would lead to the same policy of procuring organs from these categories of patients by changing the definition of death so that people in these groups were considered dead and treated accordingly. Call this 'Option 2'" (2004, p. 266; see also Siminoff, Burant, and Youngner, 2004; Schmidt, 2004; Hausman, 2004; Fost, 2004; Shewmon, 2004; Campbell, 2004).

[10] Consider, for example, Brody's arguments regarding the "sanctity of human life": "The belief in the sanctity of human life is the belief that each moment of biological life of every member of our species is of infinite value. This belief has profound implications for discussions of medical ethics; it stands in opposition, for example, to most recent discussions of death with dignity which emphasize patient autonomy and quality of life rather than the sanctity of human life. It is widely believed that traditional Judaism believes in the sanctity of human life doctrine. ... It is not my intention in this essay to argue that this viewpoint is entirely incorrect. Judaism has not traditionally stressed patient autonomy, and it has on the whole opposed suicide and euthanasia. I shall argue, however, that it is not committed to a belief in the sanctity of human life, to a belief that residual life in pain has infinite worth. Such claims totally misrepresent the traditional Jewish position and fail to bring out important aspects of it which are essential to its contemporary elaboration. It also makes it impossible for us to learn from the traditional Judaic discussions an important lesson about the structure of a balanced moral life" (Brody, 2003, pp. 229–230).

[11] The *Oxford English Dictionary* references a professor as a "public teacher or instructor of the *highest rank* in a specific faculty or branch of learning" (emphasis added, second edition, 1989 [on-line], www.dictionary.oed.com).

BIBLIOGRAPHY

——. (1968). *Black's Law Dictionary*, 4th edition, revised. St. Paul, MN: West.

Biros MH. Runge JW. Lewis RJ. Doherty C. (1998). 'Emergency medicine and the development of the Food and Drug Administration's final rule on informed consent and waiver of informed consent in emergency research circumstances,' *Academic Emergency Medicine*, 5(4), 359–68.

Brody, B.A. (2006). 'Intellectual Property and Biotechnology: The U.S. Internal Experience – Part I,' *Kennedy Institute of Ethics Journal*, 16(1), 1–37.

Brody, B.A. (2003). *Taking Issue: Pluralism and Casuistry in Bioethics*. Washington, D.C.: Georgetown University Press.

Brody, B.A., N. Dickey, S.S. Ellenberg, R.P. Heaney, R.J. Levine, R.L. O'Brien, R.B. Purtilo, and C. Weijer. (2003). 'Is the Use of Placebo Controls Ethically Permissible in Clinical Trials of Agents Intended to Reduce Fractures in Osteoporosis?' *Journal of Bone and Mineral Research*, 18(6), 110–1109.

Brody, B.A. (2002a). 'Ethical Issues in Clinical Trials in Developing Countries,' *Statistics in Medicine*, 21(19), 2853–8.

Brody, B.A. (2002b). 'Philosophical Reflections on Clinical Trials in Developing Countries,' in R. Rhodes, M. Battin, and A. Silvers (eds.). *Medicine and Social Justice*. London: Oxford University Press, pp. 197–210.

Brody, B.A. (2002c). 'The Ethical Recruitment of Subjects,' *Medical Care*, 40(4), 269–270.

Brody, B.A. (2001). 'Making Informed Consent Meaningful,' *IRB: A Review of Human Subjects Research*, 23(5), 1–5.
Brody, B.A. (2000). 'A Historical Introduction to the Requirement of Obtaining Informed Consent from Research Subjects,' in L. Doyal and J. Tobas (eds.). *Informed Consent in Medical Research*. London: BMJ Books, pp. 7–14.
Brody, B.A. (1998a). *The Ethics of Biomedical Research: An International Perspective*. NY: Oxford University Press.
Brody, B.A. (1998b). 'Research on the Vulnerable Sick,' in Jeffrey Kahn, Anna Mastroianni, and Jeremy Sugarman (eds.), *Beyond Consent: Seeking Justice in Research*. New York: Oxford University Press, pp. 32–46
Brody, B.A. (1997a). 'When are Placebo Controlled Trials No Longer Appropriate,' *Controlled Clinical Trials*, 18(6), 602–612.
Brody, B.A. (1997b). 'In case of emergency: No need for consent,' *Hastings Center Report*, 27(1), 7–12.
Brody, B.A.A. (1996). 'Conflicts of Interests and the Validity of Clinical Trials,' in Roy G. Spece, Jr., David S. Shimm, and Allen E. Buchanan (eds.). *Conflicts of Interest in Clinical Practice and Research*. NY: Oxford University Press, pp. 407–417.
Brody, B.A.A. (1995). *Ethical Issues in Drug Testing, Approval, and Pricing*. NY: Oxford University Press.
Brody, B.A.A. (1993). 'Assessing Empirical Research in Bioethics,' *Theoretical Medicine*, 14(3), 211–219.
Brody, B.A. (1992). 'Special ethical issues in the management of PVS patients,' *Law, Medicine & Health Care*, 20(1–2), 104–115.
Brody, B.A.A. (1989). 'The President's Commission: The Need to be More Philosophical,' *Journal of Medicine and Philosophy*, 14(4), 369–383.
Brody, B.A. (1988). *Life and Death Decision Making*. New York: Oxford University Press.
Brody, B.A. (1981). 'Faith healing for childhood leukemia: commentary,' *Hastings Center Report*, (February), 10–11.
Campbell, C.S. (2004). 'Harvesting the Living?: Separating Brain Death and Organ Transplantation,' *Kennedy Institute of Ethics Journal*, 14(3), 301–318.
Carnahan, S.J. (1999). 'Promoting medical research without sacrificing patient autonomy: legal and ethical issues raised by the waiver of informed consent for emergency research,' *Oklahoma Law* Review, 52(4), 565–91.
Cherry, M.J. and Zientek, D. (eds.) (2006). "Judgments at the Edge of Life and Death: Artificial Nutrition and Hydration," *Christian Bioethics*, 12(1).
Cherry, M.J. (2006). 'How should Christians make judgments at the edge of life and death,' *Christian Bioethics*, 12(1), 1–10.
Food and Drug Administration (1996). Final Rules. *Federal Register*, Oct. 2.
Fost, N. (1998). 'Waived consent for emergency research,' *American Journal of Law &* Medicine, 24(2–3), 163–83.
Fost, N. (2004). 'Reconsidering the dead donor rule: is it important that organ donors be dead?' *Kennedy Institute of Ethics Journal*, 14(3), 249–260.
Goldner, J. A. (2000) 'Dealing with conflicts of interest in biomedical research: IRB oversight as the next best solution to the abolitionist approach,' *The Journal of Law, Medicine & Ethics*, 28(4), 379–404.
Halevy, A. and Brody, B.A. (1993). 'Brain death: Reconciling definitions, criteria, and tests,' *Annals of Internal Medicine*, 119(6), 519–525.
Halevy, A. and Brody, B.A. (1996). 'A multi-institutional collaborative policy on medical futility,' *Journal of the American Medical Association*, 276, 571–574.
Halevy, A. and Brody, B.A. (1995). 'Is futility a futile concept?' *The Journal of Medicine and Philosophy*, 20, 123–144.
Halevy, A., Neal, R.C., and Brody, B.A. (1996). 'The low frequency of futility in an adult intensive care unit setting,' *Archives of Internal Medicine*, 156, 100–104.
Harris, J. (1995). 'Euthanasia and the value of life,' in J. Keown (ed.), *Euthanasia Examined*. Cambridge: Cambridge University Press, pp. 6–22.

Hausman, D.M. (2004). 'Polling and Public Policy,' *Kennedy Institute of Ethics*, 14(3), 241–247.

Karlawish, J.H. and J.B. Hall. (1996). 'The controversy over emergency research. A review of the issues and suggestions for a resolution,' *American Journal of Respiratory & Critical Care Medicine*, 153(2), 499–506.

Kipnis, K., N.M. King and R.M. Nelson. (2006). 'Trials and errors: barriers to oversight of research conducted under the emergency research consent waiver,' *IRB: A Review of Human Subjects* Research, 28(2), 16–19.

Mosley, J.B., K. O'Malley, N.J. Petersen, T.J. Menke, B.A. Brody, D.H. Kuykendall, J.C. Hollingswroth, .M. Ashton, and N.P. Wray (2002). 'A Randomized Placebo Controlled Trial of Arthoscopic Surgery for Osteoarthritis of the Knee,' *New England Journal of Medicine*, 347(2), 81–88.

President's Commission for the Study of Ethical Problems in Medicine and Biomedical and Behavioral Research (1981). *Defining Death*. Washington, D.C.: U.S. Government Printing Office.

Schmidt, T.C. (2004). 'The Ohio Study in Light of National Data and Clinical Experience,' *Kennedy Institute of Ethics Journal*, 14(3), 235–240.

Shewmon, D.A. (2004). 'The Dead Donor Rule: Lessons from Linguistics,' *Kennedy Institute of Ethics Journal*, 14(3), 277–300.

Siminoff, L.A., C. Burant, S.J. Youngner (2004). 'Death and Organ Procurement: Public Beliefs and Attitudes,' *Kennedy Institute of Ethics Journal*, 14(3), 217–234.

Society of Critical Care Medicine Ethics Committee (1994). 'Consensus Statement on the Triage of Critically Ill Patients,' *Journal of the American Medical Association*, 271 (15), 1200–1203.

Veatch, R.M. (2004). 'Abandon the Dead Donor Rule or Change the Definition of Death?' *Kennedy Institute of Ethics Journal*, 14 (3), 261–276.

SECTION I

PLURALISTIC MORAL CASUISTRY

CHAPTER 2

NOTES TOWARD A PLURALISTIC PROFESSIONAL MEDICAL ETHICS

LAURENCE B. MCCULLOUGH

Center for Medical Ethics and Health Policy, Baylor College of Medicine, Houston, Texas, U.S.A.

I. INTRODUCTION

Baruch A. Brody has made many significant and influential contributions to medical ethics and bioethics, as well as to ethics and philosophy generally. He has also played major leadership roles in medical ethics and bioethics as founding Director of the Center for Medical Ethics and Health Policy at Baylor College of Medicine in Houston, Texas, and as president of the Society for Health and Human Values, which has now merged into the American Society for Bioethics and Humanities, the field's professional organization. He is also a master teacher of medical students, residents, and fellows, as well as of undergraduate liberal arts students and graduate philosophy students. His many accomplishments as a philosopher, medical educator, mentor, builder of institutions, and leader of organizations are documented elsewhere in this volume.

Among Brody's many contributions to the medical ethics and bioethics literature and teaching, I want in this chapter to emphasize and take advantage of his major contribution to method in medical ethics and bioethics: the articulation, justification, and deployment of a pluralistic moral theory, with a particular focus on end-of-life decision making and medical care (Brody, 1998). My purpose in doing so is twofold. First, I set out the main features of Brody's pluralistic moral theory and some of its important implications for medical ethics and bioethics, especially in response to the recent criticism by Tom Beauchamp (2004) concerning an absence of ethical theory in bioethics. Second, I offer some reflections on what I take to be the main features of a pluralistic professional medical ethics, showing how such an approach to medical ethics is indebted to Brody but also differs in important ways from a general pluralistic moral theory brought to bear on moral problems in biomedical science, research, and practice, and in the organization and funding of medical care. The result is not meant to be definitive, but rather a set of notes toward a pluralistic professional medical ethics.

M.J. Cherry and A.S. Iltis (eds.), Pluralistic Casuistry, 23–36.

II. BRODY'S PLURALISTIC MORAL THEORY

In his *Life and Death Decision Making*, Brody underscores the need for a moral theory to guide and evaluate decision making in the clinical setting. He begins by taking exception to the primary focus of much of the bioethics literature on the question of who should decide. Brody points out, correctly, that addressing this question will help us to identify who has or should be accorded the authority to make controlling decisions about medical care. However, he adds, addressing this question does not provide for the resulting morally or legally authoritative individual or organization guidance about what decisions should be made. Yet, such guidance is crucial for the intellectual and moral authority of the decisions made and thus of the decision maker. Brody also makes the crucial point that a focus on the question of who should decide does not provide us with answers to questions about the justified limits on decision making authority.

If we aim to make responsible decisions about clinical cases, Brody goes on to claim, we require an adequate moral theory. In making this claim, Brody responds to skepticism that moral theories can provide us the guidance we need, because they are abstract and because their "mode of application is unclear" (Brody, 1988, p. 8). Brody then makes the ingenious move of pointing out that the path to an applicable moral theory is to take "from each of the traditional abstract moral theories a component which needs to be combined with components of other theories in a way that produces a type of model for decision making that can be applied to difficult cases" (Brody, 1988, p. 8). We require such an approach because no single-component moral theory is adequate by itself to provide comprehensively applicable guidance to decision making in concrete cases.

> We also need to recognize that each [single-component moral theory] has failed because it has recognized only one of the many legitimate moral appeals. Rather than seeing the history of moral philosophy as a history of competing theories among which we must choose, we ought to view it as a series of attempts to articulate different moral appeals, all of which will have to be combined to frame an adequate moral theory for helping us deal with difficult cases (Brody, 1988, p. 9).

Brody's point is not that a particular moral theory from the history of ethics is false. Instead, his point is that every single-component moral theory has both strengths—it gets something right about our moral lives—and weaknesses—it is incomplete, so the strategy of arguing for a canonical single-component moral theory will still leave us with those weaknesses as inadequately managed methodologic problems. These weaknesses stem precisely from the inability of a single-component theory to account for (i.e., justify and critically assess) our full range of moral intuitions, concerns, convictions, practices, and institutions. For example, Aristotle's virtue theory is to be commended, Brody says, because it insists that we incorporate consideration of the virtues into an adequate moral theory, but Aristotle's ethics is to be criticized for its failure to take account of appeals to consequences and rights.

> In short, moral philosophers over the centuries have put forward many differing moral theories. Each involves a specific type of moral appeal. The strength of each is precisely

> that the moral appeal it is invoking is a legitimate moral appeal, one that truly needs to be taken into account as we confront real cases. But the difficulty with these theories is twofold. One difficulty is internal, their failure to sufficiently articulate the moral appeals they are invoking. But the other and more important difficulty is that each theory is prone to counterexamples based explicitly on other moral appeals (Brody, 1988, p. 10).

In other words, multiple moral appeals are required because each complements the other; i.e., each successfully addresses the limitations and resulting inadequacies of explanation and justification of the other appeals in applicable guides to decision making and in critical evaluation of decision making. The success, i.e., justification, of a pluralistic moral theory turns on its ability to establish and deploy reliably the complementarity of various moral appeals (Grinnell, Bishop and McCullough, 2002).

Brody's approach to establishing the relevant moral appeals is intuitionist, but not an intuitionism of indubitable, direct grasp of moral rules and principles that are taken to be true. This variety of intuitionism produces either simple principles that do not provide adequate guidance because they do not help us deal with possible exceptions, or complex rules that do deal with exceptions but in their progressively more detailed specification become progressively less intuitively true. Brody argues, instead, for an approach that "systematizes these intuitions, explains them, and provides help in dealing with cases about which we have no intuitions" (1988, p. 11). Such a theory is confirmed by its effectiveness; it "help us deal with … hard cases in what seems at the end an intuitively satisfactory fashion" (1988, p. 12).

Brody develops, deploys, and argues for the decision making adequacy of just such a pluralistic moral theory. As is well known, Brody's pluralistic moral theory incorporates complementary moral appeals to the consequences of our actions, to rights, to respect for persons (which, emphatically, is not equivalent to respect for autonomy), to virtues, to cost-effectiveness, and to justice. Brody then shows how a pluralistic moral theory results in a "new model for the physician-patient relationship which properly incorporates all of these many different appeals", which he calls "the model of conflicting appeals" (1988, p. 72).

> The fundamental claim of the model of conflicting appeals is that both parties in the physician-patient relationship are subject to a wide variety of moral appeals which they need to satisfy. This model says that the legitimacy of the physician-patient relationship depends on the proper balancing of these conflicting moral appeals (1988, p. 75).

Brody examines and rejects, as inadequate, alternative approaches to balancing conflict moral appeals, based either on "hierarchical or lexical ordering" of them (1988, pp. 74–75), or on the "scale approach to conflict", i.e., assigning weights to conflicting moral appeals and then balancing them (1988, pp. 76–77). Instead, he advocates "the judgment approach":

> In approaching a particular case of moral conflict we must, according to this third approach [the judgment approach], identify all of the moral appeals relevant to that case and then use the theory to ascribe a significance to each. According to this judgment approach, however, this is as far as theory can take us. Having identified the various

> moral appeals that back the various proposed actions and having ascribed a significance to them in light of the theory we have developed, we are not in a position to use a common metric to derive a conclusion of what is the appropriate action. This final process, rather than being a weighting process, is a process of judgment. We look at the various appeals and their significance, and then we judge what we ought to do (1988, p. 77).

The theory provides us with considerations that we must address in assessing the significance of relevant moral appeals.

Beauchamp (2004) has recently lamented the absence of moral theory in bioethics. In particular, Beauchamp claims that "no moral philosopher has developed a detailed program or method of practical ethics supported by a general ethical theory" (2004, p. 209). He elaborates further:

> How theory can be connected to practice is a problem of greater urgency today than it was thirty years ago, when philosophers were not in touch with medical morality and virtually nothing was expected of them in this domain (2004, p. 216).

Life and Death Decision Making provides ample evidence that Brody understands this urgency and he is surely a philosopher "in touch with medical morality"—he has been immersed in the clinical setting and fully conversant with and committed to it for more than two decades. Perhaps when Beauchamp refers to theory he means the sort of abstract accounts that Brody eschews and aims to surpass. There can be little doubt that Brody's pluralistic moral theory is indeed a "detailed program or method of practical ethics that is supported by a general ethical theory". Indeed, Brody's pluralistic moral theory provides an exemplar of practical and theoretical ethics in synergy with each other, so much so that the distinction between the two becomes inapt.

III. PLURALISTIC PROFESSIONAL MEDICAL ETHICS

Having considered the main conceptual and methodologic elements of Brody's pluralistic moral theory, I turn next to the promised reflections on a pluralistic professional medical ethics. I do so, mainly, as a counterpoint to the common methodologic view that medical ethics is a subset of bioethics that, in turn, is understood to be a subset of ethics generally. On this view, the only distinctive feature of medical ethics and of bioethics is that "it relates to a particular realm of facts and concerns" (Clouser, 1978, p. 116). Medical ethics and bioethics are not special because either "embodies or appeals to some special moral principles of methodology. It [medical ethics] is applied ethics" (Clouser, 1978, p. 116). As a subset of ethics, medical ethics is to be undertaken from the moral point of view; i.e., an account of medical morality that prescinds or abstracts from social roles and role-related or role-specific obligations, and considers obligations of individuals or persons as such to each other and to other entities. Brody's approach to a pluralistic moral theory rejects a simple "applied ethics" approach but accepts medical ethics based on the moral point of view. It is at just the latter juncture that I wish to depart from it and consider what a pluralistic professional ethics involves; i.e., an approach to medical ethics that takes the social role of being a professional and its role-related obligations as the central focus of moral attention and concern.

To do so, we first need to get clear on what it means to be a professional. It is commonly thought in medical ethics and bioethics that there is Hippocratic tradition that comes down to us unbroken from ancient times (and is usually very paternalistic in nature) (Veatch, 1989; 2003). Baker has recently exploded this myth of the "Hippocratic footnote" (Baker, 1993b). In the history of Western medical ethics, the concept of medicine as a profession was first introduced, not in the ancient world, but much more recently, in the eighteenth-century enlightenments in Scotland and England, in the work of two transformative figures in the history of medical ethics, the Scottish physician-ethicist, John Gregory (1724–1773), and the English physician-ethicist, Thomas Percival (1740–1804). Gregory (1770; 1772; Haakonssen, 1997; McCullough, 1998) and Percival (1803; Haakonssen, 1997; Baker, 1993a) used the tools of ethics to reform medicine from what it then was, a largely entrepreneurial, self-interested, undertaking, into a profession worthy of the name.

Gregory and Percival articulated a three-component account of the physician as a professional. First, the physician must be competent. Second, the physician must make the protection and promotion of the patient's health-related interests the physician's primary commitment and motivation, keeping self-interest systematically secondary. Third, physicians should maintain, improve, and pass on medicine as public trust and not a private guild whose main concern is the protection of the economic, social, and other advantages of its members (McCullough, 1998; Baker and McCullough, 2008). Let me explain each component in further detail.

Gregory and Percival had to make competence the starting point for their professional medical ethics because so few of their contemporary physicians and surgeons (the two groups of practitioners were still separated from each other and also in hot competition with each other) were scientifically competent. For Gregory and Percival, scientific competence was a function of submitting to the intellectual discipline of Baconian method; medical practice and research should be based on "experience." Experience involved the careful, disciplined, and precise observation of the natural courses of diseases and the effects on disease progression of medical and surgical interventions. Appeals to one's personal experience did not count as appeals to experience properly understood, because of unavoidable bias in both reporting and analysis. Appeals to one's authority or power, of course, were intellectually corrupt and had resulted in the deaths of untold numbers of the sick. Percival called for the routine collection of data in the infirmary, or hospital, on the processes and outcomes of care, regular meetings to review these data, and application of the results of this critical evaluation to improve medical and surgical management of patients in the hospital. Readers will recognize in Gregory's and Percival's calls for a commitment to Baconian method and intellectual discipline a nascent form of what has become known as evidence-based medicine. That a revolutionary idea took widespread hold barely two centuries after its introduction represents a remarkable accomplishment from the perspective of the history of ideas.

By becoming scientifically competent physicians and surgeons could make reliable claims about health, disease, and injury, and how disease and injury could be effectively managed to restore, maintain, or improve health and resulting functional status. Baconian scientific competence would then serve as an antidote to the crisis of intellectual trust of the sick concerning physicians, surgeons, and other practitioners in eighteenth-century Britain (Porter and Porter, 1989). The sick had little or no confidence that medical practitioners' claims to intellectual authority were reliably founded. As a result, to the perils of disease and injury, the sick would have had to add the unknown and potentially lethal perils of seeking out practitioners in whom the sick had little or no intellectual confidence. It is no surprise that self-physicking, i.e., self-diagnosis and treatment, was the preferred approach to disease and injury. Its results were, unfortunately, very mixed. Scientifically competent medical practitioners were the antidote to this crisis of intellectual trust and to the clinical perils of placing oneself under the not-so-tender mercies of physicians, surgeons, or apothecaries, not to mention the irregulars (or, less charitably, quacks).

The sick in eighteenth-century Britain also experienced a moral crisis: they suspected that, more often than not, medical practitioners were more interested in "feeing" the sick, i.e., getting paid and getting rich, and in prestige and power, especially in the infirmary setting, than they were in protecting and promoting the health of the sick (Gregory, 1770; 1772; Porter and Porter, 1989). In short, practitioners were understood to be "interested" men, motivated primarily by self-interest and not by a life of service to the sick. Gregory, especially, understood that this made being sick and placing oneself, reluctantly, in the care of a medical practitioner very perilous business indeed. Practitioners committed to a life of service to the sick were the antidote to the crisis of moral trust among the sick, who could, perhaps for the first time in history of Western medicine outside religious healing orders, have confidence that their medical practitioners were mainly concerned about them, the sick, and were willing to make personal sacrifices, including financial sacrifice, in order to meet the needs of the sick.

Gregory assails the "corporation spirit" of the royal colleges of the time, which essentially used their powers of licensure to protect the entrepreneurial self-interests of their members. The *statuta moralia*, or moral rules, of the royal colleges were exclusively concerned with protecting the market share and power of college members (McCullough, 1998). Percival, however, is the first in the history of medical ethics to refer to medicine explicitly as a public trust; i.e., a social undertaking that is valuable precisely to the extent that it is based on scientific competence and involves a life of service to the sick and to society.

Gregory also understood that, once we have physicians who are professionals worthy of the name (the word 'professional' was then widely used by physicians, but in self-interested fashion by university-educated physicians to differentiate themselves from the irregulars, which was ironic, to say the least, given the absence of a required curriculum or examination), the sick become patients (McCullough, 1998). That is, once we have a profession of medicine, we have physicians, whose scientific and moral commitment to the sick transform them into patients. The professional relationship in medicine is thus a physician-patient relationship, not a patient-physician relationship as it is understood by some in contemporary bioethics (Veatch, 1989; 2003). It is worth noting that Gregory used both 'the sick' and 'patient' in his medical ethics, whereas Percival used 'patient' almost exclusively. The shift is more than linguistic; it is conceptual and introduced professional medical ethics into the Western world.

Percival to be sure, and Gregory to a lesser extent, also understood that professional ethics is not something that can be developed or maintained just by physicians. An appropriate and supportive organizational culture is also required. Percival introduced this crucial moral insight into the nature of professional medical ethics by devoting so much of his text to the identification of hospital policies and practices (Percival, 1803).

Gregory and Percival correctly understood that an account of professional medical ethics must begin with the intellectual and moral virtues of physicians. The professional virtues of physicians are traits or habits of character that direct one's attention and concern to patients, routinely dispose one to discern one's obligations to patients, and motivate one routinely to fulfill one's obligations to patients. That is, professional virtues transform the medical practitioner into a physician.

For Gregory, and therefore for professional medical ethics, the bedrock professional virtue of physicians is integrity. This virtue requires physicians to practice medicine, conduct research, manage health care organizations, and contribute to health policy by adhering uniformly to standards of intellectual and moral excellence. Intellectual excellence now means that physicians should practice medicine and conduct research on the basis of the intellectual and clinical discipline provided by evidence-based medicine. The best available evidence should be identified and carefully assessed, and then used to guide the clinical management of the conditions and problems of individual patients and to identify, evaluate, and constantly improve the components of comprehensive processes of care. Strict adherence to evidence-based medicine is, of course, required in all biomedical and clinical research and all contributions to health policy—from expert testimony in civil and criminal actions to the preparation of legislation and regulations. Achieving intellectual excellence routinely in medical practice, research, and health policy creates the basis for intellectual authority that merits and commands the respect of patients, health care organizations, payers, and society. Moral excellence means just what Gregory and Percival said it should mean: physicians should be primarily committed to the protection and promotion of the health-related interests of their patients and should, therefore, keep their own self-interests in a systematically secondary place. This means that the burden of proof is on the physician who wishes to make protection and promotion of his or her self-interest primary, and that the burden of this proof is steep.

Notice that professional integrity appeals to professional values, which include, the intellectual and moral standards that should govern the clinical judgment, decision making, and behavior of *all* physicians; i.e., professional conscience. Appeals to professional conscience are directly relevant too, because they should guide the clinical practice of all physicians; e.g., one should obtain informed consent for clinically risky invasive procedures in non-emergency situations from all competent adults. There are also appeals to individual conscience that are directly relevant to, because they should guide, clinical practice. Appeals to individual conscience are based on values, beliefs, and commitments in a physician's moral life that have their origins outside of medicine; e.g., in one's religion and other serious forms of moral reflection and ways of life (McCullough and Chervenak, 1994).

This account of professional integrity differs from Brody's, which focuses on personal values. Brody argues that integrity "calls upon health-care providers and health-care recipients to stand firm in their values" (1988, p. 37). By contrast, in professional medical ethics, professional integrity calls for physicians to stand firm both in their adherence to the intellectual and moral excellence that should be definitive of all physicians, professional conscience, and to stand firm in their adherence to intellectual and moral excellence that derives from non-medical sources in a physician's life, individual conscience. The virtue of integrity for patients is solely a function of the personal values of patients; they are not in a professional role in medical care, although they are surely protected by one.

Note that this account of the professional virtue of integrity allows us to see that a central ethical challenge in being a professional physician involves conflicts between professional and individual conscience. As a rule, individual conscience is secondary to professional conscience. Exceptions to this rule are allowed, but only consistent with fulfillment of professional obligations to protect and promote the health-related interests of patients by other physicians. Thus, the professional obligations of the physician in the informed consent process bind *all* physicians *equally* and require them to present to patients all medically reasonable alternatives for the clinical management of their condition or problem. For pregnancy, this includes the alternatives of termination and continuation of pregnancy. A physician who did not offer the alternative of termination of pregnancy (i.e., present it non-directively with an evidence-based account of its clinical benefits and risks), when a pregnancy has been diagnosed to be complicated by a fetal anomaly, would violate professional integrity. Protection of individual conscience justifies a physician who is morally opposed to performing abortions to offer to perform termination of pregnancy himself or herself, provided effective referral is made to a colleague or clinic that provides safe and effective abortions. Protection of individual conscience also prohibits residency and medical student training programs to require participation in abortions, while professional conscience requires all trainees to have the fund of knowledge and clinical skills required to diagnose and manage the clinical complications of abortion (McCullough and Chervenak, 1994). The distinctions between professional and individual conscience and between professional and individual integrity cannot reliably be made in a medical ethics based on general moral theory, because it discounts or dismisses roles and role-related obligations, but can be made in professional medical ethics.

Virtues such as integrity, compassion, self-effacement, and self-sacrifice are the core professional virtues of physicians. They create obligations of their own. For example, compassion requires physicians to recognize and respond promptly to relieve the pain, distress, and suffering of patients. Self-effacement requires physicians not to be unduly influenced by clinically irrelevant differences between themselves and their patients, such as differences in race, income, gender, religion, or sexual orientation, when these are clinically irrelevant. Self-sacrifice names itself and involves the considerable intellectual and practical task of identifying and maintaining ethically justified limits on self-sacrifice (McCullough and Chervenak, 1994).

Even when they create obligations, however, professional virtues are not sufficient by themselves to provide comprehensive action guides in clinical care and research. Here the key insight of Brody's pluralistic moral theory comes to bear powerfully. Rather than adding what could be, at worst, endless or, at best, unmanageable specification to virtues to make them do all of the work needed to fulfill the commitments to competence and protection of the patient's health-related interests, a pluralistic professional medical ethics recognizes the need for complementary action guides based on specified ethical principles such as beneficence, respect for persons, and respect for autonomy. These ethical principles, when specified to the demands of clinical practice or research, provide concrete guides to and the basis for critical assessment of clinical judgment, decision making, and behavior (McCullough and Chervenak, 1994).

For example, while the professional virtues certainly direct the physician's primary attention and commitment to the patient's health-related interests, they do not guide the physician in incorporating the patient's values and beliefs into clinical judgment, decision making, and behavior. The ethical principle of respect for autonomy, however, does so. This ethical principle also reminds the physician that the values and beliefs that patients bring to the physician-patient relationship will range far more widely than just the patient's health-related interests.

The ethical principle of beneficence directs the physician to protect and promote, on the basis of the best available evidence, the patient's health-related interests. That is, the intellectual and moral authority of physicians in the physician-patient relationship is a function of the evidence-based competencies of medicine and is limited to the health-related interests of patients. Medical competencies include the prevention and clinical management of pain, distress, suffering, disease, and injury. Obviously, disciplined judgments need to be made when these competencies cannot be carried out consistently with each other. Which competency should receive priority for implementation thus becomes a matter of judgment, taking into account not only beneficence and respect for the patient's values, beliefs, and preferences, but also, as Brody is correct to emphasize, respect for the patient as a person. The latter appeal directs the physician to protect and promote the basic health capacities of the patient. When physicians begin to approach the limits of medicine to manage disease and injury and thus to prevent death, attention to the principle of or appeal to respect for persons directs the physician's increasing attention and concern to evidence-based clinical judgments about the impact on the patient's functional status of increasing disease-related and iatrogenic disease, injury, pain, distress, and suffering. Gregory, especially, emphasizes the limits of medicine's capacities in these respects and thus does not make prolongation of life an unlimited ethical obligation of physicians. Indeed, he includes among the physician's obligations to "smooth the avenues of death" (Gregory, 1772, p. 35). In my reading of Gregory, he opens the door to physician-assisted suicide not only as not inconsistent with professional integrity but as a professional obligation in some cases. This seriously calls into question arguments against physician-assisted suicide on the basis of professional integrity (McCullough, 1998).

Note that this discussion of ethically justified limits on medical intervention makes no appeal to the goals of medicine. Claims about the goals of medicine in the medical ethics literature have two sources. One is an essentialism about medicine, identifying the goals of medicine with the final cause or perfection of medicine, a view of medicine championed especially by Pellegrino and Thomasma (1988; 1993). Whatever else Wittgenstein may or may not have accomplished regarding essentialism, he did decisively show, in his consideration of language games, that human practices or activities cannot reliably be said to have fixed essences. Aristotelian metaphysicians already had understood this centuries before, of course, so we should not regard Wittgenstein's contribution as original. Rather, we should, rightly, see it as decisive, the final historical nail in the coffin of essentialist accounts of human practices and social institutions such as medicine. It is worth noting that Pellegrino and Thomasma do not, anywhere that I can find, take up the Aristotelean-Wittgensteinian critique of essentialist accounts of human practices.

The second basis for claims about the proper goals of medicine appeals to views external to professional medical ethics of what those goals should be. Thus, for example, one might claim that it is not a proper goal of medicine to prolong life when the patient has irreversibly lost interactive capacity and thus can no longer grow and develop as a human being; e.g., when the patient has been reliably diagnosed to be in a persistent or permanent vegetative state. That is, medicine should not be understood to be vitalist, to have as one of its proper goals the preservation of life irrespective of what the patient can or cannot do with the functional status that is maintained. McCormick (1974), in his classic article, based his argument to this effect explicitly on a central tenet of Judeo-Christian medical ethics: there is no compelling obligation to keep someone alive when he or she has irreversibly lost the capacity to be in a prayerful relationship with God, which he then secularized in terms of interactive capacity. This is a reasonable claim but does not rule vitalism out as a goal of medicine; it just makes vitalism non-obligatory. There is, in the end, no non-arbitrary account of the proper goals of medicine. Happily, this is not a problem for pluralistic professional medical ethics; it requires only an account of the competencies of medicine and their evidence-based limits in order to deploy the principle of beneficence or, in Brody's terms, an appeal to clinical consequences.

A pluralistic professional medical ethics does, it appears, depart from Brody's pluralistic moral theory's appeal to cost-effectiveness and justice. Professional medical ethics prohibits physicians from practicing below an accepted, evidence-based standard of care. A rationing decision by a health care organization or payer, the effect of which would be to deny access to resources required to meet such an accepted standard of care, would have no intellectual or moral authority for physicians. Physicians should, therefore, refuse to cooperate with or implement such decisions. Resource management decisions, based on reliable evidence, to select a less resource-intensive standard of care, however, would be consistent with professional integrity, as Percival already argued (Percival, 1803). Thus, evidence-based drug selection, say for the drug of first choice for the management of newly diagnosed hypertension, undertaken in order to control formulary costs more effectively than allowing physicians unbridled freedom to select medications, would be obligatory for physicians to accept. It is when cost control becomes separated from evidence-based standards—e.g., an insurance company economically incentivizing its beneficiaries to select providers based on cost without any reference to evidence of quality—that physicians should, as a matter of professional integrity, actively and effectively resist.

Professional medical ethics thus adds a dimension to the ethical analysis of resource management decisions that current discussions of rationing in the bioethics literature fail to include, namely, the effect on professional integrity of those decisions. This failure, from the perspective of professional medical ethics, disables the bioethics literature on rationing. It should be noted that professional medical ethics does not require that there remain a profession of medicine. Professional medical ethics, however, does require that any ethics of resource management that fails to take account of professional integrity should, by policy makers and the public, be judged as inadequate and potentially predatory on a profession of medicine. Society might want to give up a profession of medicine, but it should at least know that that is what it is doing in making specific health-policy decisions about resource management that involve rationing below an evidence-based standard of care. The much-invoked "decent minimum" is defined in professional medical ethics precisely by such evidence-based standards.

Professional medical ethics also adds to Brody's pluralistic moral theory a deep concern for organizational culture and its impact on professional integrity and on clinical judgment, decision making, and behavior of physicians. Organizational culture includes the policies and practices of an organization, its expressed and actual values, and what its leadership does not tolerate and what its leadership does tolerate that it should not. Here we should take our instruction from Percival, who should be credited with being the first to understand the synergy between professional medical ethics and organizational culture. He understood clearly that introducing intellectual integrity in the form of adherence to the standards of Baconian evidence-based medicine and surgery could be accomplished in the hospital only setting by changing its culture from one in which there had previously been no accountability to one in which there would at least begin to be accountability for the quality of the processes and outcomes of medical care. It seems plain to me, as a longtime student of Percival, that he understood a key feature of professional integrity: without disciplined accountability, if only the voluntary cooperation of physicians with each other to improve quality, professional integrity will soon wither under the blandishments of economic and other conflicts of interest and patients will pay with their health and lives. Percival hoped that the intellectual discipline of monthly case conferences would transfer into private practice. However, he noted that the only form of accountability in that setting, in which physicians worked alone, was the court of individual conscience. This is a slender reed on which to build professional integrity (Sharpe, 2000). The contemporary practice of medicine, in which the vast majority of physicians practice in groups, allows for systems of accountability and therefore quality enhancement to be developed. In other words, we can create reinforcing organizational cultures of professional integrity in both the outpatient and inpatient settings. Professional medical ethics requires that we do so.

Brody develops his pluralistic moral theory, in part, to provide reliable guidance to physicians in the absence of public policy regarding resource allocation. His shrewd distinction between the requirements of cost-effectiveness in the public and in the private settings is an excellent example. There is a general lesson to be drawn from this, at least for professional medical ethics in the United States. In the United States the citizenry and the electorate, and the people whom the electorate chooses to run our representative government, have been ducking hard issues of responsible resource management for many decades. Since the best guide to the future is the past, we should not expect this political habit to change in the foreseeable future. Like Brody's pluralistic moral theory, a pluralistic medical ethics, with its emphasis on professional integrity as an unyielding moral appeal, provides a reliable moral compass for physicians, health care organizations, and payers in the midst of unsettled and even non-existent health policy. In this, I hope, a pluralistic medical ethics shares one of the great strengths of Brody's approach.

IV. CONCLUSION

In this chapter, I have sketched a pluralistic professional medical ethics. One of my goals in doing so has been to rescue medical ethics from bioethics, as understood by Clouser and the many who agree with his conceptualization of the field. As nothing special, medical ethics is not professional ethics. There is nothing special about the physician-patient relationship, i.e., a social role created by physicians committed to being professionals that, in turn, creates the social role of the patient, providing for the sick shelter from the storm of a crisis of intellectual and moral trust in health care "providers." Indeed, that so much of the bioethics and medical ethics literature shifts effortlessly, and without attention to the implications of doing so, from 'physician' to 'health care professional' to 'health care provider' to 'health care worker' speaks volumes about the de-professionalization of medical ethics by bioethics. The result is to return the patient to the fragile status of the sick, presumably armed with their autonomy to protect themselves from predatory health care providers. It is folly to think that this will be adequate; the sick had this tool already at their disposal for centuries before Gregory and Percival and it did nothing to prevent the crisis of intellectual and moral trust to which Gregory and Percival so forcefully and effectively responded.

Gregory and Percival knew all too well what such a world looks like—a practice of medicine and research dominated by, because it is driven by, the self-interests of physicians and surgeons with a resultant crisis of intellectual and moral trust. Gregory and Percival took themselves to be reforming medicine *into* a profession. By standing in the past with them, we gain a powerful critical perspective on bioethics as predatory on professional medical ethics in the hands of some of bioethics' practitioners. I do not place Brody in this group, but I do believe that we need to build on his pluralistic moral theory to articulate a pluralistic professional medical ethics.

BIBLIOGRAPHY

Baker, R.B (1993a). 'Deciphering percival's code,' in R.B. Baker, D. Porter, & R. Porter (eds.), *The Codification of Medical Morality: Historical and Cultural Studies of the Formalization of Western Medical Morality in the Eighteenth and Nineteenth Centuries: Volume One: Medical Ethics and Etiquette in the Eighteenth Century* (pp. 179–212). Dordrecht: Kluwer.

Baker, R.B. (1993b). 'Medical propriety and impropriety in the English-speaking world prior to the formalization of medical ethics: introduction,' in R.B. Baker, D. Porter, & R. Porter (eds.), *The Codification of Medical Morality: Historical and Cultural Studies of the Formalization of Western Medical Morality in the Eighteenth and Nineteenth Centuries: Volume One: Medical Ethics and Etiquette in the Eighteenth Century* (pp. 15–17). Dordrecht, The Netherlands: Kluwer Academic Publishers.

Baker, R.B., & McCullough, L.B. (2008). 'Introduction: what is the history of medical ethics?' in R.B. Baker & L.B. McCullough (eds.), *A History of Medical Ethics*. New York: Cambridge University Press, in press.

Beauchamp. T.L. (2004). 'Does ethical theory have a future in bioethics?' *Journal of Law, Medicine & Ethics,* 32, 200–217.

Brody, B.A. (1988). *Life and Death Decision Making*. New York: Oxford University Press.

Clouser, K.D. (1978). 'Bioethics,' in W.T. Reich (ed.), *Encyclopedia of Bioethics* (pp. 115–127). New York: Macmillan.

Gregory, J. (1998 [1772]). 'Observations on the duties and offices of a physician, and on the method of prosecuting enquiries in philosophy,' in L.B. McCullough (ed.), *John Gregory's Writings on Medical Ethics and Philosophy of Medicine* (pp. 93–159). Dordrecht: Kluwer Academic Publishers.

Gregory, J. (1998 [1770]). 'Lectures on the duties and qualifications of a physician,' in L.B. McCullough (ed.), *John Gregory's Writings on Medical Ethics and Philosophy of Medicine* (pp. 161–245). Dordrecht: Kluwer Academic Publishers.

Grinnell, F., Bishop J.P., & McCullough, L.B. (2002). 'Bioethical pluralism and complementarity,' *Perspectives in Biology and Medicine,* 45, 338–349.

Haakonssen, L. (1997). *Medicine and Morals in the Enlightenment: John Gregory, Thomas Percival and Benjamin Rush*. Amsterdam: Editions Rodopi.

McCormick, R.A. (1974). 'To save or let die: the dilemma of modern medicine,' *Journal of the American Medical Association,* 229, 172–176.

McCullough, L.B. (1998). *John Gregory and the Invention of Professional Medical Ethics and the Profession of Medicine*. Dordrecht: Kluwer Academic Publishers.

McCullough, L.B., & Chervenak, F.A. (1994). *Ethics in Obstetrics and Gynecology*. New York: Oxford University Press.

Pellegrino, E.D., & Thomasma, D.C. (1988). *For the Patient's Good: The Restoration of Beneficence in Health Care*. New York: Oxford University Press.

Pellegrino, E.D., & Thomasma, D.C. (1993). *The Virtues in Medical Practice*. New York: Oxford University Press.

Percival, T. (1803). *Medical Ethics, or a Code of Institutes and Precepts, Adapted to the Professional Conduct of Physicians and Surgeons*. Printed by J. Russell, for J. Johnson, St. Paul's Church Yard, & R. Bickerstaff, Strand.

Porter, D., & Porter, R. (1989). *Patient's Progress: Doctors and Doctoring in Eighteenth-Century England*. Stanford: Stanford University Press.

Sharpe, V.A. (2000). 'Behind closed doors: accountability and responsibility in health care,' *Journal of Medicine and Philosophy*, 25, 28–47.

Veatch, R.M. (1989). *Death, Dying, and the Biological Revolution: Our Last Quest for Responsibility*. New Haven: Yale University Press,

Veatch, R.M. (2003). *The Basics of Bioethics*, Second edition. Upper Saddle River: Prentice Hall.

CHAPTER 3

ETHICS AND DEEP MORAL AMBIGUITY

KEVIN WM. WILDES, S.J.
President, Loyola University New Orleans, New Orleans, LA, U.S.A.

For many people the very ideas of "ethics" and "moral ambiguity" are contradictory. Many people, religious and secular, liberal and conservative, have great clarity on ethical questions and the answers to them. And they think that those who disagree with them are wrong. The problem is, however, that many of the very clear answers to ethical problems contradict one another. So, while people may think that moral claims are clearly known, the clear pronouncements often lead to a din of confusion and counter claims.

Bioethics is often a field of inquiry that takes place in the context of public controversy. One element of these moral controversies is that people often think there is a clear answer to the problem at hand and they do not understand why other people do not see the problem in the same way. The controversies in bioethics illustrate the challenges of addressing moral issues in a morally pluralistic society. The issues are often not as simple as people might hope and there is more moral ambiguity than we might want to admit.

Bioethical controversies can be found in a variety of settings, including the research laboratory and the clinic. They are situated in the laboratory in questions of research ethics as we try to expand medical knowledge and technology (e.g., stem cell research). Bioethical controversies are found in the clinical choices surrounding patient care (e.g., how we treat patients at the end of life). With advances in scientific knowledge and medical technology, we now have choices and responsibilities in matters that for earlier generations were matters of chance. These choices also fuel public controversies as societies try to determine policy, regulation, and law for medicine and health care (e.g., how health care research and delivery are funded or how bioethical choices are regulated).

Though medical choices are often some of the most intimate choices in the lives of men and women, these choices often have a public dimension about them and become the subjects of public policy disputes. They are not simply choices that we make for ourselves but often involve others: the others we care for, the others who help deliver medical care, the others who regulate, oversee, and pay for medical care. Because of this 'public' dimension to medical choices, bioethical problems are particularly troublesome for secular societies which often are more diverse

M.J. Cherry and A.S. Iltis (eds.), Pluralistic Casuistry, 37–48.

culturally than religious societies. The diversity of cultures in secular societies often means a diversity of moral practices. One might think of the contrast between the diversity found in the United States in comparison to an Islamic state, like Iran. One might imagine how bioethical controversies would be addressed within a nation that was self consciously religious (e.g., a Christian state, an Islamic state, a Hindu state). Implicit in many of the controversies in bioethics are other challenging issues about how we think about ethics and the role of religion in resolving them.

In this essay, moral pluralism signifies more than simply different opinions or judgments about a course of action. The term moral pluralism can be used to communicate fundamental differences about how the good life is understood and envisioned. Moral pluralism can also be reflected in the different methods used to think about the issues of bioethics and to justify decisions that are made (Wildes, 2000). Different methods reflect different assumptions and views of the moral world and how the moral world works. These different methods often assume differing views about health and sickness, as well as how each should be addressed. Because of these differences in opinion and methods of ethical analysis, secular societies will face the additional question of when it is morally appropriate to regulate bioethical choices in medicine and health care.

In spite of all these differences, one often finds in bioethics a quest for simplicity and certainty. Bioethical controversies are often the subject matter for the news headlines, television news broadcasts, along with TV and radio talk shows. The choices that are made seem to have dramatic, immediate impacts on the lives of human beings. However, while we are often caught up in the most current controversy, such as stem cell research or end of life decision making, the disagreements in the field reach beyond particular topics or controversies. The issues reach to very fundamental debates about how to frame and address such controversies. Indeed, the issues of bioethics often raise fundamental questions about how we think and talk about the moral world (Wildes, 2000). Yet, people have a tendency to treat complex issues as if they were simple ones. People want to know "the" answer. People are often very confident of their views about complex issues, reducing them to very black and white concepts and choices. This simplicity leads to, or arises from, the desire for "sound bite" answers to the issues of bioethics. We want people to have something important to say in a sentence or two that we can hear on the evening news, or *Crossfire* , or *Larry King Live*. We talk about these topics casually over the water cooler or at Starbucks as we go to and from work.

The quest for simple, sound bite answers is not reserved to the media or our casual conversation partners alone. One can also find the same quest for clarity and simplicity in our desire to legislate answers to bioethical issues. But legislation is a cumbersome tool and it is difficult to capture the complexities of the moral world within the legal framework. Simplicity often leads to ideology rather than moral discourse. The desire for simplicity and clarity often leads to the reduction of complex issues to very simple forms. People often have unarticulated expectations about moral knowledge. They assume that moral knowledge is a type of knowledge that is clear, unambiguous, and can be legislated.

It is important to understand the complexity of the field if one is to appreciate the importance of the scholarship and insights of Baruch A. Brody. In many ways, Brody's work provides a sharp challenge to the desire for simplicity and sound bite answers. Brody has addressed many of the issues in the field, in areas such as research ethics, clinical ethics, and the allocation of resources. Brody not only addresses important issues, but he also examines the nature of morality as such and the fundamental philosophical questions about the moral world (see Brody, 1988; 1998; 2003).

Brody's reflections on the ontological and epistemological questions of the moral world lead him to view the moral world as pluralistic and ambiguous. Throughout his work, Brody argues that there may be times when the pluralism is irresolvable because there is "deep moral ambiguity" in the ontology of the moral world itself. It is Brody's work about morality itself and bioethics that is the focus of this essay. He asks and explores fundamental philosophical questions that are important for understanding the field and the moral problems it tries to address. In examining the different moral appeals, such as consequentialism, deontology, or virtue theory, Brody argues that the moral world is made up of different elements. How we arrange these elements will shape the way we see the moral world. His work helps to explore the complexities of moral pluralism.

While many thinkers in bioethics have addressed some of the particular problems in the field (e.g., end of life care) there are only a few who have also thought about the field of bioethics itself and the ontology of the moral world. Brody has done both. The work of Baruch Brody provides important, helpful insights into the apparent fragmentation of the moral world and the reality of pluralism. Brody's insights are rooted in views about the ontological and epistemological aspects of the moral world. In the midst of desires for clarity and simplicity, the work of Baruch Brody provides an important way of viewing the field. Rather than approach the diversity of opinions with yet another opinion, Brody begins with the assumption of diversity and he sets out to explore and understand it.

I. FRAMING MORAL PLURALISM

In the contemporary world, people often speak very casually about cultural diversity. In fact, cultural diversity is often celebrated and set as a goal for organizations to achieve. However, in celebrating cultural diversity we often overlook the relationship of cultures to morality. We fail to recognize that morality is embedded within cultures and is part of a way of life. Elsewhere I have stipulated a distinction between morality as a way of life, a first level of discourse, and ethics as a systematic reflection on morality that is a second level of discourse.[1] Morality compasses the moral practices we assume, while ethics seeks to examine those practices and moral systems. Moral practices are often embedded in cultures and a way of life. In day to day life, people assume the morality in their own particular culture to be true. In today's secular society, people often fail to understand that the desire for a multicultural society can lead to a society that is morally pluralistic. Such a diverse society will need to wrestle with questions of moral cooperation between people and communities that have different views, and such a society will have to address the appropriate *moral* role of government force and policy in a multi-cultural society.

Bioethics in secular societies must confront both the questions of cooperation and the questions about the role of government. Bioethics often finds itself beset by some of the problems and challenges of moral pluralism because health care is a moral enterprise, a co-operative venture, and has a public face. It is often forgotten that health care services are often viewed within the context of a moral framework. We understand and make our health care decisions within a moral framework. There are clearly very different views of what are morally appropriate or inappropriate courses of action for many of the questions in bioethics (e.g., treatment of patients, end of life care, abortion, assisted suicide). Moreover, the delivery of health care services often requires people, with different moral views, to work together. The problems of moral pluralism need to be addressed in secular, diverse societies if citizens in those societies are to live and work together. These issues need to be examined so that such societies can morally justify the regulation of different moral communities within such societies. And, people need to think about the questions of cooperating with others who may hold different moral views of the world.

Moral pluralism is not merely about disagreements over what one should or should not do. The pluralism is also about how we think about resolving these issues. Brody helps to address these differences and gives us a way to think about them and why they exist. There are a number of ways a "secular" society can be envisioned. For the sake of this essay, I am simply assuming that a secular society is a morally neutral society (Engelhardt, 1991, see chapter 2). There are many dimensions of life in a secular society, which are viewed in different ways by different moral communities.

There are several ways that moral pluralism has been addressed in bioethics. One way, in the face of divergent opinions and pluralism, is to say that there is only one right opinion and that all others are simply wrong and claims of pluralism are dismissed as misconceived. One finds this stance in many different moral controversies. This view of pluralism often drives the different groups in the abortion "debate." Differences of opinion sometimes reflect different understandings of facts and sometimes, in Brody's language, they represent deep moral ambiguity.

A second way to approach pluralism is a type of "soft" relativism which acknowledges the pluralism and assumes it is "ok." Here, there can be a slide from political correctness to moral relativism by way of cultural relativism. The moral relativism that emerges is based on notions of cultural diversity and the acceptance of such diversity. This view, however, also invites the use of coercive power to resolve moral differences. If differing views are "ok" and to be tolerated and respected, but we still need to work together, we must find some way to resolve the differences lest one solution be simply asserted and imposed over others.

A third approach to pluralism is to acknowledge pluralism but explain it away by arguing that there is fundamental moral agreement, but differences emerge as we specify principles to moral problems. Beauchamp and Childress are perhaps the best examples of this approach (see Beauchamp and Childress, 2001). They believe there is common ground in the four principles, but that pluralism emerges when we attempt to specify the principles. This approach to pluralism is to reduce it to sameness. The logic of such a move is to begin by acknowledging the differences but insist that the differences are more about appearance than substance. Brody sees this type of reductionism in the influential work of Beauchamp and Childress. In their book, *The Principles of Biomedical Ethics*, particularly in the earlier editions of the book, they argue that different theories of ethics will lead to "the same rules" (Beauchamp and Childress, 1979; 1983; and 1989). Later editions of the book explicitly acknowledge other ways, beyond the principles, of addressing moral questions and Beauchamp and Childress argue that these other approaches can be understood in terms of the principles.

Other thinkers in bioethics, such as H.T. Engelhardt, Jr., take a fourth approach when they argue that there is only pluralism and that morality is rooted in the life of a community (Engelhardt, 1996). He argues that it is only the practice of permission that binds "moral strangers" together. In Engelhardt's analysis, we exist either as moral friends (in the same moral community, sharing the same moral vision) or we are moral strangers. He argues for a secular morality that is based on a respect for persons and a principle of permission.

Brody offers an alternative to these different approaches to the problem of moral pluralism. He views moral pluralism as a real challenge that cannot be ignored or "explained away" as it often is in bioethics. Yet, he does not see the problem as intractable as does Engelhardt. In his work, Brody offers a *positive* account of moral pluralism. While he argues that many problems can be addressed, there are some problems that cannot be adequately addressed because of deep moral ambiguity. He does not offer a monistic account that tries to incorporate every different method and view. Nor does he argue that pluralism is the default account because there is not a monistic account.

Brody's pluralistic theory is a complex model. He does not attempt to water down moral pluralism and moral differences into "sameness". Rather, he argues for a model of pluralism rooted in the wide variety of moral appeals (Brody, 1988, p. 9). He argues that the pluralism of moral appeals is rooted in the ontology of the moral world. Furthermore, pluralism has an epistemic dimension that is tied to the importance of "judgment" and that men and women, of good will and who share common moral assumptions, will often reach different judgments about what should or should not be done. At the same time, however, he acknowledges deep moral differences (moral ambiguity), which he anchors in the nature of the moral world.

Brody's view about the nature of moral thought—and bioethics—is important as one thinks about the field and the fundamental assumptions one makes about bioethics. Brody approaches moral pluralism as part of the reality of the moral world and not simply as an obstacle to overcome. He also is intellectually comfortable in talking about deep moral ambiguity and recognizing that we will almost always have important, perhaps irresolvable differences of opinions.

II. THE PROBLEM: THE NEED FOR A MORAL THEORY

Brody argues that bioethics needs a theory of moral decision making to address the different problems that confront the field. The claim may seem obvious. People often assume that they have a theory when they confront moral issues. But, the array of moral controversies in bioethics indicates that people often deploy very different theories. Moreover, "professionals" in bioethics have developed a variety of theories for the field (Wildes, 2000). Brody recognizes the different theories and argues that the field needs a theory that accepts and understands pluralism and addresses it.

Brody's view that bioethics needs a theory seems to fly in the face of the received wisdom in the field which holds that the field does not need moral theory (Beauchamp and Childress, 1979-2001; Jonsen and Toulmin, 1988). This claim is often made in light of failure of philosophical theories, such as utilitarianism, natural law, or deontology, satisfactorily to address bioethical issues. However, Brody argues that it is precisely because of the failure of these monistic moral theories that the field needs a theory. A moral theory is needed, Brody argues, that can address the different types of moral appeals and justifications which are made in bioethics. The theory ought to help us address the questions such as where the burden of decision making falls and where are there limits to decisional authority. He argues, however, that the type of moral theory needed is not a unified theory but a theory that can account for pluralism. He argues that too often the search for a theory in bioethics has been a search for a unified theory, such as utilitarianism or deontology, and when the search for a monistic theory fails, the search for theory is abandoned.

Brody also argues that as a result of the failure to find a theory for bioethics, much of the effort in bioethics has been to address the question of who has the moral authority to make decisions (Brody, 1988, p. 6). However, Brody thinks that the field needs to contribute more than answers to the question of who should decide, even if that question can be settled. He also argues that even when it is clear who should decide, we still may want to impose limits on the decisions being made (e.g., the decisions parents may make on behalf of their children and the decisions adolescent children may make on their behalves) (Brody, 1988, p. 7). To establish such limits, we will need a substantive moral theory, not just a procedural one. In some cases we will still need to confront the question of what to do. Brody argues for the need for a theory, against the view of bioethicists such as Beauchamp and Childress or Jonsen and Toulmin. He also argues that the field needs more than an account of who has moral authority to decide a controversy. Here, he takes exception to bioethicists like Engelhardt.

The failure to find one theory for bioethics does not mean that we should abandon the search for theory. Instead, Brody thinks we need to look for a different type of theory. Rather than search for a monistic theory we should examine what a pluralistic theory would be like. He focuses not on the pluralism in moral judgment, but on the pluralism in the basic way moral theories are constructed. He develops an ontological account for the pluralism in moral theory. And he believes we can use a form of casuistry and theory construction to find ways to address many cases and problems.

III. ARGUMENTS FOR PLURALISM

Brody offers two arguments in support of his claims about moral pluralism. He characterizes one argument as a "negative" argument, centering on the failure of monistic theories to deal adequately with counter examples and exceptions. No monistic account is able to give a full account of the moral world. Beyond his negative argument, however, Brody offers a positive account of moral ambiguity.

In his negative argument, Brody argues a number of major points:

1) That monistic theories have plausibility because they appeal to some property (e.g., consequences) that is important to how we assess and judge moral actions.
2) But, for each monistic theory there are a series of actions where the rightness or wrongness of the choice of action cannot be grounded in the property singled out by the monistic theory (for example, the killing of innocents cannot be justified even when doing so would result in the good consequences of saving many lives).
3) The different monistic theories have tried to develop strategies to address these problem and Brody argues that the strategies either fail to work or they become too *ad hoc* to be consistent.
4) Pluralism is plausible because the property identified in each theory does ground the rightness of some action but each property (theory) is limited and faces challenges of counter examples (Brody, 2003, chapter one). The different monistic theories themselves offer an argument for moral pluralism as each theory identifies some plausible element of the moral world such as consequences or duties or virtue.

In light of the negative argument, Brody moves to a positive account of moral pluralism. In developing his positive argument, Brody focuses on the experience of moral ambiguity. Moral ambiguity can be manifested in several ways. Ambiguity is sometimes found in disagreements between individuals. This is often what happens in bioethics. It is sometimes in the uncertainty experienced by someone in deciding what is the right thing to do. However, the moral ambiguity that concerns Brody is neither an ambiguity about facts nor regarding the moral relevance of the facts. He is concerned with the deep moral ambiguity that rests on the ontology of the moral world.

In his positive account, Brody postulates deep moral ambiguity—both intrapersonal and interpersonal. There are many moral disagreements and often these can be accounted for by disagreements over facts (e.g., is the patient persistently vegetative), as well as disagreements about the moral relevance of the facts (e.g., when it is permissible to stop life-sustaining artificial nutrition and hydration). These type of problems hold out the possibility of resolution. But in situations of deep moral ambiguity there is no doubt about the facts or their moral relevance. Instead, there is still ambiguity about how to act. He explains moral ambiguity in light of the variety of moral appeals. One can think of the moral world as a room with different pieces of furniture which can be arranged in different ways. This leads to rooms that have similarities, but may look very different depending on which furnishings are accentuated and which are not.

But the ontological account of moral ambiguity does not give a complete account of moral ambiguity. In conjunction with the ontological account, Brody develops an epistemological account which he thinks will justify the claims of his account. His epistemological account is a form of intuitionism. In his view, intuitions are a natural cognitive capacity to form moral judgments. He ties this intuitionism about judgment to the work on theory construction in science. He proposes a method where one attempts to bring together specific moral judgments into a larger set of principles and theory. Brody deploys a moral epistemology that relies on the central role of judgment in moral knowledge. Following in the line of Aristotle, Brody argues that the nature of moral knowledge is such that it will not lead to clear and exact answers and that it is a mistake to treat moral methods as if they should lead to clear and exact answers.

IV. JUDGEMENT, AMBIGUITY, AND CASUISTRY

Brody not only examines the ontological roots of pluralism in bioethics, he also argues for an epistemological dimension to moral pluralism. He brings together the different aspects of the ontological explanation with the faculty of judgment. Moral pluralism is also explained by epistemology and an understanding of moral judgment. In his discussion of moral judgment, Brody stands in a line of thought developed by Aristotle. In his explanation of knowledge, Aristotle distinguishes between knowledge that is theoretical and knowledge that is practical. Theoretical knowledge is generalizable and independent of particular facts. Moral knowledge is practical knowledge and areas of practical knowledge will rarely have a single correct or right answer, as areas of theoretical knowledge will. Practical knowledge must take account of the actual circumstances. This does not mean, however, that any answer will do. Brody also understands the nature of moral judgment insofar as judgments are not exact deductions. If someone emphasizes one feature (e.g., consequences) over another (e.g., virtues), we are likely to get different judgments, even though sometimes we will reach the same conclusions (as Beauchamp and Childress suggest). Brody thinks that part of reaching similar conclusions will depend on getting the facts of the matter correct. He also thinks that a casuistical based reasoning can help bring judgments together.

It is at this juncture that Brody may have been too optimistic. There are different ways that one can develop a model of casuistry. One well known way is to have paradigm cases and work to match the current case or problem with one of the paradigm cases. This link would then help people determine what they should do. Brody does not propose a model of casuistry that uses paradigm cases. Rather, he links case-based reasoning with the faculty of intuition. Either approach to casuistry, in a pluralistic world, can have serious problems. Sooner or later, any casuistry will depend on the description of the case at hand. In addressing moral questions, descriptions are not neutral. The language that is used, and the terms deployed, convey some level of moral judgment. Our characterizations of cases often carry moral evaluations within the descriptions themselves. How one describes an act of physician-assisted suicide will often carry with it one's moral evaluation. Is it "killing," "murder," or an "act of mercy"? The descriptions used will often carry with them moral views.

Brody's use of casuistry is helpful for understanding his pluralistic theory. In developing different analysis and explanation of each case, Brody realizes that the differences in the case analysis sometimes rest on facts and interpretation, and sometimes on the different appeals that can be used. Often in complex cases, these different appeals do not fit neatly together. These cases and appeals lead to his views about deep moral ambiguity. However one assesses Brody's use of casuistry, his fundamental reflections on the nature of moral knowledge, as practical knowledge, are very helpful in understanding moral judgment and moral pluralism.

Moral knowledge is often 'incomplete' because it depends on the circumstances of a situation. Nevertheless, we must make decisions in light of this incompleteness. And there are different moral appeals, each of which captures true moral judgments or aspects of true moral judgments, which are not reducible to one another (consequences, duties, virtues). This will mean that our moral decision making (our knowing what to do) will be based on a model of judgment (not a weighing metaphor). It also will mean that there will be moments when people of good will, understanding all the relevant facts of a case, will reach different judgments about what ought to be done (Brody, 1988, pp. 77–79). The appeal to judgment in moral decision making and epistemology is another avenue for moral pluralism. Practical judgment involves decision making under uncertainty. So moral agents, sharing the same information and the same moral commitments, may well reach different conclusions about what should or should not be done.

V. PLURALISTIC THEORY

A problem with the language of moral pluralism, in bioethics, is that the term "pluralism" is used in a variety of ways. Brody brings disciplined reflection to the term and the ways it might be used. He helpfully distinguishes between moral disagreements and uncertainties that revolve around the facts of a case and deep moral ambiguity. He argues that disagreements and uncertainties (about the facts, or their relevance, for example) are quite common. Many disagreements regard the interpretation of the facts or the relevance of facts for a particular decision. Such cases can, in principle, be explained, understood, and resolved (Brody, 1988, p. 37). But, Brody also examines the deep, philosophical questions of ontology and epistemology that are needed for the field to operate. His contributions to specific controversies are important to the field. But, equally important are his explorations of the ontological and epistemological questions of moral pluralism.

To understand Brody's views on ontology and epistemology we can use an analogy for the moral world. Brody argues that the moral world is like a room with different pieces of furniture. Just as a room has chairs, tables, and sofas, the moral world has consequences, duties, and virtues. In a monistic system the room only has one type of furnishing. It is a room of chairs, for example, since the monistic approach to moral theory focuses on one element (e.g., consequences or duties). In a pluralistic system there are many different pieces of furniture in the room. But, how we see the room will depend on how we enter it. Do we come in from the side of the room where the chair, or the sofa, or the table located? In Brody's method, we first know the room through intuition. But intuition alone is not sufficient for Brody.[2] He argues that we need to test our intuitions by theory construction. How does our intuition about what we should do in the particular case before us fit with our over all views of the moral world? Theory construction gives us a way to develop our intuitions and test our judgments.

It is clear that Brody recognizes the difficult nature of the moral world. The pluralistic theory he develops is rooted in the ontology of the moral world. The moral world is a world that is messy as it has many different aspects, which cannot be reduced or eliminated (as often happens in bioethics). His ontological explanation helps to account for why there are so many different types of moral theory and methodologies in bioethics. Different theories and different methods focus on an aspect, or some combination of aspects, of the moral world. Some methods focus on outcomes and consequences, while other methods focus on duties and obligations, and still others focus on character and virtue. Each is a part of the furniture of the moral world.

In his positive explanation of the ontology of the moral world, Brody argues a claim about the supervenience of moral qualities. That is, that the rightness or wrongness of an action is dependent on other properties of an action (Brody, 1988, p. 32). He argues further that there are several such properties that are independent of each other. If one develops a theory of action, one can see that actions have different properties and these properties are the basis of moral qualities. Often, in modern moral philosophy, different monistic moral theories claim that there is only one property upon which the rightness or wrongness of an action depends. In Brody's pluralistic account, particular choices may have more than one property (consequences, duty, virtue). Different properties will yield different accounts of the choice.

Brody's ontology of the moral world raises important questions as to whether or not many monistic theories are, in fact, monistic (Brody, 1988, p. 33). Often, he argues, when one examines a monistic theory, especially as it addresses difficult cases or "exceptions", there are appeals to other moral properties. There is a temptation to build in other moral properties in monistic theories. Based on his examination of monistic theories and different moral appeals, Brody argues that it is a fundamental mistake for moral philosophy to assume that any one aspect captures the whole of the moral reality. He understands that the moral world is complex and cannot be reduced to a particular aspect. If one is to have completeness in addressing moral issues, we need to take account of all these aspects. From Brody's argument, one might say that many of the methods utilized in bioethics are too simple and narrow. They are incomplete.

Brody argues that there are fundamental philosophical disagreements about the nature of the moral world. In his view, a pluralistic account gives a more complete account of that pluralism and how one ought to address "exceptions." One of the challenges for any foundational account will be how to deal with exceptions. Too often, foundational or monistic accounts of moral theory deal with exceptions on an *ad hoc* basis rather than a principled one.

VI. CONCLUSIONS

Bioethics is a complex field. It covers patient care, medical research, the allocation of resources, and questions of government policy. It is a field that is further complicated in a secular, morally pluralistic society. In such societies, the field needs to navigate questions of how people, with different moral views, can cooperate in health care.

Baruch Brody's work is a rich tapestry that covers the complex field of bioethics. In reflecting on his work, one can focus on a number of different aspects. Some people will certainly reflect on Brody's work in clinical ethics or research ethics. Nevertheless, one of his great contributions to the field of bioethics has been excellent philosophy. While he has reflected and puzzled on many particular topics and questions, from research ethics to end of life care, Brody has always thought about the philosophical basis of the field. He is one of the few philosophers in bioethics who does not shy away from philosophical questions about ontology and epistemology. He has explored such questions as he has sought to develop a method for bioethics. His interest in these difficult areas stems from a deep desire to help the field be more precise, and more helpful, in addressing these very difficult issues and questions.

His development of a pluralistic theory seeks to address the reality of moral pluralism and the challenges it presents for the field of bioethics. He does not simply ignore or gloss over the differences. He has sought ways to address moral pluralism and diversity while, at the same time, acknowledging moments of "deep moral ambiguity," and that we may not be able to pull the moral word together into a nice package or theory.

While I am sure that other writers will reflect on Brody's model of casuistry, or his work on research ethics, or some other area of practical concern, it would be a mistake to overlook the questions he raises and positions he develops on moral ontology and epistemology. As a philosopher, these may well be some of his most enduring contributions to the field.

NOTES

[1] In *Moral Acquaintances*, I stipulated a distinction between morality and ethics. Morality is the level of first order discourse in which we live and make moral decisions. Morality is the moral world that we simply assume and take for granted in our everyday lives. Ethics is the level of second order discourse which steps back from our everyday assumptions and practices to examine the basic assumptions about

the moral world, such as why we regard consequences or duty as the definitive mark of morality. This second order discourse is not only the realm of philosophers or ethicists. Conflicts, at the level of morality, often lead people to this second level discussion (Wildes, p. 3).

[2] Brody's intuitionism is unlike the intuitionism of someone like W.D. Ross, who would argue that we know the right prime facie by intuition (Ross, 1930).

BIBLIOGRAPHY

Beauchamp, T. & Childress, J. (2001, 1994, 1989, 1983, 1979). *The Principles of Biomedical Ethics*. New York: Oxford University Press.

Brody, B. (1998). *The Ethics of Biomedical Research*. New York: Oxford University Press.

Brody, B. (1988). *Life and Death Decision Making*. New York: Oxford University Press.

Brody, B. (2003). *Taking Issue: Pluralism and Casuistry in Bioethics*. Washington, DC: Georgetown University Press.

Engelhardt, H.T. (1991). *Bioethics and Secular Humanism: The Search for a Common Morality*. London: SCM Press.

Engelhardt, H. T. (1996). *The Foundations of Bioethics*. New York: Oxford University Press.

Jonsen, A., & Toulmin, S. (1988). *The Abuse of Casuistry: A History of Moral Reasoning*. Los Angeles: University of California Press.

Ross, W. D. (1930). *The Right and the Good*. Oxford: Oxford University Press.

Wildes, K. (2000). *Moral Acquaintances: Methodology in Bioethics*. Notre Dame: University of Notre Dame Press.

CHAPTER 4

MORAL JUDGMENT AND THE IDEAL INTUITOR: DEALING WITH MORAL CONFUSION AND MORAL DISAGREEMENT

JANET MALEK

Department of Medical Humanities, East Carolina University, Greenville, NC, U.S.A.

In *Life and Death Decision Making*, Baruch Brody lays out his model of conflicting appeals, a moral framework for analyzing and resolving difficult ethical questions surrounding medical treatment (Brody, 1988).[1] In this chapter, I describe the model of conflicting appeals, looking closely at its keystone—moral judgment. I give reasons to question whether judgment as Brody describes it can play the role that it must play in order to make the model do what he wants it to do. I then flesh out a plausible understanding of moral judgment that plays this role successfully and that is consistent with Brody's views. In conclusion, I briefly evaluate this interpretation.

I. THE MODEL OF CONFLICTING APPEALS

The model of conflicting appeals is a decision making process that consists of three stages. It calls for a moral agent, when faced with an ethically challenging choice, to first ascertain which of several moral claims, or "appeals", are relevant to the case at hand. The agent must then assess the significance of each relevant appeal in light of the particulars of the case. Finally, the agent must make a judgment, based upon the relevant appeals and their respective significance, about the right thing to do in that case. This approach, Brody claims, is "a picture of the epistemology of morality, of how we can come to know what is right and wrong" (1988, p. 79). That is, agents can use this method to identify the morally correct option when faced with a difficult choice.

Brody identifies five ethical appeals: the appeal to the consequences of our actions, the appeal to rights, the appeal to respect for persons, the appeal to the virtues, and the appeal to cost-effectiveness and justice (pp. 17–48). Each appeal may not be applicable to every case; the decision maker must figure out which appeals are relevant to the choice he faces. And there may be ethical considerations

M.J. Cherry and A.S. Iltis (eds.), Pluralistic Casuistry, 49–64.

that are material to a particular case that are not included in Brody's list of appeals. As Brody notes, the list of appeals he has proposed is open to revision and can be easily modified without compromising the validity of the model. However, given that Brody's five appeals reflect most of the major existing lines of ethical thought, there is reason to believe that his list is a good starting point.

Once the relevant ethical appeals have been identified, the decision maker must evaluate the significance of each appeal in light of the characteristics of the case at hand. Brody also gives guidance for this step, describing conditions under which each appeal has more or less significance (pp. 79–94). For example, the significance of the appeal to consequences varies according to the consequences' likelihood of occurring, the amount of value or disvalue they generate, and the number of people they impact. A likely consequence that could have a great impact on many people has more significance in moral decision making than an unlikely consequence that would have a minor impact on only a few. Brody suggests guidelines for assessing the significance of each of the four other appeals as well. There may be factors missing from Brody's account that affect the evaluation of an appeal's importance in a particular case. Again, however, his framework can easily accommodate additions or alterations.

In the final stage of the model of conflicting appeals, the moral agent must use her assessment of the significance of the relevant appeals to arrive at a decision about what is the right thing to do. Brody does not, however, specify how these judgments should be made. He does state that moral judgments *cannot* be based on a hierarchical model of appeals, in which the appeals would be ranked by importance with the relevant appeal highest in the hierarchy trumping the others (pp. 75–79). Such a model, he claims, is implausible because the requisite hierarchy of appeals would be impossible to develop and defend. Nor, Brody argues, can judgments be based upon a process of weighing or balancing appeals. A model involving this sort of process would require the existence of a metric common to all of the appeals, that is, a measure by which the significance of the appeals could be compared. Brody argues that no common metric exists among the appeals to make this sort of weighing or balancing possible. Essentially, Brody holds that the process used to arrive at a moral judgment cannot be described by a function or algorithm, but is instead a product of human intuition. A decision maker must look at the picture painted by the appeals holistically and simply make a judgment about what is the morally correct choice. This idea of intuitive moral judgment is the keystone of Brody's approach to moral decision making.

According to Brody, judgments about individual cases can serve as a foundation for the formation of a generalized moral theory. He claims that, "the data about which we theorize are [our] initial intuitions. The goal is to find a theory that systematizes these intuitions, explains, them, and provides help in dealing with cases about which we have no intuitions" (p. 13). In other words, judgments about individual cases can be compared, contrasted and eventually combined into a coherent moral theory. Cases with similar features should produce similar moral judgments, which then serve as a guide for future cases with those same features.

Brody categorizes his approach as pluralistic casuistry. The account is pluralistic in that it "accepts the legitimacy of a wide variety of very different moral appeals" (p. 9). The five appeals represent many different types of ethical considerations, all of which do significant work in this model. The model is also casuistic because it is the intuitive judgments about individual cases that form the foundation of the approach. The generalizations that eventually constitute the moral theory are derived from intuitions about these individual cases.

II. MORAL JUDGMENT AND OBJECTIVE MORAL TRUTH

The model of conflicting appeals is a plausible account of how thoughtful moral agents make difficult moral decisions. It seems likely that many agents use a process similar to the one Brody describes to make these kinds of choices. They take a variety of considerations into account and then make a judgment based not on hierarchies of appeals or on specifiable algorithms, but on their intuitions about the case. One important reason for the model's strong appeal is its descriptive strength.

Brody, however, holds that the model of conflicting appeals is more than a description of what agents do when confronted with a moral problem. As noted above, he claims that the approach "is a picture of... how we can come to know what is right and wrong" (p. 79). On this view, the model of conflicting appeals can serve as a guide to help agents make ethical decisions; in other words, it has prescriptive power as well as descriptive power. But this seems to be only partially true. The first two stages of Brody's approach have prescriptive elements. The first stage gives a moral agent guidance about what types of things need to be considered during moral decision making, providing a contentful foundation upon which the agent can begin moral decision making. In the second stage, the model helps the agent determine how significant each of those considerations should be in that decision, and so also has prescriptive power.

However, this approach gives no guidance to the confused moral agent about how she should navigate the third and most important stage of the decision making process—how she should make a judgment about the right thing to do. Without such guidance, the prescriptive power of the model is significantly compromised. Brody acknowledges this characteristic of his model. He admits that "one wants the moral theory to provide an answer. The judgment approach is not comforting in dealing with moral conflict precisely because it insists that moral theory cannot do this" (p. 78). The model of conflicting appeals, therefore, does not produce a definitive solution to every moral problem. However, Brody claims, this characteristic is a strength of the model because it recognizes and helps to explain the existence of "deep moral ambiguities." Brody, therefore, seems to embrace rather than be embarrassed by this feature of the model of conflicting appeals.

Is this a tenable position? Brody wants his framework to give guidance that allows moral agents to discern a morally correct choice, but also holds that a moral theory cannot algorithmically provide answers about moral conflicts. In order to maintain both of these stances, he is forced into one of two possible corners. He must hold either that (1) any judgment a moral agent arrives at constitutes a morally acceptable decision, or (2) there are some type of constraints on the judgment process that ensure that the agent makes a morally acceptable decision.

Brody clearly rejects the relativistic position of option (1). He holds that "a moral judgment about a particular action being right is true if the action in question has the supervenient property of rightness, a property that depends on the nonmoral properties of the action . . . the property is not supervenient on sociological properties of how the action is viewed by individuals or by society" (p. 12). In other words, some actions have the property of rightness and others do not. As a result, moral judgments about the rightness of those actions can be true or false. A judgment is only true if it identifies as right the choice that is, in fact, the right thing to do. Because Brody holds that some judgments successfully pick out morally correct options and others do not, he would be unwilling to accept option (1). Further evidence that Brody would not tolerate the relativistic approach is found in the role that judgments play in his account of moral theory formation. If conflicting judgments were allowed into the process of theory formation, it would be difficult or impossible to systematize those judgments into a coherent moral theory. Brody must, therefore, accept something along the lines of option (2).

Option (2) suggests that there are constraints on the judgment process that ensure that a decision maker identifies the right action as right. Such constraints could be either externally imposed, taking the form of rules that the judgment process must conform to in order to be true, or internally constructed, taking the form of cognitive limitations that inevitably lead an agent to a correct judgment. Brody argues against several types of external constraints. As noted earlier, he claims that judgments cannot be based upon a hierarchical order of appeals, because no one appeal consistently takes precedence over others, and cannot be arrived at by weighing or balancing the appeals because the appeals are incommensurate. There are, however, other types of external constraints that could serve the necessary function. Perhaps a judgment can only be known to be true when it is consistent with other judgments. But there may not be judgments that are similar enough to use as comparisons. Further, the set of judgments could be consistent but false. Alternatively, judgments could have to be logically provable in order to be true. However, the model of conflicting appeals does not provide any moral premises to begin with, making it impossible to create sound arguments. Finally, true judgments could be identified by their correlation with or proximity to moral truth. This approach is a nonstarter because it is clearly circular, requiring that an agent know the very thing she is trying to ascertain in order to evaluate the truth of her judgment. None of these types of external constraints, therefore, seems promising.

The possibility that there are internal constraints that lead agents to true moral judgments, however, is one that Brody embraces. A closer look at Brody's work reveals that he may have an account of such constraints, although he does not explicitly acknowledge the central role that that account plays. In *Life and Death Decision Making*, Brody claims that "we have a fundamental cognitive capacity which enables us to recognize the moral value of individuals, actions, and social arrangements" (p. 12). This claim, however, is almost stated as an aside, without elaboration or an explanation about its importance to the model of conflicting appeals. Fortunately, in a separate paper, "Intuitions and Objective Moral Knowledge," Brody deals with this question in more detail (Brody, 1979). Here he holds that, "to a large degree, our fundamental moral intuitions about particular actions, agents, and institutions are forced upon us by our moral cognitive faculties" and that there are "constraints imposed upon our formation of initial judgments, both empirical and moral, by our cognitive mechanisms, by our nature" (p. 455). In other words, the constraints that are necessary in order to avoid the relativistic option are not external like those suggested above, but instead are part of human nature, putting limits on the judgments at which moral agents can arrive.

The idea of cognitive constraints is, therefore, central to the understanding of moral judgment needed to make the model of conflicting appeals work. If such constraints exist, the model may have the prescriptive abilities that Brody wants it to have; that is, it may provide moral agents with a way to reliably identify morally appropriate choices. Currently, there is no direct empirical test for the existence of such constraints on cognitive processes, nor is there an instrument to measure or characterize them. We are, therefore, forced to try to ascertain this information indirectly.

Several philosophers have developed arguments purporting to demonstrate the existence of the type of cognitive constraints that Brody needs. Noam Chomsky has suggested that it would be impossible for children to learn all of the complicated principles of morality as quickly as they do simply through experience. They do not encounter enough sufficiently rich cases for them to understand morality at the level that they do without additional internal guidance (Chomsky, 1988). Alan Fiske has taken a different approach, based upon evidence that the same four relational models can be found in the social interactions of diverse cultures. He infers from this commonality among cultures that, "these structures cannot be products of the particular conditions of each disparate domain or individual experience, as researchers have generally assumed. These modes of organizing social life must be endogenous products of the human mind" (1992, p. 690). In other words, the significant similarities among human interaction in many different contexts can only be explained by the existence of some type of cognitive constraints. A third account defending the concept of cognitive constraints has been proposed by Michael Ruse and Edward Wilson. Their naturalist theory ties the evolution of morality to the evolution of the human species, holding that, "[internal moral premises] are immanent in the unique programmes of the brain that originated during evolution ... The constraints on [human mental] development are the sources of our strongest feelings of right and wrong, and they are powerful enough to serve as a foundation for ethical codes" (1986, p. 194). Cognitive constraints are, like other human traits, determined at least in part by our genes.[2]

There are objections that can be made to these arguments that relate to the specific premises of each. However, in order to avoid a long digression into the merits and difficulties of each account, I will focus on two questions that trouble all three defenses of the idea of cognitive constraints. First, if such constraints exist, how is *intrapersonal* moral confusion possible? If moral agents have cognitive constraints of the kind the model of conflicting appeals requires, it seems that those constraints would, as Brody suggests, force a particular intuition upon them. Second, if such constraints exist, how is *interpersonal* moral disagreement possible? Cognitive constraints powerful enough to avoid the relativistic option should lead all (or at least most) moral agents to the same conclusions. In the remaining pages, I will look at these two problems and attempt to defend Brody's theory against them.

III. INTRAPERSONAL MORAL CONFUSION

The first potential problem for the idea that moral agents have cognitive constraints that shape their moral judgments is the existence of intrapersonal moral confusion—when an agent is uncertain what the right thing to do is in a given situation. For example, consider the following case about physician-assisted suicide.

> Mr. M, a 28 year-old single man with no children was diagnosed with Ewing's Sarcoma six years ago. The cancer was found in his pelvis and was promptly and aggressively treated. At first, Mr. M was responding well to the treatments, but eighteen months ago the cancer was found in his lungs and seems to be continuing to spread. His oncologist, Dr. D, estimates that he has about 3 months to live. Mr. M is in a lot of pain that can only be managed by significant quantities of pain medication, which Mr. M dislikes because they make him "fuzzy". Mr. M asks Dr. D if there is a combination of the drugs he is currently taking that would be lethal and, if not, if she would be willing to prescribe something that would be. Dr. D is not sure what she should do.

Let us assume that Dr. D wants to use the model of conflicting appeals to make a decision about this difficult case. She considers all of the relevant appeals and their relative significance. She thinks about the consequences of giving Mr. M the information or the means to end his own life—the significant pain he will certainly experience until a natural death and the likelihood that he would die sooner if given the opportunity. She also thinks about the consequences that this option would have for her and her ability to practice medicine. She takes into account her duty to respect Mr. M's apparently competent choice. She considers how to act most compassionately and whether giving Mr. M the opportunity to end his life would threaten her integrity. And after all of this deliberation, it seems quite likely that Dr. D still will not know what to do. There are at least two different kinds of confusion that Dr. D may be experiencing.

For one, Dr. D could have conflicting intuitions; that is, have strong inclinations toward two different judgments. She could have the intuition, based upon thorough consideration of the relevant appeals, that she should prescribe a lethal combination of medication with the knowledge that Mr. M may use it to end his life. But Dr. D could, at the same time, have the intuition that she should not prescribe such medication. This position may be logically incoherent, but it is not uncommon for agents to hold contradictory beliefs. In fact, these contradictory intuitions are the source of her moral confusion. This type of intrapersonal uncertainty speaks against the possibility of cognitive constraints, because it seems that such constraints would lead an agent down one path to a particular moral judgment rather than in two contradictory directions.

Another possible source of Dr. D's moral confusion could be the lack of any intuition about a situation at all. In other words, Dr. D may consider the morality of a physician assisting in a patient's suicide in a particular case, and be unable to make a judgment about what is the right and wrong thing to do in that case because she has no inclination either way. Cognitive constraints that put boundaries on the process of moral judgment would provide the guidance that an agent needs to arrive at a judgment, precluding the possibility of this sort of uncertainty. So the existence of these two kinds of moral confusion seems to, at least at first glance, tell against the existence of cognitive constraints.

An objector could argue that intrapersonal moral confusion is not actually a problem for the idea that there are cognitive constraints on moral judgment because I have overestimated the power of these constraints. One could claim that such constraints may exist but not have the strength that I have assumed—that the existence of such constraints might not actually *force* an individual to arrive at a particular judgment. The constraints could, therefore, exist but be unable to prevent the sort of confusion described above. However, it seems that cognitive constraints fitting this weaker description would be unable to reliably lead moral agents to knowledge about the right thing to do in a given situation, an ability that Brody needs them to have in order for the model of conflicting appeals to be prescriptive.

The existence of intrapersonal moral confusion, therefore, may cast doubt on the idea that there are cognitive constraints of the strong type that Brody needs. However, there are numerous possible explanations for the existence of such confusion, some of which do not challenge the idea that our judgments are guided by cognitive constraints.

First, there could be more than one objectively right option in a given case. That is, there could be two or more different choices that are equally and maximally morally correct. One could object that this solution would be a logical impossibility in some cases. For example, it would be contradictory for it to be objectively true both that "Dr. D should not provide Mr. M with the information or means to take his life," and that "Dr. D should provide Mr. M with the information or means to take his life." Even so, some have argued that such moral dilemmas do exist and that agents can be faced with them in the real world (Sinnott-Armstrong, 1988). It seems possible that Brody could be open to this option. He clearly holds that objective moral truths exist, but does not specify that there can be only one for any given case. Further, this explanation would be compatible with the idea that we have cognitive constraints, because such constraints would lead to confusion if there were two morally correct choices. However, if they do exist, cases with more than one morally correct option are rarities in the real world. Therefore, this cannot be a comprehensive explanation for intrapersonal moral confusion.

A second explanation for an agent's moral uncertainty could be that there is ambiguity or a lack of information about the cases. Ambiguity about important facts could cause the generation of different intuitions depending upon the assumptions made in light of that ambiguity, or could prevent the generation of any intuitions at all. Further, without all of the relevant information, it could be difficult for a moral agent to reach a judgment about the case. In many cases in the real world, uncertainty about the appropriate course of action can be resolved when clarification and additional information are provided. Such cases are not evidence against the existence of cognitive constraints on human judgment because the confusion resolves with the additional information. However, there are other cases in which an agent knows all of the relevant information, but still experiences moral confusion.

Finally, intrapersonal moral confusion could occur when a moral agent is uncertain which ethical appeals are relevant or how much significance each of those appeals should be given in the case at hand. He could come to several preliminary conclusions depending on which appeals are included or could be unable to arrive at any conclusion at all. Similarly, if the moral agent is not sure about the significance that each appeal should have, she could have different intuitions based upon the various possible significances of the appeals, or could have no intuition about the correct course of action. This explanation is also not necessarily damaging to Brody's claim about cognitive processes, and could account for some of the cases of intrapersonal moral confusion.

It seems likely that many cases of moral confusion can be explained by one of these three accounts. In such cases, moral confusion is not evidence against the idea that there are cognitive constraints on our moral judgment. If all cases of moral confusion fall under one of these explanations, then this kind of confusion is perfectly compatible with the idea of cognitive constraints. However, in difficult cases like the one described above, an agent could be perfectly informed and could accurately assess the relevance and significance of the appeals and still be unsure about the right thing to do. The existence of such cases cannot be empirically proven, but it seems unlikely that the many reports of moral confusion can all be accounted for by one of these explanations. If these cases do exist, they would be evidence against the existence of cognitive constraints of the strong kind that Brody needs in order for his theory to serve as a method of moral decision making.

IV. INTERPERSONAL MORAL DISAGREEMENT

The second kind of conflict that could be problematic for the existence of cognitive constraints is interpersonal moral disagreement. In such cases, moral agents come to different judgments about what is the morally right thing to do in a particular case. The existence of such moral disagreement can be confirmed by experience (and by the fact that ethicists can find employment). Let us return to the fictional but plausible case described above. Dr. D's uncertainty about whether or not she should provide Mr. M with the means to end his life drives her to consult two of her colleagues, Drs. Y and N. She tells them in detail about Mr. M and his request. After some consideration, Dr. Y concludes with certainty that Dr. D should grant Mr. M's request. Dr. N feels certain that she should not. If moral judgments are guided by cognitive constraints, how is it possible that Drs. Y and N arrive at different conclusions? There are several possible explanations for this not uncommon phenomenon.

One possibility, as discussed in the previous section, is that there is more than one right thing for Dr. D to do in the case at hand. If there is more than one morally right option, Drs. Y and N could each use the model of conflicting appeals to arrive at a morally correct decision. They could each feel certain that their own judgment is right because both judgments are correct, even though they are different. This possible explanation does not undermine the idea of cognitive constraints. And, as suggested above, because it admits the existence of objective moral truths, Brody might be willing to accept it. However, the number of cases in which there are multiple morally correct resolutions is probably quite small.

Another possible explanation of interpersonal moral disagreement is that there is a single objectively right thing for Dr. D to do, but either Dr. Y or Dr. N (or both of them) failed to identify that course of action due to error in the decision making process. There are at least two different types of errors a moral agent could make: errors about the nonmoral properties of the case and errors about its moral properties.

The agents' failure to make the same moral judgment could be due to disagreement about one or more of the nonmoral properties of the case. If Drs. Y and N do not have identical information about Dr. D's case, they could easily come to different conclusions about it. One of them could be missing a vital piece of information or could be misinformed about some fact of the case; perhaps Dr. N heard Dr. D say that Mr. M has 13 months to live rather than 3. Further, even if they have the same information about the case, one of them could have a misunderstanding about what some of that information means. For example, Mr. M's reports of feeling "fuzzy" on painkillers could mean that he is nearly incapacitated to Dr. Y, but just mean that he is a little slow to process thoughts to Dr. N.

Alternatively, moral agents could make errors in evaluating the case's moral properties, that is, in assessing the relevance and significance of the various ethical appeals. An agent who recognizes relevant appeals as relevant is likely to arrive at a different moral judgment than an agent who fails to do so. Similarly, disagreement could occur when one of the agents gives too much or not enough significance to a particular appeal. For example, Dr. Y's judgment could be affected if he does not realize that Mr. M's claim to have his wishes respected could be less important because his pain could be compromising his ability to make decisions. Or, if Dr. N fails to properly account for the low quality of Mr. M's remaining life, he could put too much emphasis on the negative consequence of Mr. M's death.

It seems clear that errors, either about cases' moral or nonmoral properties, generate some instances of interpersonal moral disagreement. Such cases do not challenge the idea of cognitive constraints, because disagreement could occur in these cases even if such constraints exist. However, it does not seem likely that the above explanations can account for all of the cases of interpersonal disagreement. Cases of moral disagreement that do not fall under one of these explanations cannot be empirically proven to exist, but our experience suggests that they do.

V. GENUINE MORAL DISAGREEMENT

Let us assume (as I believe Brody would) that there is a single morally correct resolution to the case above and stipulate that Drs. Y and N agree on the facts of the case and have not made errors in their assessment of the appeals. Given these guidelines, it seems that there is genuine disagreement in the judgments made by Drs. Y and N. How could such substantive interpersonal moral disagreement be compatible with the existence of cognitive constraints? In the following paragraphs, I will try to develop a plausible response to this question that is compatible with Brody's philosophy.

The genuine moral disagreement in this case could exist because Drs. Y and N have different sets of cognitive constraints. That is, the boundaries that guide the two agents' cognitive processes might not be identical. Dr. Y might have a tendency to respect autonomy while Dr. N is predisposed to value the preservation of life. This could explain why the two come to different judgments even when they agree on the facts and have identical assessments of the significance of the appeals.

In cases in which there is widespread agreement among moral agents and only a minority of dissenters, Brody might find this account acceptable. For example, if 99.9% of people agree with Dr. Y and only 0.1% of people agree with Dr. N, it may be plausible to say that those people in the 0.1% minority do not have the same cognitive constraints that everyone else has. Just as some people are born with a genetic mutation or are exposed to an environmental toxin that causes cancer, some people are born with or acquire a cognitive "mutation" that causes the formation of different cognitive constraints. In such a case, different people would have different constraints, but the "mutant" ones can be qualified as abnormal. They, therefore, do not shed doubt on the overall ability of cognitive constraints to lead to objective moral truths just as the existence of cancerous cells does not shed doubt on the ability of normal cells to promote health. Brody claims that, "some [intuitive judgments] can be rejected to save a theory which otherwise explains and systematizes most of the judgments" (Brody, 1979, p. 454), suggesting that he might be willing to take this approach to deal with cases like these.

In at least some cases of interpersonal moral disagreement, however, there is a deep division in the judgments of moral agents. In such cases, a substantial proportion of people has arrived at each of two or more moral judgments and it is not possible to determine which are the judgments derived from "normal" cognitive constraints. While there are probably more cases in which a clear majority shares a moral judgment and only a small minority does not, cases in which there is deep division among judgments clearly exist. For example, it seems likely that the judgments of both Dr. Y and Dr. N would be supported by a significant proportion of moral agents. Perhaps this disagreement can be explained by the fact that there exist two (or more) different sets of cognitive constraints in the population of moral agents, neither (or none) of which can be disregarded as "abnormal".

While this explanation in isolation does not sound implausible, it is problematic in the context of the model of conflicting appeals because if it were accurate, cognitive constraints would not necessarily lead to objectively true moral judgments. Unless the case at hand is one of the few in which there is more than one morally correct choice, some judgments derived from cognitive constraints are correct and others are not. If there is more than one set of cognitive constraints, which of these should be privileged? Without a method of determining which judgments are true and which are not (which we do not have, thus the need for the model of conflicting appeals), it is not possible to identify objective moral truths. There is no way to differentiate between sets of cognitive constraints so that it is clear which one produces correct moral judgments. If the model of conflicting appeals admits that Dr. Y and Dr. N have different but equally "normal" cognitive constraints that lead them to different judgments about what Dr. D should do, it cannot be informative about which judgment is true. As a result, this explanation for interpersonal moral disagreement is not one that Brody's model can accommodate. Cognitive constraints must be universal (or nearly so) in order to make the model of conflicting appeals prescriptive.

Although he does not argue for it in this context, Brody embraces the idea of a universal set of cognitive constraints. He states that the "assumption of the constancy of our cognitive nature across individuals... seems reasonable enough providing that this cognitive nature is physiologically rooted and genetically determined. It would seem highly unlikely that members of a species would not then share this cognitive nature" (1979, p. 455). In other words, if our cognitive processes are derived from our physical beings, it is probable that all humans have similar cognitive processes, just as we have similar bodies. Brody admits this is an assumption, but clearly finds it plausible. Assuming, then, that there is a single set of cognitive constraints on our intuitive judgments and that there is a single morally correct judgment in most cases, how does Brody account for interpersonal moral disagreements that are not caused by errors?

Perhaps each moral agent begins with the same set of cognitive constraints, but those constraints are altered during the course of that agent's life. In other words, the cognitive processes that guide intuitive judgments could be the same in all people initially, but could change or be obscured over time by life experiences. As an analogy, envision two stained glass windows, created in the same pattern of wrought iron, but filled in by different artists with different patterns of colored glass. It seems possible that all humans have a basic cognitive structure that is colored in by life experiences and the influences of families, friends, and society in a way that is unique to each individual. For example, if Dr. Y had watched his grandfather die a slow and painful death from cancer, his cognitive constraints are likely to be affected such that he disvalues such a death more than someone who has not had that experience. Similarly, if Dr. N's grandmother recovered from the brink of death from cancer to live another happy ten years, his cognitive constraints are likely to bend to make room for hope under even the direst circumstances. Drs. Y and N may have initially had identical cognitive constraints, but the difference in their life experiences could cause them to make different judgments about Dr. D's case.

I believe that Brody would be likely to find this type of explanation for moral disagreement acceptable. It allows for the existence of objective moral truth. It gives an account of the constraints on moral judgment needed to avoid the relativistic option, which he rejects. Further, this account complies with his assumption that there is a single cognitive process that guides moral agents' judgment. At the same time, this explanation successfully shows how it is possible for substantial interpersonal moral disagreement to exist even if all of these claims are true. As a result, if Brody were to accept this account, he could deal more effectively with objections to his account that are based upon such disagreement.

Further, this account leaves room for the possibility that the model of conflicting appeals could be developed into a fully prescriptive moral theory. All that is necessary to achieve this goal is a method of figuring out what judgments would be made by someone with cognitive constraints as they were before being distorted by life's experiences and influences. If such a method were developed, it should be possible to deduce what the correct moral judgment would be for any given case.

VI. ISOLATING COGNITIVE CONSTRAINTS

There are at least three candidate methods for figuring out to which judgments cognitive constraints would guide moral agents. One approach would be to use the level of intersubjective agreement about a given moral judgment about a particular case as an indicator of that judgment's proximity to the true judgment. In other words, the more people who share a particular moral judgment about what is the right thing to do in a given case, the more likely that judgment is to reflect our constraints. One could argue that, given the many different directions in which people's cognitive processes can change, the most common moral judgments reflect the cognitive constraints rather than a consistent alteration of cognitive processes.

A second possible approach to this problem, also based upon intersubjectivity, would be to analyze a wide variety of moral judgments made by moral agents and to attempt to identify any common denominators—similarities among aspects of the moral judgments—among them. These common denominators could take the form of common values, principles, or virtues. They are likely to be indicative of the nature of the cognitive constraints because they are shared by most, or at least many moral agents. The fundamental nature of such denominators could plausibly reflect the fundamental nature of cognitive constraints and could be informative about judgments that would be derived from those constraints.

There are, however, two serious objections to these intersubjective methods of finding the judgments to which our cognitive constraints would lead. First, even in cases in which there is near consensus about a moral judgment, the high level of intersubjectivity of that judgment could easily be brought about by strong social norms or mores. Such social influences would be able to alter the judgments of many moral agents in similar ways, so that those judgments appear to reflect cognitive constraints but in actuality reflect only a common alteration of those constraints. There are many examples of intersubjective agreement among moral agents' judgments that now seem clearly false, such as infamous judgments about slavery and about female circumcision. A second problem with these methods is the difficulty of finding intersubjective agreement or common denominators when there are deep divisions among moral agents' judgments. The method is not likely to give a good indication of the judgment that would be derived from cognitive constraints without substantial intersubjective agreement, which often does not exist.

A third candidate for a method of identifying the moral judgments to which cognitive constraints would lead takes a different approach. Rather than attempting to derive these judgments from the judgments made by *actual* moral agents, this approach uses the judgments of a *hypothetical* moral agent: an "ideal intuitor." Similar to an "ideal observer" (Firth, 1952), an ideal intuitor would be an agent who is in a special position to make judgments that would be entirely determined by the basic framework of cognitive constraints.[3] This position would be defined by the possession of a certain set of qualities. Firth (1952, pp. 333–345) and Carson (1984, pp. 56–80) have developed separate accounts of the particular qualities that an ideal observer must have. Firth claims that an ideal observer must be omniscient, omnipercipient, disinterested, dispassionate, consistent, and otherwise "normal". Carson holds that an ideal observer must be a human being, be fully informed about nonmoral facts, be uninfluenced by non-ideal observers, and have full knowledge of relevant moral principles. The ideal observer's judgments and attitudes must not involve emotional displacement or self-deception and need not be impartial, disinterested, dispassionate, or normal. It is plausible to consider these sets of qualities as descriptive of an ideal intuitor even though they are used in the context of describing an ideal observer. An analysis of these and other possible qualities of an ideal intuitor would be tangential to my main point: it is sufficient to claim that some combination of qualities would be sufficient to create an ideal intuitor whose moral judgments would be entirely determined by her natural cognitive constraints. She would need to have the intelligence and imagination to appropriately identify and assess the ethical appeals relevant to any possible case. She would have to be free of life experiences and social influences that could color her judgments. It seems plausible that she would need to be impartial. It is worth noting, however, that the more qualities that are specified as characteristics of an ideal intuitor, the greater the distance between the hypothetical ideal intuitor and average moral agents becomes.

Moral judgments made by an ideal intuitor would be those to which natural cognitive constraints would lead because those constraints are unadulterated in the ideal intuitor. In order to figure out what those judgments would be, then, it is necessary for a moral agent to put himself in the shoes of an ideal intuitor. He will need to ask himself, "What judgment would an ideal intuitor make in this case?". The addition of this piece to the model of conflicting appeals adds a prescriptive element to the model's third step. A moral agent is led to a correct moral judgment by identifying the judgment that an ideal intuitor would make in a given case. As a result, the idea of an ideal intuitor solves both of the challenges to the use of cognitive constraints in the model of conflicting appeals identified above—the problem of intrapersonal moral confusion and the problem of interpersonal moral disagreement. This construct provides guidance to the confused moral agent, at least to the extent to which an individual would be able to put himself in the ideal intuitor's shoes. So Dr. D could, therefore, use the modified model of conflicting appeals to help decide whether she should provide Mr. M with the means to take his own life. And it makes it possible to privilege some moral judgments over others, so that the interpersonal moral disagreement between Drs. Y and N can be resolved using this approach. In addition, the construct of the ideal observer is conceptually compatible with the framework Brody proposes in the model of conflicting appeals.

The use of the ideal intuitor avoids the problems of the intersubjective methods of identifying the moral judgments to which cognitive constraints would lead. However, this approach also has drawbacks. First, one could object to this proposal as some have to Firth's concept of the ideal observer (Brandt, 1955, p. 408; Carson, 1984, p. 5) and argue that two ideal intuitors could arrive at different conclusions about what is the right thing to do in a given situation. If this is the case, then the construct of the ideal intuitor has gotten us no closer to giving the model of conflicting appeals prescriptive power; that is, to leading an agent to the single correct moral judgment. Russ Shafer-Landau lays out the problem generated by the possibility that ideal intuitors could disagree in the following argument:

1. "If there are objective moral facts, then there can be no intractable moral disagreement among ideal moral judges.
2. There can be such disagreement.
3. Therefore, there are no objective moral facts" (1994, p. 332).

In other words, if ideal intuitors reliably pick out true moral judgments, then either ideal intuitors must agree or there are no such things as true moral judgments. As is clear from this chapter's analysis, Brody needs moral agents to be able to pick out true moral judgments, and the concept of the ideal intuitor helps them to do so. Further, as discussed earlier, Brody holds that there are objectively true moral judgments. As a result, the approach that he would surely take would be to deny premise (2) above, rather than to accept the conclusion (3). He would need to hold that true ideal intuitors would consistently make identical moral judgments about the cases they consider. Firth, also, takes this position, stating, "I cannot believe that there is any convincing ethnological or psychological evidence that two ideal observers could have conflicting moral experience with respect to the same act" (1955, p. 416). This seems to be a plausible position, at least under some definition of the ideal intuitor. The more qualities that are built into that definition, the more likely it is that ideal intuitors would agree. At some point, it seems likely that the characteristics of ideal intuitors would be similar enough to all but guarantee that they would arrive at the same conclusion about the morally correct decision to make for a particular case. However, empirical evidence would be required to confirm this hypothesis.

This point, however, brings attention to the second limitation of the ideal intuitor addition to the model of conflicting appeals. A true ideal intuitor is only a hypothetical entity; the cognitive constraints of any actual person with the capacity to form moral judgments would be affected by her relationships and life experiences. And the more "ideal" the ideal intuitor is conceived to be, the greater the difference between normal human beings and the intuitor becomes. As a result, the value of this construct is dependent on the degree to which human beings can put themselves in the shoes of the ideal intuitor. The closer that an agent can get to perceiving a case as an ideal intuitor would, the more likely it is that he will arrive at a true moral judgment. This means that the prescriptive power of the model of conflicting appeals is limited by the ability of the moral agent to successfully make the required mental shift. It is possible that no actual agent would be fully able to see a case from the perspective of the ideal intuitor and that many actual agents would have difficulty finding a viewpoint that is sufficiently similar to the one required to make morally correct decisions. This is not an insignificant drawback to this modified model of conflicting appeals. Nonetheless, the addition of the ideal intuitor provides a solid theoretical foundation for the model as well as a practical approach to guide agents toward true moral judgments, something that the model of conflicting appeals is unable to do without this piece.

VII. TWO IMPORTANT OBJECTIONS

The ideal intuitor construct, therefore, sounds like a plausible addition to Brody's model of conflicting appeals. Significantly, it seems to be consistent with Brody's broader philosophy. It is a sound way to develop the model of conflicting appeals prescriptively while working within the structure Brody has provided. However, I have serious concerns about two critical assumptions that this additional piece relies upon.

First, do cognitive constraints exist? Brody claims that they do, but does not argue in support of this claim. To my knowledge there is no empirical evidence to corroborate the existence of such constraints. Brody could argue in response that although it is impossible to observe cognitive constraints directly, evidence of their existence can be found in the effects that they have on human behavior and judgment. While this could be true for some types of cognitive constraints, such as those that govern human emotion or reasoning, it is unclear how the effects of moral cognitive constraints could be measured or shown to exist even indirectly. The most plausible measure of these effects would be the morality of the judgments made by moral agents. However, an argument that attempts to demonstrate the existence of these constraints based upon the fact that they identify morally correct judgments immediately becomes circular.

Second, and more importantly, even if cognitive constraints do exist, is there any reason to believe that those constraints lead us to moral truths? There seems to be no necessary connection between cognitive constraints, even if they are innate, and objective moral truths. This connection is critical to Brody's model. Brody clearly assumes, but never explicitly states, that cognitive constraints would lead to moral truths. However, I cannot find any good reason to believe that a moral agent who accurately deduced the judgment of an ideal intuitor should feel confident that he has identified the morally correct course of action. The existence of this connection is appealing, but unsubstantiated.

Given the fundamental nature of these two doubts, I cannot, in the end, find even this modified version of Brody's approach persuasive from a prescriptive standpoint. Nonetheless, I believe the model of conflicting appeals is a valuable descriptive account. It accurately describes what thoughtful moral agents do in searching for moral truths, if such things exist.

NOTES

[1] Although most of the cases Brody considers relate to decisions that could make the difference between life and death, his framework can be applied in a much wider variety of situations within the medical context, as well as to moral decision making outside of that context.

[2] All of these are arguments for the existence of *innate* cognitive constraints—constraints that human beings are born with—as opposed to constraints that are *acquired* from an outside source. Such innate constraints seem to be what Brody had in mind when describing his account. However, neither Brody's account nor my analysis of it is dependent upon this distinction.

[3] Firth uses the concept of the ideal observer in the following context: "statements of the form 'x is P,' in which P is some particular ethical predicate, [would be] identical in meaning with statements of the form: 'any ideal observer would react to x in such and such a way under such and such conditions' " (1952, p. 321). This is slightly but importantly different from the version of the ideal intuitor I am suggesting, which would take a form something like: "we know that 'x is P,' in which P is some particular ethical predicate, is a correct moral judgment when an ideal intuitor judges that x is P."

BIBLIOGRAPHY

Brandt, R.B. (1955). 'The definition of an "ideal observer" theory in ethics,' *Philosophy and Phenomenological Research*, 15, 407–413.

Brody, B. (1979). 'Intuitions and objective moral knowledge,' *The Monist*, 62, 446–456.

Brody, B. (1988). *Life and Death Decision Making*. New York: Oxford University Press.

Carson, T.L. (1984). *The Status of Morality*. Dordrecht: D. Reidel Publishing Company.

Chomsky, N. (1988). *Language and Problems of Knowledge*. Cambridge: The MIT Press.

Firth, R. (1952). 'Ethical absolutism and the ideal observer,' *Philosophy and Phenomenological Research*, 12, 317–345.

Firth, R. (1955). 'Reply to Professor Brandt,' *Philosophy and Phenomenological Research*, 15, 414–421.

Fiske, A.P. (1992). 'The four elementary forms of sociality: framework for a unified theory of social relations,' *Psychological Review*, 99, 689–723.

Ruse, M. & Wilson, E.O. (1986). 'Moral philosophy and applied science,'*Philosophy*, 61, 173–192.

Schafer-Landau, R. (1994). 'Ethical disagreement, ethical objectivism and moral indeterminacy,' *Philosophy and Phenomenological Research*, 54, 331–344.

Sinnott-Armstrong, W. (1988). *Moral Dilemmas*. Oxford: Basil Blackwell.

CHAPTER 5

CASUISTRY NATURALIZED*

MAUREEN KELLEY

Treuman Katz Center for Pediatric Bioethics, University of Washington School of Medicine, Seattle, Washington, U.S.A.

Widely adopted in medical school curricula and in clinical consultation practice, the theory of applied ethics known as "moral casuistry" has enjoyed renewed popularity with the rise of clinical bioethics in the last five decades. The two most systematic accounts of contemporary casuistry as a theory of practical morality in bioethics can be found in the work of Jonsen and Toulmin, and Baruch Brody (Brody, 1988; Jonsen and Toulmin, 1988). Carson Strong has also offered important defenses and refinements to the theory (Strong, 1999; 2000). Like its ancient and medieval theoretical ancestors, contemporary casuistry attempts to offer a reliable, practical process for moral decision-making that seems tailor made for the practice of medicine. Faced with a moral quandary or decision the casuist will reflect on the nonmoral and moral features of the case at hand and compare these features to a paradigm case, one where there is stable social consensus about the right course of action. General ethical norms emerge from families of cases to guide moral reasoning over new or more ambiguous cases. Trained reflection on the features of new cases may then lead us to adjust, refine, or better specify the general norms via the mechanism of a reflective equilibrium, seeking the appropriate balance between general moral norms and concrete cases or decisions (Rawls, 1971, pp. 48–51).

Contemporary casuistry as developed in particular by Baruch Brody has two characteristic features. First, it relies on a pluralistic theory of value in the tradition of early twentieth century moralist W.D. Ross. Experience with the vast variety of real moral cases in the law, politics, and everyday moral decision-making reveals the relevance of multiple core moral appeals. Entire theories and supporting arguments have been offered in defense of each of these appeals, which in the history of moral philosophy include appeals to consequences, rights, virtues, and concerns of justice.

---

*I would like to thank Baruch Brody for his intellectual generosity and dedicated mentoring throughout my years of graduate study and during my first two years as a faculty member at Baylor College of Medicine. Every time I consult on a case in the hospital or puzzle through a philosophical problem with my own students I carry his advice and many valuable lessons along with me. I would also like to thank Dien Ho for helpful comments on an earlier draft of this paper, and Lisa Rasmussen and Ben Hippen for insightful discussions on the topic.

M.J. Cherry and A.S. Iltis (eds.), Pluralistic Casuistry, 65–82.

Each monistic (or single-valued) theory encounters the problem of exceptions: no single theory fully captures our intuitions about right action in all cases. How then do we know when to make exceptions to the central principles of the theory and on what grounds, if not within the theory? In rights theory, for example, even philosophers like Robert Nozick, who in his earlier work defended an absolute conception of rights as negative side-constraints, allowed that protecting an individual's right to life or property may be sacrificed if it means preventing devastating and massive moral consequences, such as a holocaust (1974, pp. 28–30). Brody, and other moral pluralists take this as evidence that morality is likely more complex rather then less so. As Bernard Williams asked somewhat rhetorically, "If there is such a thing as the truth about the subject matter of ethics...why is there any expectation that it should be simple? In particular, why should it be conceptually simple, using only one or two ethical concepts, such as *duty* or *good state of affairs*, rather than many?" (1985, p. 17). The second distinguishing feature of Brody's brand of casuistry is the central role granted to intuitionism as the means to moral knowledge (Brody, 1979). Here moral knowledge includes intuitions about the basic moral appeals, knowledge of the grouping and analysis of families of moral cases, and knowledge or wisdom about striking the appropriate balance between the appeals when they come into conflict in particular cases. An important feature of Brody's intuitionism is its humility; knowledge about new cases and first impressions ought to be tentative and open to discussion and further reflection. Even knowledge about time-tested pluralistic moral appeals should be open to revision, though these intuitions, because they have stood the test of time and public deliberation, are considered more stable.

The theory's value pluralism, its central role for moral judgment, and its empirical starting point in real cases in everyday decision-making seem to make the theory particularly suitable for decision-making in bioethics, given its domain in the health professions and sciences. Here we find a natural affinity between the practice of casuistry and the everyday practice of medicine and science. Where the idea of a "standard of care" in clinical medicine emerges from a professional consensus about how families of cases ought to be clinically managed, so too the idea of a paradigm case in medical ethics represents a more settled area of consensus on how best ethically to manage a case. Just as clinicians struggle to fit anomalous disorders within stable concepts of disease and existing standard of care for disease management, clinical bioethicists on the casuistry model make tentative judgments about novel moral quandaries by comparing moral features of the new cases with more familiar and settled paradigm cases.

Somewhat surprising, given these natural affinities with actual clinical practice and its widespread use in the clinical teaching and consulting, casuistry has been under fire from sociology and from within bioethics for failing to fit the way people and clinicians typically make moral decisions.[1] If correct, this is a particularly devastating charge given the aims of the theory. The rebirth of contemporary casuistry in bioethics with its concern with the particularity of individual cases and patients marked an attempt to bring the subject, the case, and lived experience, back into ethical deliberation and action. If indeed it turns out to be a poor guide for action in real choices, as recent critics claim, this would not only be a devastating result for casuistry but for any theory of applied ethics sharing the critically objectionable features of intuitionism, pluralism, and moral judgment in a social context. Here, I will consider the merits of these recent objections and I will suggest a possible response and a way forward for casuists (and by implication for other theories of applied ethics). I argue that casuistry can offer more effective moral guidance in moral conflict only if it can offer a more explicit empirical account of the epistemic skills necessary for striking a moral balance where values conflict or case interpretations remain unclear. Such an account would also require better understanding of the social and psychological biases affecting moral choices, and the perspectives of marginalized groups. Joining the empirical turn in ethics, I will argue that the next step in a defense of casuistry is to reject the standard intuitionism and reflective equilibrium central to the moral epistemology of the theory and accept instead a naturalized account of moral judgment and its limitations in practical moral decision-making, one where both judgment itself is informed by empirical study, and where the truth of particular decisions is determined by social deliberation within the constraints of empirical inquiry.

I. THE EMPIRICAL TURN IN ETHICS

A robust theory of practical morality will not only answer the question: what ought I do? Practical morality is a balancing act. And so we ought to expect a good theory of practical morality to give guidance on the sort of person who balances competing values well, who strikes an equilibrium between theory and practice, who appreciates values gained and values lost when acting in a moral dilemma or in a difficult conflict. Such an individual exemplifies important epistemic virtues, virtues of moral perception. Of course these virtues are not acquired or honed in a vacuum. Moral reasoning is a deeply social practice, and so such a theory should also give attention to the sorts of institutions and relationships that reinforce moral norms and facilitate moral learning. Largely due to the influence of the social sciences, and the work in epistemology of W. V. Quine, we are seeing a return in moral philosophy to an explicit recognition of the practical, empirical, and social nature of moral reasoning, a more widespread endorsement of ethical naturalism, a starting point taken for granted by Aristotle, David Hume and John Stuart Mill. Naturalists in ethics offer an explicit defense of ethics in the natural world, as the proper subject of empirical science, especially the social sciences and cognitive psychology.[2] Quine himself was skeptical that a science of morality is possible (Quine, 1979, pp. 471–80), but the subsequent work of Owen Flanagan has convincingly shown that a Quinean naturalized philosophy commits one to naturalism in ethics and that the methodological problems Quine attributed uniquely to moral inquiry are in fact equally problematic in scientific methodology (Flanagan, 1982, pp. 56–74). The interesting question remaining is: how thoroughgoing a naturalist ought we to be?

Varieties of ethical naturalism can perhaps best be distinguished by the degree to which the proponents wish to press the "double-aspect" theory of ethics, as opposed to reducing normative values to natural or descriptive facts about human beings and the world. The theory of ethics defended by Owen Flanagan maintains a distinction between the factual and normative features of ethical discourse. Daniel Dennett, like Flanagan, offers an evolutionary account of moral norms, or "memes," that emerge and evolve over time in social contexts, subject to social and genetic evolutionary forces not unlike genes (Dennett, 1995, pp. 453–510). While Dennett and other Darwinian naturalists resist the idea that a genealogical story about the rise of normative rules, principles, and values entails a thoroughgoing reduction of value to fact, the ardency with which naturalists mind the gap between fact and value in the face of the reductive criticism offers a useful litmus test for the brand of naturalism on offer.[3] On the double-aspect accounts, ethical questions cannot be entirely reduced to empirical questions, but the social sciences and cognitive psychology are viewed as important tools in answering questions about moral motivation, barriers to moral behavior, and the acquisition of moral knowledge.

For the purposes of argument, I will assume only that the weakest variety of ethical naturalism is true, that on the most straightforward interpretation of Kant's principle 'ought implies can' ethics must be concerned with what humans are capable of doing, our physical and psychological capacities and limitations when it comes to prescribing moral reasoning and action. Some have suggested that this form of ethical naturalism is so weak as to be uninteresting. And that ethical naturalism is not really naturalism at all if it does not require the derivation from empirical facts to "oughts" or normative conclusions.[4] This seems false on both counts. According to basic naturalism, social moral deliberation is necessarily constrained by and informed by social science, neuroscience, cognitive psychology and the other scientific inquiries into animal and physical nature. The empirical turn seeks to explain ethical motivation, capabilities, interests, inclinations, and dispositions, patterns of behavior, barriers to moral understanding and agreement. It does not follow from such explanations of human moral nature that ethics is *explained* away.[5] This basic form of naturalism captures a very basic, and in fact quite ancient, sensibility about the nature of morality. As Owen Flanagan has put it delightfully in a reference to Bernard Williams' work on shame, "the Greek concept of shame tied morality to where it should be tied—to being seen naked, in the eyes of others" (1996, p. 19). With the Greeks and later with Hume, Smith, and Hutchinson, and Mill the moral "ought" and moral sentiments situated moral reasoning, organically, in the thick of social relationships, practices, and institutions. If we understand the purpose of moral theory to provide guidance in actual moral decisions in social life, the empirical turn provides a crucial correction to moral theory of the modern era. Bare ethical principles cannot successfully guide real choices if they fail to account for the real nature of human moral motivation, psychological biases in moral knowledge, and the social forces affecting moral decision-making in a social context.

This return to a form of ethical naturalism comes on the heels of the post-modern turn in ethics. Post-modern critics of early twentieth century moral theory raised the veil on the intuitionism and bias in moral and political theory, and brought due attention to overlooked groups, perspectives, and ways of reasoning. Post-modernism brought power to the fore of discussions about morality and politics in a way we have not seen since Hobbes and Machiavelli. While the movement is often perceived as delivering up negative, destructive criticism, discussions of moral theory today are more exciting and relevant than ever, largely due to the radical critiques of post-modernism and the sincere attempts to meet these criticisms. Training a more radical eye on empirically grounded work in ethics remains crucial, since mere appeal to descriptive accounts of human behavior, cognitive biases, and social forces can lead us unintentionally to acquiesce in such descriptive realities. Without the critical normative voice, we lose the "ought," the progressive voice in our moral and political debates that is essential to improving the human condition.

Bioethics as a sub-discipline of ethical theory and a cross-discipline of the social and life sciences, law, and literature, has taken the post-modern and empirical turns, with both still in progress. Some of the most robust contemporary discussions of practical morality have occurred in the field of bioethics, where the daily decisions of health professionals provide an ideal context for observing and reflecting on the nature of everyday moral reasoning in a variety of situations, involving diverse people and institutions. Oddly enough, we have not seen an explicit defense of ethical naturalism in bioethical theory. Given the close ties between clinical bioethics and biomedical sciences, perhaps most clinical bioethicists take some version of ethical naturalism as an obvious starting-point, not in need of defense. We are certainly seeing an increase in the number of empirical studies on bioethics in the social sciences, especially in the area of research ethics. It remains surprising, though, that in the debates surrounding bioethical theory and arguments regarding competing theories of applied ethics, we do not find explicit discussions of ethical naturalism as an important measure of theoretical strength or weakness. In what follows, I hope to bring just these sorts of questions to bear on the strengths and weaknesses of one of the most widely used theories of practical morality in bioethics: moral casuistry.

II. CHALLENGES TO CASUISTRY

Serious challenges to contemporary casuistry in bioethics can be broken down into three distinct objections: normative, psychological, and sociological critiques. What unites these objections is a common concern about the apparent lack of fit between this theory of applied ethics and the real world or ordinary decision-makers. Such a failure of fit ought to be of concern to those in the business of applied ethics, since it likely implies a failure of the theory to resolve concrete cases, the central aim of applied ethics. Here I will present the objections as raised specifically against theories of moral casuistry. In the next section, I propose a road map for research in applied ethics that will enable contemporary casuistry (and other theories of applied ethics) to meet these challenges.

A. The Normative Critique

Casuistry, along with other theories of applied ethics, has been criticized for failing to offer a principled rule or reliable mechanism for deciding between conflicting interpretations of cases and, more specifically, conflicting moral appeals within such case interpretations. Without such a principle or reliable mechanism the theory has to presuppose the very moral content it is supposed to provide (Engelhardt, 1996, pp. 42–45; Beauchamp and Childress, 1994, pp. 96–97). Such disagreements enter at two different levels: at the level of paradigm cases and within particular case interpretations.

First, reasonable people may disagree about whether a case counts as a paradigm case against which other cases ought to be measured. Should physicians take the Nancy Cruzan case as the paradigm for resolving end of life disputes when the patient's wishes are unknown, or should the more recent Terri Schiavo case be the new paradigm case? Was the latter a mistaken departure from the appropriate balance between a patient's wishes, a family's wishes, patient well being and dignity, and patient's rights, or is it a new interpretation of the appropriate balance? There is not clear consensus on this case and so bioethicists, physicians, and the public alike point to the paradigm case that supports their predetermined sense of what constitutes the appropriate balance between the conflicting values in such heart-breaking cases.

Second, such disagreements enter at a deeper level, at the level of irreducible moral appeals, such as the appeals to rights, consequences, moral virtue, or justice. As critics have charged, casuistry offers no systematic or principled way to assign priorities to the various moral appeals and such priorities are necessary to resolve value conflict in particular cases. Take the difficult cases involving pediatric assent to research. The rule of thumb in pediatric research bioethics in the United States since 1995 has been to include the assent (or input) from older children and adolescents, in addition to obtaining the consent of the parents (Committee on Bioethics, 1995, pp. 314–317; Bartholome, 1996, pp. 981–982). However, in practice, when the well being of the child or adolescent is at stake, and a coherent, competent, articulate twelve-year-old refuses life-saving treatment, the standard of care in pediatric ethics is to override the minor's refusal in most cases. The developing rights of the minor are then trumped by concerns about the patient's well being, and the parents' rights and obligations toward a child. If multiple moral appeals, such as those to consequences and rights, clash in a particular case, how should we decide what weight to give the various values at stake? When there is an evolving debate within the relevant specialty, as there is in pediatrics, casuistry gives little guidance. This would seem to relegate casuistry's guiding power to settled areas of bioethics. So, even when there is clear consensus on a paradigm or family of paradigm cases, conflict may also arise when deciding how best to balance the conflicting values in the new case, since paradigm cases underdetermine what one ought to do in new cases.

B. The Psychological Critique

The normative critique is a familiar one in discussions of applied ethics and bioethical theory. Less familiar is an objection that I will call the psychological critique, though it is a variation on longstanding concerns about the reliability of intuitions in ethical judgment. Since casuistry centers on a moral epistemology of intuitionism, it is open to the standard objections about bias in moral judgment, forms of self-deception about moral truth, and the social factors which influence our intuitions, especially the relationship between power and consensus. Recent advances in cognitive psychology and social psychology confirm and deepen these longstanding concerns about the reliability of intuitions in moral judgment, especially in social contexts involving collective decision-making.

In 1957, Festinger's groundbreaking work on cognitive dissonance offered a theory to explain the human tendency to reconcile conflicting beliefs to reduce the anxiety and tension of internal and socially perceived inconsistency. Since then a great deal of exciting work has been done to confirm many of Festinger's initial insights and to explore similar phenomena in social decision-making (Festinger, 1957; Elliot and Devine, 1994; Matz and Wood, 2005). Recent work by Wood, for example, explores the degree to which disagreement with others produces arousal and discomfort, or "dissonance stress". This distress can arise from the mere anticipation of joining a group wherein one will hold the minority view or belief, and it can also trigger a change in belief accompanied by an ex-post rationalization of the veracity of the beliefs one held before one joined the group. In general, cognitive dissonance theory demonstrates preferences for agreement over disagreement, the latter creating an unpleasant or uncomfortable state. This state motivates information processing and other mechanisms for change (Matz and Wood, 2005, p. 23). Related research on social influence reveals complex affective and cognitive mechanisms underlying changes in attitudes or beliefs. These attitudes are embedded in social relations and may include social influences arising, for example, from one's membership in the minority or majority in a group, particularly groups with which one may identify more deeply (Wood, 2000). That such behaviors seem independent of the content of particularly held beliefs has significant implications for our confidence in moral judgments in a social context. On the weakest interpretation of the data, ethical theorists should at least be concerned about the relationship between intrapersonal pressures to reconcile moral belief with interpersonal influences, where the latter may have little to do with truth and more to do with how brute disagreement itself creates cognitive dissonance which can in turn encourage belief change without argument or evidence. Returning to casuistry in bioethics, we should ask, how reliable are my intuitions about the proper balancing of values in a withdrawal of treatment case when I am surrounded by other members of an ethics consult team who collectively endorse the decision in Cruzan as the relevant paradigm case? How reliable are the dominant intuitions in the group? And are the latter not what we would ordinarily refer to as "consensus"? Research on cognitive dissonance gives empirical bite to longstanding concerns about the epistemic reliability of intuitionism as a basis for both theorizing and practical moral judgment.

A second related concern with the epistemic reliability of the casuistic method is the problem of bias. One such bias that is particularly relevant to moral decision-making in a social professional context is masked social category bias. Norton et al., have demonstrated a sophisticated tendency to use a perversion of ethical casuistry, understood instead as "specious reasoning in the service of justifying questionable behavior", to mask the real reasons motivating the subject's choice (Norton et al., 2004, p. 817).[6] In this study, the subject was asked to offer objective criteria he would rely on when choosing a job candidate. The subject was then asked to screen candidates and make a choice. Afterwards, the subject was asked to explain the reasons behind the final choice. As it turned out, subjects were making the choice based on prior judgments about a job candidate, in this case, mere knowledge of a candidate's gender. The subjects then searched for reasons to support the biased preference (other than the objectionable reasons of gender). By doing so, the subject could maintain the semblance of objective decision-making based on publicly shared reasons and avoid public criticism and internal feelings of guilt by masking the biased choice (Norton et al., p. 817, following on the work of Pyszcynski and Greenberg, 1987). While the data are not extensive, the study conducted raises a concern worth taking seriously in applied ethics. If it is to contend among the competitors for theories of public justification of practical ethical choices in real cases, casuistry needs to offer a way of safeguarding against this perversion of ethical casuistry, the tendency to rely on cases to support an ex post rationalization of a prior decision or pre-judgment. A clinician makes an all-things-considered best judgment about how best to resolve a case, and then appeals to consensus or a paradigm case to support the decision publicly after the decision has already been made. Thus, the public reasons offered to explain one's choice do not necessarily reflect the real reasons for one's decision. Casuistry does not offer a way to get behind the veil of public reason.

This problem of bias and the concerns raised by cognitive dissonance research suggest there is a significant gap between how real people reason morally and how moral theories assume people reason or are capable of reasoning. Again, if a theory of practical morality has hopes of genuinely guiding ethical decisions, it is hard to imagine how it can succeed if it idealizes the objectivity of persons making the moral judgments, making inferences from initial intuitions to more refined intuitions about cases, and moving from even stable intuitions to resolutions about what to do in particular cases. Each phase in the path from tentative first intuitions to moral action depends on the unbiased, reliable judgments of persons most often made in social contexts, where such judgments have been shown to be subject to bias and social influence.

C. The Sociological Critique

A more recent, third, objection to casuistry stems from a longstanding insight from sociology and is closely related to the post-modern critiques of moral and political theory. "Such critiques have pointed to the universalizing and abstract tendencies of bioethics as limiting the field's ability to engage the lived worlds of diversely constituted and situated social groups, particularly those of the marginalized" (Kelly, 2003, p. 2277). If true, casuistry would fail to capture the true particularity of the human condition in moral choice.

The concern is that the social and political process of identifying paradigm cases and the appeal to reflective equilibrium and consensus may overlook marginalized patient groups and perspectives, or exclude certain groups or ways of thinking. A certain degree of abstraction and generalization is required in any theory, ethical and otherwise. For a theory that is intended to be practical and action-guiding there must be procedures for moving from general principles to concrete recommendations. Casuistry offers just such a procedure in its system of case-based reasoning. However, as discussed above, the procedure relies significantly on professional and social consensus about paradigm cases. In addition to the normative concerns about disagreement, there are sociological errors that can occur in the attribution of moral consensus about cases.

First is an *intragroup error*: one may mistakenly lump members of such groups together, attributing views or characteristics to persons by virtue of group membership, which individual members may not ascribe to. The debate about community consultation in research ethics illustrates this error well. In the context of HIV/AIDS research in traditional communities, where individual consent would violate community norms and traditions of patriarchal or community driven decision-making, a research protocol that takes the consent of the tribal chief or dominate group as representative of all members, or the decision of the husband not to bring his wife to be tested for HIV to be representative of the wife's wishes, commits the intragroup error. Tendency to commit this error in general reasoning about groups raises concerns about the accuracy of attribution in epistemic judgments of those analyzing and recommending bioethics policy on issues such as population-based genetic research or behavioral studies. Such inaccuracies may lead to interventions or policies that do not fairly represent all members of the group, or more seriously, that may violate the rights of more vulnerable and less powerful members of the group.

This leads us to the second type of error that can occur in case-based reasoning: the *exclusion error*. Certain groups may not be large enough or medically/scientifically significant enough to be noticed or included in public moral debates affecting them. Bioethics itself offers an interesting example of the exclusion error. Consider the dominance of rule-centered approaches to bioethics in the United States (whether rights-based, consequentialist, or the pluralistic principles approach offered by Beauchamp and Childress). In one sense, the ethics of care movement can be seen as an important correction to intragroup error, attributing a consensus about the nature of ethics and moral dealings in the world to "American bioethicists." Discussions about an "ethics of care," whether as a substitution for the rule-centered approach or a theoretical complement, highlighted a way of relating to patients that was quite familiar to the profession of nursing and social work, but not part of the modern culture of medicine. Medicine, as the dominant profession, influenced consensus about the relevant ethical norms of medical ethics. The more characteristically feminist, relationship-centered approach to ethics had been overlooked, in part because nursing and social work did not have a dominant voice in the earlier public discussions and consensus formation in medical ethics.

Recent attention to rural health represents a similar correction of a substantive moral oversight in bioethics. Sociologist, Susan Kelly, has argued that bioethics has given little or no voice to rural residents and the practice of rural medicine (Kelly, 2003, p. 2278).[7] Her own research and programs like the National Rural Bioethics Project at the University Montana-Missoula are contributing important insights into different ways of thinking about patients, the physician-patient-family relationship, informed consent, and even such basic notions as time and space. Studies like this at the international level will be central to the development of global bioethics that is more inclusive of the rich varieties of moral life in less studied regions and communities.

Intragroup and exclusion errors are morally problematic for two reasons. First, the errors signal a breakdown in fair and equal representation of all members in important public discussions and consensus formation, and so represent a failure of justice. Second, the errors signal a breakdown in valuable social moral epistemology. That is, a break down in the social institutions and processes necessary to the exploration of moral truth. In the discovery and construction of important moral truths a society, professions, and other social groups are less likely to arrive at a robust truth without the input of overlooked members. Updating John Stuart Mill's insight in *On Liberty* with what we know from modern sociology, the error is significant because it impedes truly *public* moral exploration of the difficult questions of our time. Casuistry, like its competing theories of practical morality in the professions, depends on moral expertise. This expertise is obtained by training and public deliberation in a profession that is not free from exclusionary practices, even if unintentional.[8] The examples of community consultation, ethics of care, and rural medicine illustrate both the great injustice of the silenced voice and the great potential gains of inclusion.

III. CASUISTRY NATURALIZED

Can contemporary casuistry respond to these serious objections? In order to offer a thorough response to each objection, I think a new research program is called for, not just for those defending moral casuistry but also for anyone hoping to defend applied ethics as an action-guiding enterprise. Here I will focus on casuistry's response and recommend a road map for future normative and empirical work, including the questions that need to be answered in order to give a satisfactory response to the above objections, and more constructively, point to exciting new explorations in bioethics and other areas of applied moral philosophy. The proposed approach takes its starting point in the basic sense of ethical naturalism described earlier: that at minimum applied moral philosophy must take into account epistemic and psychological abilities and limitations among those making moral judgments and prescriptions. We need to know the errors of our ways. And better understanding the propensities for misjudgment lends itself to a more robust and action-guiding positive program in applied ethics.

A. Response to the Normative Critique: Casuistry and Moral Compromise

Can the normative critique of casuistry be overcome? Is there a way reliably to resolve value conflict within the resources of the theory? Casuistry is not alone in struggling with this serious objection. All pluralistic theories inherit the criticisms first levied against W. D. Ross. In a system comprised of prima facie values, with no priority or weight given to any one value, how are we to decide between the values when they come into conflict? Casuists, including Brody, explicitly reject the idea of a meta-principle or lexical ranking of competing moral appeals on the grounds that both attempts fail to capture the complex nature of specific cases. In real cases, the strengths and relevance of competing moral appeals can vary depending on the circumstances. Appeals to a single metric, meta-value, or set ranking fail to capture these moral phenomena. Brody openly endorses an alternative he finds troubling: that we rely instead on deliberation and judgment to seek guidance in value conflict (1988, pp. 76–79). The alternative is troubling for three reasons, according to Brody: (1) it offers individuals little certainty and guidance in particular cases, (2) it offers no more certainty or guidance to groups who disagree on the appropriate balance of conflicting values, and (3) there is no theoretical guard against those who wish merely to rationalize a pre-judgment about the case (pp. 77–78). As I have argued, empirical data in cognitive and social psychology raise even deeper and more widespread concerns about the reliability of moral intuitionism, but a possible answer to the normative critique rests within these criticisms. The intuitionism shared by contemporary casuistry and other theories of practical morality assumes an outmoded moral epistemology that calls for revision in light of recent discoveries about human cognitive function. The intuitionism of old assumes the possibility of direct access to objective moral knowledge via direct perception and reflection. This epistemology has been revised in contemporary political theory to include a central role for the types of social institutions that contribute to reliable moral knowledge and the stability of norms and practices. (Buchanan, 2004, p. 95-130). Such institutions, with the right features (such as transparency, for example), provide the checks and balances needed to dampen the effects of biased or faulty moral judgments. A further revision is needed to fill out the details and nature of such bias and error in social decision-making to give us better empirical tools for evaluating competing institutions and arrangements. In addition to the normative and empirical inquiry of institutional design we need also a moral psychology and epistemology that is responsive to information from the social sciences, cognitive science, and neuroscience.[9] Such research is essential to understanding the best approaches to training and expertise in the fields of applied ethics. Within such an empirical framework, a satisfactory response to the problem of normative disagreement should not abandon value pluralism for the reasons given by W.D. Ross and others working in that tradition. Rather, it should embrace the degree to which everyday moral decision-making requires difficult moral compromises over irreducible values. The empirical approach captures an inescapable fact about social moral life: given our own imperfections and genuine value pluralism, where efforts to achieve consensus fail, stable compromises are the best we can hope for in serious moral disagreements.

One of casuistry's strengths, particularly the theory offered by Brody, is its foundation in value pluralism. The domain of applied ethics is most often the public sphere: hospitals, businesses, and other institutions that bring together people from varied moral backgrounds. Pluralism more accurately reflects the complex nature of human moral interactions and by doing so has a better chance of offering relevant guidance in real moral problems. Brody himself presents the limits of the theory as a virtue. Pluralistic casuistry under-determines solutions to concrete cases, but by doing so, remains humble in the face of moral ambiguity. I would like to push this insight even further by offering two additional reasons for remaining committed to a pluralistic theory with moral judgment at its center. First, a strength of the theory that has not been heralded nearly enough is its recognition of the *moral residual.* In the spirit of W. D. Ross, an account of prima facie moral principles that resists the desire to rank or reduce truly conflicting and distinct moral appeals is the only sort of theoretical approach that can show proper homage to the moral loss incurred when any one value is overridden or counter-balanced. That is, the moral judgment approach recognizes the hardships and losses of moral compromise.[10] By explicitly recognizing the messy features of the moral life among others, this model of applied ethics gives proper homage to what is lost in serious moral disagreement: a moral residual that meta-principles and monistic theories pride themselves in wiping away. Second, by acknowledging genuine moral ambiguity, under-determination in specific cases, and human fallibility, this approach encourages moral learning and makes moral debate, reflection, and deliberation essential to the enterprise of moral judgment and action.

What we are as yet missing is the empirical research needed to deepen our understanding of intrapersonal and interpersonal moral compromise and the barriers to and best conditions for successful moral deliberation and moral learning. One of the significant advantages of the naturalist turn in ethics is that it strengthens our empirical understanding of the barriers to conflict resolution, sources of and corrections for bias in moral judgment, and the social influences at play in public moral deliberation. The empirical objections that I canvassed in the critical section above should not be taken as a *reductio* on the pluralistic casuistry offered by Brody. Instead, it should be seen as a road map for future work. Initial data on the complexity of human moral psychology may make any appeal to moral judgment seem more than troubling. Notice, though, that all of practical morality comes back to moral judgment, and usually judgment in a social context. This is not a fatal blow to the possibility of practical moral judgments. This is a brute reality about moral life. We should acknowledge the uncertainties and our human cognitive imperfections and set out better to understand our own limitations and strengths. Expanding our empirical knowledge about the nature of disagreement and moral judgment will bring us closer to more effective solutions to moral conflict, while still honoring what is lost when such solutions are imposed and even accepted.

Primary among the solutions I will discuss in the next section is a more detailed account of the training needed for moral judgments and expertise.

B. Response to Psychological and Sociological Critiques: Social Moral Psychology

Should the complex psychological factors at play in social moral decisions shake our confidence in any theory of applied ethics centered on intuitionism, since such intuitions are proven much less reliable than previously assumed by ethical theorists? Yes, we should be shaken, but not defeated. I have instead proposed that such research offers exciting and promising direction for a deeper understanding of moral judgment among others. Psychological and sociological biases do not discriminate between Kantians and pluralistic casuists. Any theory of applied, practical morality where principles or moral appeals must be applied to specific cases must contend with what naturalized ethics is learning from cognitive psychology and neuroscience. Even those trained to discern cases and recommend action are subject to psychological and social factors that may skew or bias particular decisions. One of the advantages of a naturalized casuistry is that it allows us to correct for judgment bias by building in a central role for trained moral judgments and moral learning.

Where should this research be focused? Two things are still lacking in our present understanding of practical moral judgment: (1) a better understanding of the psychological barriers to, and capabilities for, reliable moral judgment, and (2) an account of the institutions that best support the practice of social moral deliberation, including data on the institutional features that are most responsive to the psychological limitations revealed in the first research arm.

Returning to the literature on cognitive dissonance theory, for a moment, we can see a promising path for future research on the first question. We already know most people have strong preferences for agreement over disagreement, and the latter creates an unpleasant or uncomfortable state. This state then seems to motivate information processing and other mechanisms for change (Wood, 2000). So, how might we harness the motivational boost for agreement and still protect sincerity of belief, for example? What factors or conditions increase the desire to forge ties with others and build support, and what factors tend to encourage dissociation from the group? What are the benefits of dissociation, if any? We still need studies that explicitly control for moral vs. nonmoral content in agreements. Are the pressures to agree stronger or weaker between the two types of conflicts? Is disagreement more or less distressing when it is moral in nature, and why?

Further empirical research can also help address the concerns raised from a sociological perspective, and begin to address the second gap in our current understanding about social moral psychology. Sociological research can help identify group prejudices, power relations, and overlooked segments of society or a profession. The sociological approach offers tools for addressing problems that philosophy on its own has trouble accounting for, or even noticing. By focusing on systems, social patterns, and social institutions, sociology and even cognitive social psychology offer an important complement to existing philosophical work on justice and human rights. In moral and political philosophy, Allen Buchanan has recently directed philosophers' attention to this direction of research (2002). One of the central questions posed by social moral epistemology is this: How do we identify genuine epistemic authorities and avoid factual errors in moral beliefs? This suggests an important direction for research and discussion in bioethics: Given the central role of medical and scientific institutions, professional organizations, human rights and other health advocacy organizations, and universities, in moral reasoning, can we better define the features of such institutions that encourage or discourage robust moral deliberation? Because of the biases in individual decision-making in social settings, institutional safeguards are essential to correcting for bias and offering reliable mechanisms for ethical decision-making.

A central piece of the institutional research project will be further to explore the institutions that best support the teaching of applied ethics. For a moral theory to be truly action guiding we have to ask: Is it reasonable to expect that people can develop this fine-tuned sense of moral balance and can it reasonably be taught to others? If not, the theory is open to charges of elitism or intellectualism: the theory is useful only for exceptionally trained moral experts, such as clergy or other spiritual advisors, or specially trained ethicists. Parents cannot readily rely on its guidance to raise good children and professionals cannot be trained to internalize the skills needed to make reliable moral judgments. This invokes age-old questions about moral education. From Plato and Aristotle to Hutcheson and Hume, moral philosophers have wrestled with the problem of practical moral teaching and the role of natural abilities in moral deliberation and action. Such questions were lost for a time in modern and contemporary moral philosophy, but the naturalist turn is bringing them back into view. Casuistry, like any theory of applied ethics, is subject to the cognitive biases and errors of agents insofar as it seeks to guide everyday moral reasoners. Does it rely on psychological features that are only present by nature in a select few? Which features should we care most about and attempt to teach, even if most of us do not come by them naturally? Can morality be taught, and to whom, and how hard is it to do so? What is the extent and nature of the psychological variation between people and how does cognitive variation affect decision-making styles? Important advances in developmental psychology have begun to answer such questions in pediatric and adolescent populations. Similar research into cultural and sex-based differences is also instructive.

From practical experience in fields like clinical bioethics, legal arbitration, and political diplomacy, we have a wealth of experience on successful decision-making strategies in contexts of conflict and disagreement. Social scientists, with the normative guidance of moral philosophers, can better develop models for reliable moral deliberation in social contexts by investigating and conducting controlled experiments on what we might call *prima facie* virtues of moral perception and deliberation. Such a list can help shape the hypotheses for studies in experimental economics and cognitive psychology. For example, let me propose a preliminary list of *prima facie* virtues for the "good casuist", based on experience and ongoing debate in clinical bioethics consultation:

Courage. To what degree does someone who advises on moral decision-making and conflict need a developed sense of courage and is this something that can be cultivated and taught? Jumping into the fray of moral disagreement, and remaining there until a resolution is reached, takes a certain kind of bravery. It would be interesting to further develop our understanding of the role for courage in moral contexts. For example, if the literature on cognitive dissonance reveals anything, it is the widespread and profound discomfort felt when we find ourselves at odds with those around us, or being pressured into a view we do not hold. Acting on our moral beliefs or acting on behalf of those who cannot speak for themselves, will often require a willingness to risk losing one's own moral interests, not being able to keep one's promises to others, losing face, or being misunderstood. It would be important to investigate related questions about how moral paralysis or indecision arises: what prompts the "deer in headlights" or Buridan's Ass effects in social conflict? When, if ever, is social isolation and dissociation protective for individuals and groups? And to what degree can we reasonable expect people or groups to brave the costly possibilities of sustained conflict or conversion to an opposing moral view?

Mental Flexibility and Integrity. It would also be important to investigate the cognitive flexibility suitable for contexts of heated or serious moral disagreement. With a starting point in complexity and human fallibility the astute navigator of practical morality would need to be open to novel solutions and approaches, and open to compromise, but within what limits? When is someone ill suited for the pressures of moral disagreement over big stakes issues that warrant careful deliberation and not political expediency? On the other extreme, what psychological or social conditions contribute to a *lack* of flexibility—dogmatism or absolutism in ethical discussions? How hard-wired are we to be impatient with others or with circumstances outside of our control, to lose sight of the balanced picture, and all-things-considered judgments? And to what degree can such temperaments be cultivated in one's upbringing or professional training? The counterbalance to mental flexibility is some sense of what can never be given away, an ability to preserve spirit of moral appeals even when making necessary exceptions.

Emotional Sensitivity. A theory of practical morality that fails to recognize an important emotive, affective component in moral judgment would be anemic. Hume taught us this important lesson. Especially for those who are frequently involved in tough moral arbitration, diplomacy, or other highly sensitive social situations, emotional sensitivity seems an essential virtue. In team consultations, for example, it is often helpful to have someone involved who is aware, not only of the conflicting moral interests at stake, but of the fears, anxieties, jealousies, and hopes of those who hold those interests. Someone who asks: What will it mean emotionally for the Jehovah's Witness parents to live with the decision to allow their daughter to be transfused? Can we try to appreciate and sympathize with the decision these parents face? Again, we would want to investigate the optimal balance between such sensitivity on the one hand, and being emotionally tone deaf. Even Hume recognized the value of cooler emotions and distanced perception of an otherwise charged event. But in light of the brute need for moral compromise discussed earlier, an important feature of trained moral perception would be to appreciate that even a rational decision to compromise likely carries a residual sense of loss for the persons accepting the compromise. And again, how difficult is it to master this sensitivity in practice? Might it be a gift of personality rather than a virtue that can be taught? Even if this is the case, a better understanding of the affective features in moral reasoning can give us a better idea about who ought to be recruited for such important positions within institutions where moral knowledge, learning, and conflict resolution take place. This is only a suggested starting point for further empirical research, targeting the questions most central to social moral epistemology and moral judgments in social contexts. A great deal of work remains to be done but the prospects are exciting.

IV. CONCLUSION

In order to be a viable, robust theory of practical bioethics, I have argued, contemporary casuistry needs an empirically grounded understanding of the psychological mechanisms central to moral judgment. This should include an account of the social conditions that make such decision-making robust and reliable, and any such account should include an understanding of the psychological biases that need to be overcome in individual and group decision-making. If casuists join the broader empirical turn in ethics, it will bring two rewards: it will give contemporary casuists a way to respond to otherwise serious criticisms regarding action-guidance in moral conflict, and it will reinforce the pedagogical utility of the theory as an extremely valuable teaching tool in contexts, like hospitals and medical schools, where practical morality needs to be taught, and taught well. Finally, the proposed turn is a natural one for casuistry in more ways than one. A detailed exploration of the social scientific questions raised here is a natural next step for a theory whose ancient roots were firmly fixed in the experience and practice of everyday moral judgment. As such the contemporary empirical turn captures the original spirit of this valuable theory of practical morality.

NOTES

[1] See, for example, Annette Braunack-Mayer, "Casuistry as Bioethical Method: An Empirical Perspective" (2001); Carl Elliott, "Where Ethics Comes From and What to Do About It" (1992); B. Hoffmaster, "Can Ethnography Save the Life of Medical Ethics?" (1992); Loretta Kopelman, "Case Method and Casuistry: The Problem of Bias" (1994).

[2] See for example, Owen Flanagan, *Varieties of Moral Personality: Ethics and Psychological Realism* (1991); Allan Gibbard, *Wise Choices, Apt Feelings* (1990); Mark Johnson, *Moral Imagination: Implications of Cognitive Science for Ethics* (1993); Alvin Goldman, "Ethics and Cognitive Science" (1993); Allen Buchanan, "Social Moral Epistemology" (2002); Michael Slote, "Ethics Naturalized" (1992).

[3] One ought to distinguish between the brand of naturalism that Dennett defends and the brand he is committed to. Tamler Sommers and Alex Rosenberg have argued that Dennett's Darwinian naturalism

commits him to moral nihilism: the view that moral concepts, beliefs, and statements don't merely refer to natural facts in the world but are, rather, false and meaningless. See, Tamler Sommers and Alex Rosenberg, "Darwin's Nihilistic Idea: Evolution and the Meaningless of Life," (2003), p. 655.

[4] See again, Tamler Sommers and Alexander Rosenberg, "Darwin's Nihilistic Idea: Evolution and the Meaningless of Life," (2003), p. 658.

[5] As Sommers and Rosenberg argue, "If all apparently purposive processes, states, events, and conditions are in reality the operation of a purely mechanical substrate neutral algorithm, then as far explanatory tasks go, the only values we need attribute to biological systems are instrumental ones. An evolutionary account of moral belief will not only explain ethics but it will explain it away." (2003), p. 661.

[6] This offers further empirical support for concerns raised by Loretta Kopelman, "Case Method and Casuistry: The Problem of Bias" (1994).

[7] For another critique of the standard understanding of the autonomous subject in bioethics, from the standpoint of rural medicine, see Dien Ho, "Informed Consent in Rural Settings: An Eastern Kentucky Case Study", draft.

[8] For a helpful survey of this issue see Lisa M. Rasmussen, ed., *Ethics Expertise: History, Contemporary Perspectives, and Applications* (2005).

[9] As mentioned previously, I think the objections raised in section II can be raised against any theory of practical morality that grants a central role to intuitions or moral judgments about principles, cases, norms, rules, or competing specifications (see Richardson, 1990; 2000). And so the offered response, to naturalize the moral psychology behind Brody's pluralistic casuistry, is also intended as a preliminary response to the similar problems in competing theories of applied ethics. Given the spirit of this volume, I will not expand on that argument here.

[10] For a full defense of this theory, see Maureen Kelley, *The Nature and Limits of Moral Compromise*, manuscript in draft.

BIBLIOGRAPHY

American Academy of Pediatrics, Committee on Bioethics (1995). 'Informed consent, paternal permission, and assent in pediatric practice,' *Pediatrics*, 95(2), 314–7.

Beauchamp, T.L. & Childress, J.F. (1994). *Principles of Bioethics*, Fourth Edition. New York: Oxford University Press.

Bartholome, W.G. (1995). 'Informed consent, paternal permission, and assent in pediatric practice,' *Pediatrics*, 96(5 pt 1), 981–982.

Brody, B.A. (1988). *Life and Death Decision Making*. New York: Oxford University Press.

Brody, B.A. (1979). 'Intuitions and objective moral knowledge,' *The Monist*, 62(4), 446–456.

Buchanan, A. (2002), 'Social moral epistemology,' *Social Philosophy and Policy*, 19(2), 126–152.

Buchanan, A. (2004). 'Political liberalism and social epistemology,' *Philosophy & Public Affairs*, 32(2), 95–130.

Braunack-Mayer, A. (2001). 'Casuistry as bioethical method: an empirical perspective,' *Social Science and Medicine*, 53, 71-81.

Dennett, D. (1995). *Darwin's Dangerous Idea*. New York: Simon and Schuster.

Elliott, C. (1992). 'Where ethics comes from and what to do about it,' *Hastings Center Report*, 22, 28–35.

Engelhardt, H.T. (1996). *Foundations of Bioethics*, Second Edition. New York: Oxford University Press.

Festinger, L. (1957). *A Theory of Cognitive Dissonance*. Stanford: Stanford University Press.

Flanagan, O. (1982). 'Quinean ethics,' *Ethics*, 93(1), 56–74.

Flanagan, O. (1996). 'Ethics naturalized: ethics as human ecology,' in. L. May, M. Friedman, A. Clark (Eds.), *Mind and Morals: Essays on Ethics and Cognitive Science*. Cambridge: MIT Press.

Hoffmaster, B. (1992). 'Can ethnography save the life of medical ethics?' *Social Science and Medicine*, 35, 1421–1431.

Jonsen, A. R., & Toulmin, S.E. (1988). *The Abuse of Casuistry: A History of Moral Reasoning*. Berkeley: University of California Press.

Kelly, S.E. (2003). 'Bioethics and rural health: theorizing place, space, and subjects,' *Social Science and Medicine*, 56, 2277–2288.

Kopelman, L.M. (1994). 'Case method and casuistry: the problem of bias,' *Theoretical Medicine*, 15, 21–37.

Matz, D.C., & Wood, W. (2005). 'Cognitive dissonance in groups: the consequences of disagreement,' *Journal of Personality and Social Psychology*, 88(1), 22–37.

Norton, M. I., Darley, J.M., & Vandello J.A. (2004). 'Casuistry and Social Category Bias,' *Journal of Personality and Social Psychology*, 87(6), 817–831.

Nozick, R. (1974). *Anarchy, State, and Utopia*. Basic Books: New York.

Pyszcynski, T. & Greenberg, J. (1987). 'Toward an integration of cognitive and motivational perspectives on social inference: a biased hypothesis-testing model,' in L. Berkowitz (Ed.), *Advances in Experimental Social Psychology*, vol. 20. San Diego, CA: Academic Press pp. 297–340.

Quine, W. V. (1979). 'On the nature of moral values,' *Critical Inquiry*, 5, 471–480.

Rawls, J. (1971). *A Theory of Justice*. Cambridge: Harvard University Press.

Richardson, H.S. (2000). 'Specifying, balancing, and interpreting bioethics principles,' *The Journal of Medicine and Philosophy* 25(3), 285–307.

Richardson, H.S. (1990). 'Specifying norms as a way to resolve concrete ethical problems, *Philosophy and Public Affairs* 19, 279–310.

Ross, W.D. (1930). *The Right and the Good*. Oxford: Clarendon Press.

Sommers, T. & Rosenberg, A. (2003). 'Darwin's nihilistic idea: evolution and the meaningless of life,' *Biology and Philosophy*, 18(5), 653–668.

Strong, C. (1999). 'Critiques of casuistry and why they are mistaken,' *Theoretical Medicine*, 20(5), 395-411.

Strong, C. (2000). 'Specified principlism versus casuistry,' *Journal of Medicine and Philosophy*, 25(3), 323–341.

Williams, B. (1985). *Ethics and the Limits of Philosophy*. Cambridge: Harvard University Press.

Wood, W. (2000). 'Attitude change: persuasion and social influence,' *Annual Review of Psychology*, 51, 539–570.

SECTION II

JEWISH MEDICAL ETHICS

CHAPTER 6

INTUITIONISM, DIVINE COMMANDS, AND NATURAL LAW

B. ANDREW LUSTIG

Department of Religion, Davidson College, Davidson, NC, U.S.A.

I. INTRODUCTION

With characteristic erudition, Baruch Brody has made distinguished contributions to many areas in philosophy, ethics, and religious studies—including philosophy of science, philosophy of law, metaphysics, theoretical ethics, and applied ethics, among the latter especially bioethics, medical ethics, and health policy. Other contributors to this volume are ably equipped to explore Brody's focused work on specific philosophical topics. As a scholar who works primarily in the discipline of religious studies, I propose to do something a bit less definitive here. I will consider in brief compass two interrelated issues that have not been at the heart of Brody's inquiry, but that have been raised intriguingly, although only in passing, in several of his writings devoted primarily to other topics. I will consider the question of whether one can identify in Brody's work in philosophical and religious ethics at least the gleanings of what might be called "natural law" categories. Because Brody's own comments about natural law have been limited, my discussion may appear to be largely speculative, but I believe that it will prove instructive. First, I will consider certain similarities and differences between Brody's method in philosophical ethics and his approach to religious ethics more generally and Jewish ethics in particular. In his work in philosophical ethics, Brody develops a moral epistemology that is recognizably intuitionist, but is clearly intuitionism of a distinctive sort. He describes his substantive moral method as a form of pluralistic casuistry, but it again emerges as casuistry of a distinctive sort when compared with other versions of pluralistic theory, largely because of Brody's distinctive version of intuitionism. Both aspects of Brody's ethical perspective—a case-driven intuitionist moral epistemology and a substantive commitment to moral pluralism—would appear to distinguish it sharply from both classical and revised approaches to natural law reasoning; nonetheless, some attention to specific points of comparison and contrast may be useful.

Second, in Brody's general analysis of approaches in religious ethics, he has defended the cogency of certain features of religiously based morality against standard criticisms of divine-command theory. In his approach to the use of

M.J. Cherry and A.S. Iltis (eds.), Pluralistic Casuistry, 85–97.

specifically Jewish materials in ethical discussion, Brody expresses skepticism about efforts to identify certain universally binding duties (the so-called Noahide commands) as, in effect, a Jewish version of natural law reasoning. In light of his expressed skepticism about that sort of identification, I will contrast Brody's position with the far more positive identification by Jewish scholar David Novak of certain natural law features to be found in Jewish thought. While Novak acknowledges the presence of certain divine command elements in Jewish morality, he also argues for the plausibility of seeing some aspects of the Noahide commands in natural law terms. I will place Novak in conversation with Brody for purposes of both contrast and comparison. In my conclusion, I will review certain features of the Noahide commandments—ones which Novak describes in natural law terms—as the context for possible rapprochement between standard natural law reasoning and casuistry and Brody's own philosophical intuitionism. While clear theoretical differences between standard accounts and Brody's approach will be noted, the possibilities for significant overlap between Brody's perspective and other versions of practical reasoning will also be identified.

II. BRODY'S PHILOSOPHICAL MORALITY

A. Intuitionism

Brody's own self-identified theory is a pluralistic form of casuistic intuitionism. A word about each of these three elements is very much in order as a prelude to subsequent comparisons between Brody's approach and other perspectives. As Brody suggests, most versions of intuitionism make claims about certain "indubitable and evident intuitions of the truth or falsehood of purported moral rules" (2003, p. 45). That is the case with the intuitionism of Whewell and Ross, as well as with the principlism developed by Beauchamp and Childress that, while emphasizing so-called mid-level principles rather than moral rules, in effect transfers Ross's characterization of moral intuitions in relation to moral rules to the obviousness and theoretical incorrigibility of its four basic principles of non-maleficence, beneficence, respect for persons (autonomy), and justice (Beauchamp and Childress, 2001). Brody argues that two key features of his approach immunize it from standard objections to intuitionism in moral theory. First, according to Brody,

> the fundamental moral intuitions are judgments about the rightness or wrongness of particular actions, the justice or injustice of particular social arrangements, the blameworthiness of particular individuals, etc. These judgments are neither evident nor indubitable. What they are are tentative judgments which are based upon our observations of these particular individuals, actions, or arrangements, but which go beyond what is observed or what can be deductively or inductively inferred from what is observed (and do so without the aid of any moral theory (Brody, 2003, pp. 45–46).

These primary intuitions, then, are moral judgments about particular cases rather than judgments about moral rules or principles. Second, based on these "first intuitions" as the basis for moral theory, we then seek to "find a theory which systematizes these intuitions, explains them, and provides us with moral judgments about cases for which we have no intuitions" (2003, p. 46). Third, and quite importantly, in the process of systematizing our intuitions according to theory, the intuitions that provide basic data for subsequent systematization are themselves corrigible; according to Brody, we may well modify or even reject our initial judgments in the process of theory formation. The theoretical generalizations that result from systematizing our intuitions are subject to various sorts of challenge, either "because they don't allow for generalizations that we intuitively judge should be allowed for or because they come into conflict with other generalizations" (2003, p. 50). As Brody observes, that may mean that in the process of systematization we will "formulate more complex and non-intuitive generalizations to replace the earlier ones" (p. 50). But the need for such correction does not, of itself, dislodge the primacy of intuitions about particular cases as a matter of moral epistemology; indeed, the corrigibility of such intuitions emerges as a strength of Brody's approach relative to other accounts of intuitionism.

B. Brody's Pluralistic Casuistry

Brody's moral epistemology is, as we have seen above, a distinctive version of intuitionism. His account of substantive morality incorporates a pluralistic moral theory. Whether pluralistic theories are best viewed metaethically in ontological, epistemological, or logical terms, two basic claims undergird all such perspectives—one about supervenience, the other about the irreducible plurality of basic normative appeals. Supervenience, Brody says, involves the "claim that the rightness or wrongness of actions (or the knowledge that we can draw about that rightness or wrongness) is dependent upon other [non-moral] properties of those actions (or our knowledge of those other properties or the premises we accept about those other properties)" (2003, p. 32). Pluralism involves the claim "that there are several such properties and that they are independent of each other." On the one hand, Brody does not defend pluralism by "a demonstration of the fundamental truths of some pluralistic theory" (p. 39) in a manner akin to such attempts for various monistic accounts of morality. He eschews that effort because, as he confesses, he is "skeptical about the whole idea of demonstrating the truth of the fundamental claims of some moral theory" (p. 39). On the other hand, as I have suggested above, Brody's method of intuitionism moves from primary data about specific actions (rather than general moral claims) to the formation of moral theory in a process he finds analogous to that of theory formation in scientific practice. Thus, Brody observes that in both science and ethics "a scientific generalization or theory need not account for all of the data in its area in order to be successful.... [and] ... I believe that the same thing happens in the process of morally theory formation" (p. 47). Therefore, "the data about which we theorize are ... initial intuitions", while "[t]he goal is to find a theory that systematizes these intuitions, explains them, and provides help in dealing with cases about which we have no intuitions" (1988, p. 13).

In the context of Brody's intuitionism, his method of casuistry about specific actions and arrangements is substantively and procedurally pluralist; i.e., "one that accepts the legitimacy of a wide variety of very different moral appeals", including the appeal to consequences, to rights, to respect for persons, to virtues, and to cost-effectiveness and justice (1988, p. 9). Unlike monistic appeals, Brody proposes that "we should regard each of these [monistic] theories as correct in advocating a moral appeal whose use is certainly legitimate but limited". Therefore, we will "need to incorporate all of these moral appeals, each treated as an independent appeal whose validity is not based on its derivability from some other moral appeal, into a pluralistic moral theory" (p. 9). The difficulties of standard monistic accounts to deal with significant instances of deep moral ambiguity are well-known, and I will not rehearse them here. But the inability of standard monistic approaches to account for such deep ambiguity provides at least negative evidence for their systemic inadequacy, while, according to Brody, such ambiguity is positive evidence for the preferability of pluralism. In the presence of two independent right-making properties (A and B) of two possible actions, if action 1 exhibits A and action 2 exhibits B, "there is room in ... simple pluralism for moral uncertainty or disagreement based upon different judgments about the moral priority of A rather than B" (2003, p. 38). The power of Brody's pluralism is in its acknowledgement of such substantive complexity, while his action-based intuitionism allows for a dialectic between the initial data of intuitions and subsequent theory formation that better comports with the presence of genuine ambiguity in moral choices and actions.

C. Brody's Intuitionism and Natural Law Perspectives

Natural law accounts have traditionally provided axiologies either in terms of certain natural inclinations which provide the basis for subsequent deliberations of practical reason, or, more recently, in terms of certain basic and incommensurable forms of human flourishing that provide *per se nota* first principles of practical reasoning. Traditional Scholastic accounts of natural law proceeded from an explicitly theological context of reflection; natural law is the "eternal law" as expressed in the human creature. For Thomas, natural law is properly understood within a robust teleology and according to the central function of "right reason" (*recta ratio*) within that metaphysic. For Thomas, both theoretical and practical reasoning are functions that are most distinctively human.[1] In the moral realm, right reason emerges as that practical principle which, in effect, harmonizes the various ends appropriate to the different spheres of individual and group existence.

More recently, so-called "revised natural law" accounts have been developed which are expressly philosophical in their orientation. John Finnis is a primary exemplar of this shift. In a number of works, Finnis sets forth a natural law morality binding upon humans *qua* humans as a matter of the logic of practical reason itself, independent of any even implicit reliance on a theological context for its ultimate justification.

According to Finnis, there are two sorts of core natural law principles. First, there are "basic practical principles which indicate the basic forms of human flourishing as goods to be pursued ... and which are ... used by everyone who considers what to do, however unsound his conclusions" (1983, p. 23). Second, there are

> basic methodological requirements of practical reasonableness (itself one of the basic forms of human flourishing) which distinguish sound from unsound practical thinking and which, when all brought to bear, provide the criteria for distinguishing between acts that (always or in particular circumstances) are reasonable-all-things-considered (and not merely relative to a particular purpose) and acts that are unreasonable-all-things-considered, i.e., between ways of acting that are morally right or morally wrong (1983, p. 23).

With regard to the first sort of principle described above, Finnis contends that the basic goods in his axiology become general *values*, rather than merely hypothetical means to preference satisfaction, in the deliberations of practical reason itself as specified by the second set of methodological requirements. In his account of that process, Finnis compares practical reason's first principles to those of theoretical reason, very much in the tradition of Aristotle's own discussion of *theoria* and *phronesis* (1980, pp. 31–32, 64-69). The point of theoretical reason is to answer questions about the constitution of states of affairs or of objects. Practical reason, on the other hand, is directly concerned with human *action*. It includes, of course, those descriptive components that one must know in order to act purposefully, but practical reason is distinguished by its prescriptive mode: one reasons practically to determine what should be the case. Both activities are, by definition, rationally intelligible. While since the time of David Hume, the different activities and objects of theoretical and practical reasoning have been captured by short-hand distinctions between the realms of fact and value, Finnis moves beyond such distinctions in his own discussion by arguing for a meaningful similarity of basic epistemological form at the level of first epistemic principles. In doing so, he develops two key points: first, that classical distinctions between theoretical and practical reason indeed obtain, but second, that the functioning of first principles in both realms allows us to draw meaningful analogies. For example, in discussing the first principles of theoretical reason, Finnis observes that

> [I]n general, all theoretical disciplines rest implicitly upon epistemic principles, or norms of theoretical rationality, which are undemonstrated, indemonstrable, but self-evident in a manner strongly analogous to the self-evidence ascribed by Aquinas to the basic principles of practical reasonableness (1980, p. 32).

The point of such remarks is not to buttress the case for relativism in the theoretical realm. Rather, it is to recognize the usefulness, reliability, and adequacy of scientific principles and models, which attain coherence, indeed "functional objectivity," despite their failure to achieve the status of strict analytic necessity. But the success of the analogy Finnis seeks to draw between *theoria* and *phronesis* rests upon his account of the first principles of practical reason as the second term. On the one hand, Finnis agrees with the thrust of Hume's objection to illicit inferences from fact to value (Hume, 1968, p. 43). On the other hand, Finnis concludes that the logical truth of Hume's point "itself in no way entails or authorizes Hume's conclusion that the distinctions between 'vice' and 'virtue' are not perceived by reason" (1980, p. 37). For Finnis, as for Aristotle and Aquinas, the two functions of the intellect are relevantly similar in the obviousness of their first principles, while nonetheless distinguished by their different *teloi*. When the subject matter is ethics, Finnis describes the hybrid character of practical reason, wherein he seeks to locate the sense of natural obligation that reflection upon the basic goods engenders:

> [E]thics is a genuinely theoretical pursuit (i.e., it contains elements of description). But ethics is also precisely and primarily ('formally') practical because the object one has in mind in doing ethics is precisely my realizing in my actions the real and true goods attainable by a human being and thus my participating in those goods. Notice: ethics is not practical merely by having as its subject matter human action (*praxis*). Large parts of history and psychology and of anthropology have human *praxis* as their subject matter; but these pursuits are not practical. No: ethics is practical because my choosing and acting and living in a certain sort of way ... is not a ... side-effect of success in the intellectual enterprise; rather, it is *the very objective primarily envisaged* as well as the subject matter about which I hope to be able to affirm true propositions. (1983, p. 3).

With such fundamental features of Scholastic and revised natural law as background, I can now offer a brief comparison and contrast with the core features of Brody's philosophical perspective outlined above. As we have seen, Brody's primary intuitionism does *not* proceed from generalities about principles or rules or, by extension, general forms of human flourishing. To be sure, systematization among and across our primary intuitions is a necessary feature in developing subsequent moral theory, but particular intuitions about actions and cases remain the source of subsequent systemic adjustment. Clearly, then, Brody's philosophical commitment to the primacy of intuitions about particular cases is dissimilar to the general axiology exemplified by both Aquinas and Finnis. Both Scholastic and revised natural law accounts proceed from general features, practical reasoning, and axiology as primary; only in the subsequent process of practical reasoning do insights and particular applications emerge. This precisely reverses Brody's intuitionist epistemology; hence, in this regard, the differences between natural law accounts and Brody's philosophical approach appear quite significant.

At a later point, however, there may emerge significant agreement in judgments about particular cases between Brody's pluralistic casuistry and natural law-based casuistry. That agreement could occur at two quite different points. There may be agreement between Brody's primary intuitions about particular cases and the secondary judgments in natural law reasoning made as subsequent applications of primary general principles. Alternatively, there may be agreement between Brody's application of theoretically articulated "principles" that systematize (and perhaps modify or even discard) intuitions about primary cases and the sometimes variable application of primary natural law directives (what Thomas called the way of "determining," and what others have called "seconday" or "tertiary" directives of natural law). Still, it is difficult to generalize about the likelihood of agreement between Brody's pluralistic casuistry and natural law approaches. As

Albert Jonsen and Stephen Toulmin acknowledge, traditional natural law-based accounts led to casuistic development and applications from more transparent, perhaps even paradigmatic, cases to other far less obvious cases, with the degree of "certainty" or universality attached to the latter judgments thereby lessened (1990). To that extent, perhaps, the "corrigibility" of Brody's primary intuitions in subsequent theory-building finds something of a functional parallel in such subsequent extensions of natural-law reasoning. Nevertheless, while the apparent "obviousness" imputed to the primary principles of practical reason suggests a parallel with Brody's primary intuitions, Brody's theoretical openness to the subsequent corrigibility of his initial data appears to find no parallel in natural law accounts, whether Scholastic or revised. Indeed, despite the salience of much recent criticism of natural law for its static and invariant character, the target of such criticism has been the so-called secondary directives or applications, rather than of the primary principles themselves. Thus, even if we might anticipate the likelihood of concrete agreement on particular cases between Brody's intuitionism and natural law approaches, Brody's openness to corrigibility of his primary data of intuitions suggests a flexibility in theory formation and moral judgment that may be less characteristic of natural law approaches. That obvious difference between Brody's perspective and all renditions of natural law should not be gainsaid.

D. Brody's Discussions of Jewish Ethics: The Status of the Noahide Commandments

I turn now from Brody's philosophical approach to his analysis of religious arguments in ethics, and more specifically, to his analysis of the use of Jewish materials in ethical argument. In a discussion of the use of halakhic materials in medical ethics, Brody considers three ways that such materials may be employed: first, as a "source for ideas" that can be "defended independently of their origins"; second, as a source for Jewish believers specifically bound by halakhic law; and third, as a source for ethical prescriptions for society in general (2003, pp. 266–267). Brody's comments on the third usage of halakhic materials are especially relevant to the status of the so-called Noahide commandments, deemed to be binding on both Jews and non-Jews. The "seven commandments of the children of Noah" are as follows:

> (1) the requirement to establish a judicial system in society (*denim*); (2) the prohibition of blasphemy (*birkat ha-shem*); (3) the prohibition of idolatry (*avodah zarah*); (4) the prohibition of wanton destruction of human life (*shefikhut damin*); (5) the prohibition of adultery, incest, homosexuality, and bestiality (*gillui arayot*); (6) the prohibition of robbery (*gezel*); (7) the prohibition of eating a limb torn from a living animal (*ever min ha-hai*) (Novak, 1998, p. 149, quoting T. Avodah Zarah 8.4; B. Sanhedrin 56a).

Brody expresses reservations about characterizing these Noahide commands as a Jewish version of natural law conclusions. He acknowledges that some would interpret the Noahide commandments as "the basic principles which natural reason would accept governing the relations between humans and God, interhuman relations, and man and nature" (2003, p. 269). Brody appears unpersuaded by such interpretations for three reasons. First, he cites Maimonides concerning the "true motives for the observance of these commandments", viz., because God has commanded them, not because they are innately reasonable. Second, important details concerning the interpretations and application of Noahide commandments are derived from biblically or Talmudically based arguments rather than from considerations of "pure reason." Third, not all of the commandments (especially the so-called "law of mixed species") comport with "natural law categories" based on considerations of practical reason (2003, p. 269). Let us consider each of these arguments in greater detail.

In quoting Maimonides regarding the question of motivation, Brody's judgment about the best way to characterize Jewish ethics appears to comport with a more general argument that he has offered concerning divine-command theory as a meaningful basis for some forms of religious morality. Brody has taken issue with a number of philosophical critiques of the cogency of divine-command accounts, which have considered classical arguments first enunciated in Plato's *Euthyphro*, 9d-11b. As classically formulated, the basic issue concerned the primacy of the divine will as the basis of moral judgments. Are morally good acts willed by God because they are morally good, or are they morally good because they are willed by God? The horns of that apparent dilemma are well known in the philosophical literature. Whichever way a theist seeks to resolve the dilemma, problems arise. If one affirms the priority of independent criteria of rightness and wrongness, the cogency of divine command theory seems thereby to be undermined. If one affirms the primacy of the divine will, other sorts of issues arise, especially so-called problems of arbitrariness and of abhorrent demands.

It may be relevant to our subsequent discussion to consider, in brief, several elements of Brody's discussion of standard critiques of divine command theory and his own more positive construal of that approach. According to Brody, there are a number of dubious premises at work in the standard critiques of divine command theory, and he provides a careful analysis of the inadequacy of several such arguments that I will not pursue here. But even more fundamentally, Brody raises the issue of whether or not the fact that God is our creator is, in and of itself, of moral relevance. Here he draws an important analogy between God as creator and obligations to parents:

> Consider, for a moment, our special obligation to obey the wishes of our parents. Why do we have that obligation? Isn't it because they created us? And since this is so, we seem to have an obligation, in at least some cases, to follow their wishes. So, *x*'s being our creator can be part of a reason for doing (refraining from doing) an action A if the other part is that that is *x*'s wish. And if this is so in the case of our parents, why shouldn't it also be so in the case of God (Brody, 1974, pp. 594–95)?

In light of such special obligations owed to God as creator, Brody's general conclusions about the (at least limited) plausibility of divine command theory appear warranted. Moreover, they may well provide an important backdrop to his own skepticism about interpretations of the Noahide commandments as a Jewish version of natural law.

Nonetheless, given Brody's philosophical commitment to intuitionism as a generally shared moral epistemology—seemingly by believers and non-believers alike—two issues deserve further scrutiny. Maimonides, as cited by Brody, emphasizes obedience to divine command as the primary motive for ascribing righteousness to the Gentile. Beyond the questions of motivation and the ascription of righteousness, however, the question remains whether independent substantive grounds can be identified in traditions of Jewish reflection that might justify the characterization of the Noahide commands (partially or in their entirety) as a Jewish version of natural law. Moreover, in light of Brody's general philosophical position, are there any possibilities for "teasing out" similarities between Brody's philosophically developed moral theory and the elements identified by others as "natural law" aspects to the Noahide commands? That is, does an affirmative answer to the question of whether there are some natural law elements in the Jewish discussion of the Noahide commands provide Brody a clearer way to link his general philosophical account of moral judgment to his discussion of the warrants for halakhic judgments about universally binding obligations?

E. Natural Law in Jewish Thought?

As we saw above, Brody expresses reservations about the description of the Noahide commandments as a Jewish version of natural law. This is the case, it would seem, because the primary *motivation* for obedience by either Jews or Gentiles, according to Maimonides, is obedience to the commands of God. As David Novak observes, that God's commandments are to be obeyed, on that basis, constitutes a "reason, not just a rule": "The will of the creator is to be obeyed … because to disobey it is to deny the truth of creation, which is that in all areas of existence, including our moral life, 'He has made us not we ourselves' (Psalms 100:3)" (1998, pp. 64–65).

However, in contrast to Brody, Novak also traces a self-consciously rationalist strain in the Jewish tradition, wherein "the commandments are to be obeyed because to obey them is to attain what the wisely beneficent creator has intended as good for us" (1998, p. 65). Hence the classical question, as posed by Plato's *Euthyphro*, arises again in this context and seems relevant to judgments about the plausibility of interpreting the Noahide commandments in natural law terms. This would especially seem to be the case when considering commandments that govern interhuman relations. In a careful survey and analysis of the relevant literature, Novak traces the conceptual distinction introduced in the ninth century between "revealed commandments" and "rational commandments," which he deems "an authentic development of earlier rabbinic sources". Novak quotes an early rabbinic source as follows:

> 'And my ordinances you shall practice' (Leviticus 18:4)': these are matters written in the Torah which even if they had not been written there, reason would have required that they be written … Some examples: laws prohibiting robbery, laws prohibiting

> incest, adultery, homosexuality and bestiality, laws prohibiting idolatry and blasphemy, and laws prohibiting shedding human blood. Even if they had not been written [in the Torah], reason … would have required that they be written (1998, p. 73, quoting *Sifra*: Aharei-Mot, ed. Weiss, 86 a. See B. Yoma, 67b).

Obviously, not all of the above ordinances are "ethical" in the modern sense; they include explicitly religious duties as well as most of the Noahide commandments governing interhuman relations. However, for purposes of present discussion, I will focus on the latter.

As Novak observes, "The debates over natural law in the Jewish tradition have inevitably become debates over the meaning of the Noahide commandments" (1998, p. 149). The reasons for this focus are clear enough. First, if one analyzes those Noahide commands that are "ethical" in the modern sense (that is, as involving interhuman duties of other-regardingness), broader warrants for such universally binding ethical obligations might be more likely to be identified than explicitly religious warrants at work in the halakhic materials. Second, in medieval rabbinical discussions of the Noahide commands, such laws were viewed in terms quite similar to medieval perspectives on the *ius gentium* in Roman law, "namely, points in common between *ius civile* (for which the Jewish analogue is the Torah: Scripture and Rabbinic Tradition) and the actual practices of the gentiles at hand." In their discussions of the Noahide laws, the rabbis, according to Novak, seek to "derive them from actual scriptural statements that seem to be about humankind *per se*" (1998, p. 151) However, in Novak's judgment such verses are not, *strictu sensu*, the "sources of these laws, but … used as allusions (*asmakhta*) at best" (p. 151). Novak therefore concludes that "the Rabbis were engaged more in speculating about the overall teaching of Scripture and its analogues in the outside world than they were engaged in strictly legal exegesis when they were developing the doctrine of the Noahide law" (p. 151). And it is the "speculative character of rabbinic thought" here that, for Novak, "strengthens the case of those who have argued for natural law in Judaism, and who have located natural law primarily in the doctrine of Noahide law" (p. 151). The Noahide laws that emerge from speculations are "basically imagined (that is, abstracted from generalizations) rather than strictly derived from specific authoritative texts … Noahide law functions more as a system of principles than it does as an actual body of rules. This is brought out by the very generality of these seven laws; they are more like categories than like immediately prescriptive precepts" (p. 151).

Here, Novak's interpretation of the legitimacy of describing the rabbinical speculations in natural law terms distinguishes his judgment from Brody's more skeptical assessment. However, the natural law elements that Novak identifies in rabbinical discussions here are of a quite limited sort because Novak differentiates two quite different approaches to natural law. The first approach, as seen in Aquinas and later Scholasticism, involves straightforward teleological speculation; i.e., reflection "on what the ends of law are and how natural law precepts are the proper means to fulfill them" (Novak, 1998, pp. 125 ff., p. 151). "Nature", in this approach, is "an all-intelligent encompassing whole, each of whose parts is a good attracting intelligent human action" (Novak, 1998, p. 152). The second approach, which is the one Novak defends, is negative rather than positive in its orientation. Here, natural law theory moves from "the inherent negative limits of the human condition" and, therefore, sees law as "the way of practically affirming the truth of that limitation of a finite creature [as] apprehended by its intelligence" (Novak, 1998, p. 152).

However, since Novak and Brody agree that that halakhic law in Judaism is primarily inspired by the requirements of the Torah—that is, law formulated and developed within a foundationally covenantal perspective—why and how does Novak nonetheless affirm a substantive place for natural law reasoning in Jewish ethics? As he observes about the Noahide commands,

> six are negative and only one positive. Moreover, the positive one, namely, the requirement to establish a judicial system, is actually for the sake of the adjudication of cases involving the other six, negative precepts. In other words, it is a procedure that functions as a means to the actual social enforcement of the others (1998, p. 154).

In light of their essentially negative character, Novak concludes that "the data from the rabbinic sources are more supportive" of a minimalist perspective concerning Noahide law than a more positive characterization (1998, p. 156).

Nonetheless, even that minimalist perspective is at odds with the views of many recent commentators and Novak, therefore, draws a second important distinction between nonrational and rational commands in rabbinical discussions. In his argument, Novak critiques the work of American Orthodox theologian Marvin Fox. Fox has been an outspoken opponent of those who would identify natural law strains in the Jewish tradition. For example, in commenting on the well-known passage (B. Yoma 67b, quoted in Novak, 1992, p. 25) that if laws against idolatry, murder, robbery, and so forth had not been written in the Torah, "reason would have required that they be written", Fox concludes the following: "What is asserted is only that, having been commanded to avoid these prohibited acts, we can now see after the fact, that these prohibitions are useful and desirable" (quoted in Novak, 1992, p. 25). Novak rejects Fox's conclusion as an adequate characterization of the Jewish moral tradition because Fox fails to distinguish nonrational commands, such as the proscription of eating pork, from the separate category of rational commandments discussed in many rabbinical sources. Paradigmatic of the latter is the proscription of murder. Novak argues, contra Fox, that the proscription of murder "does not introduce us to a new experience whose meaning is only subsequently inferred; rather, it is itself inferred from an experience the Rabbis considered to be universal" (1992, p. 26). Moreover, the "phenomenological sequence" in each type of command is distinctive. In non-rational commands, one moves from commandment to experience to "inference of secondary meaning." In rational commands, one moves from experience to the inference of primary meaning and only then to the commandment "whose reason is the continued duration of that which the experience showed to be good" (p. 26). Hence, relative to the proscription of murder, "the fact that ... [it] is also a divine commandment indicates that it has a covenantal meaning *over and above* its ordinary human meaning", whereas the non-rational commands "have only covenantal meaning, at least for us in this world" (p. 27).

Novak's discussion is far more extensive than his pointed critique of Fox's position, but his distinction between non-rational and rational commands provides the general basis for his affirmation of natural law elements in Jewish thought. As we have seen, Brody expresses reservations about depicting the Noahide commands as a Jewish version of natural law. However, there may be greater room for rapprochement between Novak and Brody than their different interpretations suggest. First, Novak's version of natural law is largely negative and minimalist, while Brody's skepticism appears to be directed at more expansive and robustly teleological versions of natural law. Second, while not specified in terms of Brody's intuitions about particular actions and judgments, Novak's emphasis on the primacy of experience in the phenomenological sequence of the rational commands may generate significant agreement about the bindingess of certain apparently universally held proscriptions. To be sure, Novak does not develop the epistemological basis of his claims about such universal experiences, but I see nothing in principle to deny the possibility of recasting Novak's claims about the rational commands in terms of Brody's version of intuitionism. (Much would depend, of course, upon the matter of whether Novak sees such primary experiences as corrigible upon further reflection, in keeping with Brody's own theoretical dialectic.) Finally, in seeking broader bases for the rational commands in expressly religious discourse, possibilities for fruitful exchange between philosophical and religious accounts may emerge—not by ignoring the obvious distinctions between their broad justificatory contexts, but by seeking to identify common features of experience and reasoning between philosophical and religious perspectives.

III. CONCLUSION

I close with a series of summary observations about Brody's approaches to philosophical and religious ethics, along with the questions they invite for ongoing discussion.

First, in his judgments about Jewish ethics, Brody appears to emphasize its divine-command character. In light of Novak's distinction between non-rational and rational commands, do greater possibilities emerge for independent (or at least separable) appeals in Jewish ethics than Brody's emphasis on divine-command suggests? If so, how might these broader philosophical and methodological elements interact with divine-command elements in religious ethics?

Second, Brody distinguishes his account of intuitionism from other versions of that epistemology, as well as from natural law perspectives. Despite such differences, are the intuitions upon which Brody relies likely to find affinities with the concrete specifications of more general norms developed in natural law accounts? Despite obvious differences in the way that the primary data for subsequent reflection are specified, might there be significant convergence at the "mid-level" of theoretical application? I would suggest that we might expect that to be the case, given the rich tradition of casuistry developed in natural law contexts.

Finally, Brody's reservations about describing Noahide commandments as a Jewish version of natural law stand in contrast with Novak's conclusions. If Novak's "minimalist" reading of natural law elements in Judaism is defensible does it suggest the plausibility of some shared warrants for universally binding claims between expressly religious approaches motivated by obedience to a creator God and philosophically grounded accounts of interhuman duties?

NOTE

[1] *ST* (London: Burnes, Oates and Washbourne, Ltd., 1929 [Dominicans' Edition], I-II, 1:1" "Man differs from irrational animals in this, that he is master of his actions. Wherefore those actions alone are properly called human of which man is master. Now man is master of his actions through his reason and will, whence free will is defined as the faculty and will of reason. Therefore, those actions are properly called human which proceed from a deliberate will."

BIBLIOGRAPHY

Beauchamp, T. & Childress, J. (2001). *Principles of Biomedical Ethics*, Fifth edition. New York: Oxford University Press.

Brody, B. (1988). *Life and Death Decision Making*. New York: Oxford University Press.

Brody, B. (1974). 'Morality and religion reconsidered' in B. Brody (ed.), *Readings in the Philosophy of Religion*. Englewood Cliffs, New Jersey: Prentice Hall, pp. 592–603.

Brody, B. (2003). *Taking Issue: Pluralism and Casuistry in Bioethics*. Washington, D.C.: Georgetown University Press.

Finnis, J. (1980). *Natural Law and Natural Rights*. Oxford: Clarendon Press.

Finnis, J.:(1983). *Fundamentals of Ethics*. Oxford: Clarendon Press.

Hume, D. (1968). 'A treatise on human nature,' in H. Aiken (ed.), *Hume's Moral and Political Philosophy*. New York: Hafner Press.

Jonsen. A. & Toulmin, S. (1990). *The Abuse of Casuistry: A History of Moral Reasoning*, Reprint edition. Berkeley: University of California Press.

Novak, D. (1992). *Jewish Social Ethics*. New York: Oxford University Press.

Novak, D. (1998). *Natural Law in Judaism*. Cambridge: Cambridge University Press.

CHAPTER 7

IN CASE: CONTINGENCY AND PARTICULARITY IN BIOETHICS

DISCURSIVE METHOD AND CLINICAL MIDRASH BRIEF NOTES ON A LESSON FROM MY TEACHER BARUCH BRODY

LAURIE ZOLOTH

Director, Center for Bioethics, Science and Society, Northwestern University, Feinberg School of Medicine, Chicago, IL, U.S.A.

I. OPENING AND SAYING: IN BIOETHICS

When you work at the major regional pediatric medical center, you can watch the evening news and see the case coming. It is summer, and the first drowning case is on, in a blur of ambulance lights flashing on the walls of a small suburban apartment house, and a very small body on the stretcher. The reporter interviews a stunned young mother who tells the camera that she wants to beg other parents to simply move away from places with pools. "You just cannot make it safe," she tells us. "I am talking now to save your child," she says.

There were fences, closed doors, a careful grandmother. But nothing keeps us perfectly safe. I turn from the screen to my oldest son, the grown-up. "I am going to see that one," I say. "God, I hope she lives," he says. "Look how little she is." The TV is on to the next drama, framing the world as a series of hungers, losses, desires, with commercial breaks for solutions that you can buy. I am thinking—"or I hope—that she is swiftly dead".

A month later, when the call comes for a consult, I remember the name from the TV story, and it is the very worst of options. The little girl is not quite dead and not quite alive, but comatose, unstable (an individual that Baruch Brody has named a "not dead, but no longer a person"), her case deepening into a tragic finality. Her father now wants to ask the doctors about when they should stop the life support, the tubes that are feeding her. And, the doctors want to ask the ethics committee.

We meet in the hospital cafeteria conference room. It is too hot in here, there are no windows, and the walls are thin enough that you can hear the laughter and calls of other kids in the hall outside, alive, alive as we listen to the story. We are quiet, waiting for the last committee members, the parents politely sitting, and the

M.J. Cherry and A.S. Iltis (eds.), Pluralistic Casuistry, 99–108.

laughter of other people's children rises outside in bursts, and they keep their polite faces. When we hear them laugh, I see the attending look at me, and shudder.

Not everyone can come, it was short notice and, it being summer, the season of swimming pools, after all, people are away. Sister Mary Margaret is here, and Lilly Jolson, and Catherine Key representing the community of parents, one with her very severely disabled child, the other with her story of her child's fitful cancer recovery: they come to our committee meeting, wearing their stories like letters on their skin. There are two doctors, Roger, the serious assistant head of the NICU, and Len, the hematologist; the committee social worker, Bettina, is here, and the risk manager, Lawrence, and the new lawyer, Pam, edgy, and fiddling with her pen as she walks in the door.

The child's health care team speaks first: of the hopelessness of the case, of the full 10 minutes before the child was found, the full 43 minutes before her heart beat again, the abnormality of the evoked potentials, the EEG, the CT scans. It is early in the course of things, only 30 days. But still. The rehabilitation team sees little hope. The baby girl is as limp as a rag doll, and this is not a good sign.

The parents speak. Father first. "I have spent the night in thought and prayer," he says. "And I know that it's time my baby girl is with my Lord. I think that she is there now, and that we need to let her body go there too. But I know that this has got to be a shared decision, and I know that my wife is not there yet with this decision. So we came here to talk about it." He is a large, white, young guy, with a sweet, soft, round face, and a round belly, with big hands. He reaches for his wife and takes her hands in his. She looks enough like him to be his sister: blond, round-faced, round bodied. They are both tagged with the little cheery paper name tags from security ("Hi! My Name is...."). They have both written their first names carefully, like good students in fourth grade.

"We don't want to be having to make this decision," she says. "We don't want to be here at all. But we are, and we just need to figure out what is the best thing. For Emily." I remember her brown eyes from TV.

Here is the problem. The parents, even now, just want to be sure that Emily will not wake up. "Give us anything you know," says the father, opening his big hands to us.

But we cannot be sure, not in the way they mean. We cannot know anything, really. We know, I tell him, that most of the time, babies this sick, with brains this damaged, either die, or live just like this, curled on the bed, wake and sleep, wake and sleep, with no more sweet smile, or favorite toy, or knowing that it is morning. While we can do lots of miraculous stuff with the extraordinary machines, we cannot do miracles. Emily has been without oxygen to the cells of her brain far longer than the four minutes that it takes for the neurons to begin to die. This is called anoxic brain damage, and it causes widespread and global neuronal death. The brain cells become necrotic, liquefying, then calcify and the brain structures are destroyed; the fine architecture that holds thoughts, nursery rhythms and dreams in place disappears. We can see this on the brain scans already and the neurologist shows the family a scan of a brain in the final stages of this process. "This is where she is now, and this is where we think she is going." We all look at the shadows and light he holds up. I think to myself: where she is going, the oddity of the phrase.

We cannot fix everyone that is broken in the children's intensive care unit. This summer, there will be many children, drowned, silent there. So the main thing we do need to know is how to stop. When to not do what we can do, and why it is that we can think this way, that it is permissible to stop the machines that are feeding his girl.

I tell him that it is OK to decide to let go, that we do this here, and in pediatric hospitals all over the country, today, other parents are making the same decision, and that we know that it is out of love. He asks us if we know about their family, about the other children, who are five and three. We do. He tells us that if God intends to make a miracle, then God will do that whatever we do. "We are Christians, he tells us, and we know we will see Emily again, even if we let her go."

Then the committee talks about this. We have two options, to continue as we are, which means supporting the child, but not doing anything more aggressive, no antibiotics, no re-intubation, and no code, or to withdraw nutrition and hydration which we know will surely end her life.

The rehabilitation team raises the persistent problem: if they are going to do rehabilitation to maximize any slim chance that she may have, then they will need the ICU to be aggressive in exactly the way we agree is not a good idea. If we do not, we may be making her chances less, and contributing to worsening her condition. This usually stops us but not in this case. No one is convinced she has any chance at all.

And then the ICU team begins to worry out loud about Baby Doe laws, which are regulations passed by Congress in the Reagan Administration, when some raised concerns that doctors would withdraw feeding tubes on babies that were disabled. Don't we have to feed her?

I was trained for this part, and here, one learns the words to say. One of my teachers, Dr. Baruch Brody, would describe this as the illumination of one of the moral appeals that will be needed to think clearly about the case. What are the legal arguments? I point out that not only does California not use these regulations, which were enacted and enforced state by state, but also that even under their most restrictive interpretation, Emily, in her irreversible coma, would be one of the exceptions. One by one, the doctors and nurses begin to speak honestly about how withdrawing nutrition and hydration feels different from the withdrawal of a ventilator. It is my job to listen and to explain the law and the ethics that surround the law, the reasons that it has become reasonable and defensible for doctors to stop aggressive treatments and allow patients to die. I tell them the way we think about this now, and about the Cruzan case. We have withdrawn nutrition and hydration a handful of times in this unit. It is hard each time. Even with the ethical and legal arguments, it still feels "like starvation," the doctor tells us. The nurses are split over the decision. Some think the child is suffering, and I remind them that even the part of her brain that can receive pain signals is destroyed, and they tell me that it does not matter: her body is being invaded, they must draw blood for testing every day, and they must do spinal taps when her fever rises.

For the bioethics committee, there is another story going on; hence, this story is "intertextual" in this sense: two months ago, we were asked whether to stop life support on another blond two-year-old, this one beaten into a coma by her father, and left to strangle in her own blood. The ICU attending and the rehab team had said the same things: grim prognosis, terrible shadow and lines on the various tests, and no hope of recovery. But that baby's family was less sophisticated, and not religious. They wanted everything done; we called this denial, but we would not stop nutrition and hydration without specific family consent. We agreed to wait a bit, frustrating the doctors. A week later, the baby woke up, just like her grandma said she would, just like the neurologist swore she would not, and a few months later I had taken to visiting her and watching her practice walking in the physical therapy room. She was well enough to wave "bye-bye" to her ethicist, me, who would have agreed it was a moral act to stop her care.

Ethical decisions at the end of life are largely about limits: ours, the families, and the limits of possibility. Because the ICU teams look like certainty itself, all bright lights and big machines, a secret language, an air of having seen it all, parents come to expect that we can know it all. Yet, that is the problem for the clinical ethicist: not being "know-it-alls" is in part what the ethics committee must remember to be. It is our work, in part to uncover and lift up the weakness of uncertainty honestly to ourselves, and finally to the family, offering them our solidarity if we cannot offer our certainty. It is our work, to remember the Talmudic idea: perhaps the opposite is also the case.

This is why we are often uneasy about having families at ethics committee meetings. We want a "thick description," but it is hard for some clinicians to display the depth of our uncertainty, the possibility of error. We argue to stop for the day, and reconvene tomorrow—it is a hard case, and the committee wants more time.

In the room after the meeting, we second guess. I think that these parents are the clearest that I have ever heard. I am so moved by the gesture of the man's big hand reaching to his wife. Everyone else thinks I am a sucker. "You can't know what is really going on, Dr. Zoloth," they say.

I am a sucker. And I have been wrong before, and we all know it. Someone else thinks they are "too together," and thinks it is weird. She was on TV, what was that about? Why isn't there anger? "I wonder," says one of the community members softly, "what I looked like when I was here. What they said about me."

What gives us the warrant to be here? To say any of this? We are such strangers, powerful and powerless, intimate and not. The parents are in that most American moment, of feeling this all is like a thing you see on TV, of waiting for the moment when it will not be real, when they can pick Emily up and take her home. How did it get to be this late?

And we are both noble, and getting paid to be here. The moral gesture of medicine and nursing at the social moment of the death of a child is one of the most complex moral gestures possible in our society. Bioethics, witnessing, the compass in our hands, compounds the gesture. We are trying to think of everything, to be both in the case (in-case) and not, and what brings us to this moment, not to another, to the task of the outsider, not to the center, is a mere contingency. I leave the meeting and walk over to the child's room with her nurse and her parents.

At the bedside, we watch her astonishing beauty: she is the size and shape of my youngest, the damp baby curls, the round face of her mother, the pink palms and the small curled shell of her hands. "Such a beauty-girl," I say. Not the ethicist, or maybe yes the ethicist for a moment. They show me the pictures, and tell me how this, their youngest child, wakes up earliest, talking and laughing, coming into their bed. They show me the pictures, taken two weeks ago: the cousin's wedding, all dressed up Emily, Emily in the garden, Emily and the other kids, with her teddy bear that I see lays still next to the still Emily in the bed. We speak of the day of the accident: just gone for a second, now where could those kids be? The way the 3-year-old was sitting quietly in the water, his sister floating nearby.

They will do everything right: the therapy, the support groups with the other parents, the church casseroles will come and come, and so will we. We will tell them that it is a hard, but fine, moral decision to allow the child to die. We will make it peaceful, in her own corner of the unit, a quiet place, lots of holding, the other children there to say good-bye. We will offer hospice and death at home. We will give pain meds and comfort care, even when we understand that she will not be in pain. We will offer such things for ourselves, for a world that we have built in which the most vulnerable is beloved, is treated with all we have, hospitality, which is pathetically little, and a great deal.

After I speak to them, I go back. I tell them to take their time, because it is their time, their baby's death, not ours. I tell them to ask to see the articles that we read. So they know that we have no secrets from them. And I worry that this will not make the ICU staff so happy. They will tell them that they cannot understand.

I worry the whole way home that we are making a mistake. We will not ever know if we proceed, if that is the case. Lily has asked us to think about how disabled is too disabled. I tell the committee about the studies that show that 80% of severely disabled people report they have a good life, and would really like to be resuscitated should they need to be, but that only 17% of emergency personnel think of a severely disabled person as having a "good quality life." The gap between what nurses and doctors think, and what professionals think, is historically quite wide. But I ask Lily, who cares for her own child, now 15, after his near fatal car crash, who she wants to make the decision. "Me," she says, "and I pray that they are the good ones." She looks up. "Just like they do." We agree, and tell the team: you ought to do what this family is asking you. Let them let go.

I am supposed to write up the case and, not disconnectedly, the billing right away. This is considered responsible practice as a consultant. I do not do either. I have missed dinner; the kids are in bed, still damp from the bath, with their clean teeth, and night-clothes on. I crawl into bed with them, and my own baby girl crawls happily on to my lap. I pull her close and stroke her arm; her baby skin feels exactly like Emily's arm, there is no difference from the outside. I cannot bear that

I have left them for a minute, when the large, possible world surrounds us, mutable in its entirety when you look away for just a moment. I cannot believe the depths of the miracles we live with every day.

We are given, of course, the whole of the life of the dying child. Standing at the bedside, it is the life of the stranger who we have the great privilege, the great ob1igation, and the great permission to touch.

This month has brought us several families in the odd quiet, the odd twilight that lives in the corners of the ICU. One family with the inevitability dying child, who we help go home to live, with full aggressive support until the first six months of Tet are over and it is an auspicious time to die. One family, who will care for their persistently comatose child all of their lives, if that is what God has given them for their portion. One family, who insist that the tumor is just a birth mark, and resist our efforts to remove it. A Jehovah Witness family with a boy who is ten, severely retarded since birth, who they want to allow to bleed to death with his hemophilia. In all these cases, there were medical and nursing—staff who felt they strongly knew what we should do, or who felt ready to allocate resources. In each case, we heard the stories, listened, told them our signposts for reflection.

II. OUT POSTS: WHY BEING "IN-CASE" IS CRITICAL

Baruch Brody, whose work inspires this chapter and my larger reflections on the necessity of the case and of contingency (the "in-case-ness" of bioethics) has argued for the need for a "totality of concern" that only a casuistry account of bioethics will allow the fullest range of proper ethical choices to families that face dilemmas in the clinical setting.

Ethics reminds us of the necessity of our minimal standards. We must control pain, we must clearly hear and honor the voice of the family, we must allow ourselves disenchantment with our own technology. What I want to add is that we must also understand the maximal standards, the ceaseless responsibility that we have for the family and the child. This relationship asks from us a totality that, in the words of Emmanuel Levinas, breaks open the world for us (see "Enigma and Phenomenon", "Substitution", and "Transcendence and Height", 1996). Lost in the numbing ordinariness of managed care medicine, of routine, of the way we follow orders, lost in a passionate crusade of modernity against death, the encounter with the dying child ought to stop us, with full force every time. We answer not merely with the honesty of the informed consent relationship, but with the deepest and most total honesty I can imagine—one borne along by both the realization about our uncertainty, and our commitment to learning, always, from the case itself, from its particularities, including the ways that it differs from the common, even reasonable, generalizations about principles of bioethics.

For Baruch Brody, this idea is understood as rooted in "a pluralistic account of morality—supported by and in turn supporting, a radical casuistry" as the only one capable of allowing a "moral universe" (2003, p. 1).

My long and detailed telling of the case—including the counter case in the middle—and the allusion to the other articles that will be given, the cases borne by the community members of the ethics committee, the reality that the ethicist herself bears a story—serves two functions. First, it allows us to ground our theories in empirical reality, but not the quantitative accounting method. It allows the moral complexities of cases to unfold and to "in-case" us in the world of competing moral appeals. Second, it allows us to understand that even if all the facts in a case that can be known (and allowing for the uncertainty of clinical medicine) are understood, and even if all agree, as Brody notes, about the moral relevance of the facts, that there are still serious moral differences (what Brody has called "deep moral ambiguity") in the case. In thinking of this problem, I will make one more claim, from the theory of Levinas. This is that even if the facts are coherent, the moral relevance clear, and the move ambiguous, the act will still be claimed—and yet, even then (as in this case) there is more to be said. There is the trace of the alternate choice. At times, this is only a jet trail or print of action, considered but untaken, lost in wind, chance, or history; but at other times, the "moral trace" will exist, standing in for us as a sense of our incompleteness. This is perhaps an idea more theological than philosophical, but here, too, Brody is a teacher, for his moral philosophy is never far from the Jewish tradition itself, in which the moral actor is never completely alone. Brody's argument that there is no final scale or metrics for judging moral action that will work in all cases is a fine one. He notes: "I have instead advocated that all we can do is make reasonable judgments, or non-mechanical conclusions, about which appeals take precedence in a given case" (2003, p. 3). It was Brody who stressed work in justice, impressing a then graduate student (this author) with his argument that families in the clinical realm were more than adjuncts to the ill child, but persons with obligations to other children outside our view, and with the capacity to organize and order their lives for those children as well as the ill one we knew.[1] A part of our attention to the case in clinical ethics is the reality of the diversion of our proper gaze toward the context of the family.

Finally, that the case itself matters and is particular, which means that similar cases will not have a stock answer. Several years ago, a young legally trained colleague, feeling she understood the casuistic method and hoping to learn from it, hoped, in her first year as a clinical ethics consultant, to create a handy database. She reasoned that each case she heard could be summarized into a paragraph, something easily accessible on line. Then, she told us confidently, one could type in key words: PVS, 2-year-old, near-drowning, and get all the similar cases and "see how we had decided." What she would have is then a record of how a committee had acted and perhaps a record of what sort of moral arguments were important—what she would not have, as Brody reminds us, is an answer about what is right to do in the case at hand (2003, p. 5). The particularities of the case and the partialities of the family—what will be asked, what can be given which appeal is central, why this is so—all will vary. There is no easy answer for each case should disturb us. The stakes are death as it bares its stark face and we turn toward the finality of the task.

III. DISCURSIVE METHOD AND CLINICAL MIDRASH

I am arguing we can do this best with an emphasis is on cases, but the limits on the idea of cases as entirely paradigmatic is an important correction, both to the idea of an "actuarial bioethics" and to narrative ethics as well. For while the case account does have aspects of a "production" or a "fiction," it is the work of the clinical ethicist to understand and elucidate the competing moral appeals at the core of the narrative. It is the task, also, of the clinical ethicist, to create the possibility for intertextuality by drawing our attention to other cases, not in a simplistic way, but so as to deepen the complexity. Cases without deep moral ambiguity will leave no mark on us, no trace of humility. Cases that yield to simple clarity or retelling (the so-called "errors in communication") are not of interest in this sense, for the problem of ethics is how to apply the appeals that all agree are important yet not prima facie. As the case in clinical ethics is not the case in family law or corporate litigation; the narrative in clinical ethics is not the literature of medicine. It is not about the story of the interesting other, but about how the other comes to challenge our course of action and how we are called to respond to her in her suffering. Medical ethics in the clinical sense is fundamentally about the actual, tangible, high stakes choices and chances people must take as a part of the search to heal illness, pain, and injury. Medicine is not only the location for our theory, it is the stuff of the theory itself, and its limits, choices, and dilemmas are not merely cultural manifestations; it is the place where we must write down our opinions in the actual charts of patients who will live and die depending on the words our hands will draw. Without this moral seriousness (and without this praxis I might add) the grace we promise is cheap indeed. In this way, I would argue that one appropriate form of moral literature it can draw from is the very much older (far older than novels or short fiction) forms found in religious narrative, such as the midrashic tradition of the Jewish Talmud, the Gospels, and the Koran.

The midrashic literature (one in which Brody has the most scholarly familiarity) is the narrative stories that illuminate and expand the legal and linguistic discourse in the Talmud, which is itself a re-consideration of the ways that a particular generation of rabbinic scholars considered the laws of the Hebrew Scriptures. While the halachah, or legal decisions, are considered binding, the Midrash, or narrative, which surround them textually draw attention to their complexity, their discrepancies, or the way that emerging generations of rabbis keep tell of other cases, bring new cases for review against the stories, and link the story to the Scriptural story and to history itself to ground the laws in context and authority. Brody, along with others from the Jewish tradition (Zoloth, 1999), is proposing the idea that the linguistically constructed universe of the Talmud is intended to structure the sorts of discourse that are necessary to order a moral life. He reminds us that these narratives are intended to reflect on a law understood as a divine gift that enables people properly to organize their personal, social, and spiritual lives. Order in this sense is both reflective and verifiable, in that the act of the creation and discipline implied is a covenant with the Lawgiver and with the others in the similarly obligated community. It is, I believe, implied in Brody's work, that medical professionals and the patients we serve, if we moral philosophers intend to be allowed the privilege of participation, are also moral actors in a certain moral covenant as well. Thus, it is the cases that we are marked by the bonds, or ligneous, that tie us into the story as well. We are indeed roped in, and it is proper that we are, for what makes the story resolutely a moral one is that it presents the ethicist with the same moral challenge as the subject of the tale, for whom, as in this case, moral luck has run out.

The midrashic literature is tied to the teller of the tale by a series of devices. First, that each story is credited to a teacher, and told in his name[2] (as I do in this reflection on Brody). Next, each narrative is tied to a moral idea or argument, and finally, the entire discussion happens across centuries, continents and cultural shifts, allowing entrance into a set of reasons that are both detailed and oddly unique, and held to be a likely and reasonable source of wisdom. More research on this theory will be needed to explore how to develop this idea further. One way is to demonstrate how a coherent, clinical Midrash may emerge in the field. Clinical Midrash will be structured around precisely the same elements of classic midrashic narrative: how is this case unlike the other we are presented with, and how is it the same? If we draw a precept to explain our decision in the case before us, and the prior case we remember creates some tension or contradiction, how can we understand this? Is the opposite argument from our own also valid? What about the story told by the minority voice, the one whose ideas lost out in the precedencial case? What of paradoxical or miracle stories (which clearly are a feature of many Scriptural accounts and then in turn inform and confound our clinical narratives)?

IV. CONCLUSION

The very alert reader will have noted a small internal joke in the first page of the article: I call the account of the case I tell you "opening and saying," which is the classic introduction used for how the discourse between student and teacher begins in much of Jewish literature.[3] It would be remiss of this author not to note my indebtedness to Baruch Brody in the field of Jewish thought, and in particular, in Modern Orthodoxy. His work is instrumental in the project of understanding how a particular religion, albeit one with a strong claim to universal ideas about human law, order, justice and necessity, can function as a source for guidance in the clinical realm. Brody recalls for us that Orthodox Jewish bioethics can fail unless it rejects dogmatic accounts, or simplistic formulations. He returns us to the texts—even the difficult texts—that contradict widely (misunderstood) ideas: that suicide is always prohibited, or that treatment can never be halted, or that resources allocation has no place in end of life decision making. By allowing access to the themes in the tradition he teaches us two things. First, that any reasonable, scholarly reading of the tradition must include a plurality of views on bioethics, and second, that the Jewish tradition may advance some important arguments not found in secular reasoning, most importantly, about the issue of justice and allocation.

Finally, a short note of personal praise and gratitude to my teacher for the many lessons he taught me and my family about how properly to behave as an ethical being. My eldest son (now a bearded Orthodox rabbi with two children), on hearing I was writing this short set of reflections, reminded me of the first such lesson. When Professor Brody first was invited to eat at our home, and he, an eager 13-year-old cook, enthusiastically shook red chili pepper instead of paprika over the quiche he had made for our first world famous bioethicist. Professor Brody ate it with brave equanimity of spirit, an actual curiosity about the event, and a warm sense of humor that allowed everyone's dignity to remain in place. In a difficult world filled with error and incompleteness, normally only having our reason, moral intuition, and our faith as our key defenses, it is of inestimable value to also have a calm, funny, brave, and curious colleague leading the way.

NOTES

[1] Brody, Baruch, "Clinical Ethics" in the Kaiser Permanente Leadership Ethics Summit, in the years 1989. 1990, 2000.

[2] Rarely hers, but with some notable exceptions, as in the midrashim in which Beruiah plays a large role.

[3] See, for example, Daniel Matt in his translation of the *Zohar*, in which he uses this traditional formulation in the introduction of all his translations.

BIBLIOGRAPHY

Brody, B.A. (2003). *Taking Issue: Pluralism and Casuistry in Bioethics*. Washington, D.C.: Georgetown University Press.

Chambers, T. (1999). *The Fiction of Bioethics*. New York: Routledge.

Levinas, E. (1996). 'Enigma and phenomenon,' in A. Peperzak, S. Critchley & R. Bernasconi (eds.), *Emmanuel Levinas: Basic Philosophical Writings*. Bloomington: Indiana University Press, pp. 65–77.

Levinas, E. (1996). 'Substitution,' in A. Peperzak, S. Critchley & R. Bernasconi (eds.), *Emmanuel Levinas: Basic Philosophical Writings*. Bloomington: Indiana University Press, pp. 80–95.

Levinas, E. (1996). 'Transcendence and height,' in A. Peperzak, S. Critchley & R. Bernasconi (eds.), *Emmanuel Levinas: Basic Philosophical Writings*. Bloomington: Indiana University Press, pp. 11–37.

Zoloth, L. (1999). 'Reading like a girl: gender and text in Jewish bioethics,' *Judaism*, 48(190), 165–174.

Matt, D. (2003). *The Zohar: Pritzker Edition*. Stanford: Stanford University Press.

CHAPTER 8

THE EUTHYPHRO'S DILEMMA RECONSIDERED: A VARIATION ON A THEME FROM BRODY ON HALAKHIC METHOD

H. TRISTRAM ENGELHARDT, JR.

Department of Philosophy, Rice University, Houston, Texas and Department of Medicine, Baylor College of Medicine, Houston, Texas

I. INTRODUCTION: BEYOND MORALITY AND THEOLOGY

Baruch Brody's extensive scholarly work possesses an impressive scope: his publications range from bioethics to the philosophy of science, from the philosophy of law to the philosophy of religion. The last area of reflection provides this essay's point of departure: an article from Brody's corpus that explores the relation (as well as the lack thereof) between secular bioethics and medical-moral norms grounded in halakhic reflection (Brody, 1983). On a number of levels, Brody's study heuristically engages foundational themes concerning the relationship of morality, moral philosophy, metaphysics, epistemology, and theology. In this article, he emphasizes three ways in which halakhic reflection can bear on bioethics:[1] (1) "the use of halakhic material [serves] as a source for ideas about medical ethics which can be defended independently of their origins", (2) "the use of halakhic material [serves] as a basis for mandating certain forms of behavior for members of the Jewish faith who are perceived as bound by Jewish Law", and (3) "the use of halakhic material [serves] as the basis for claims about the Jewish view on disputed topics in medical ethics" (Brody, 1983, pp. 318, 319). It is the third of these that he shows to be problematic. As Brody demonstrates, the difficulty with the last area of use of halakhic materials is that it confuses what the law of Moses requires of Jews with what the covenant with Noah requires of Gentiles.

This essay's goal is to generalize one of the concerns Brody addresses in developing his arguments regarding the second and third points. Namely, Brody emphasizes that halakhah "distinguishes between the obligations, positive and negative, fulfillment of which is required of the Jewish people, and on the other hand obligations, positive and negative, fulfillment of which is required of all people" (Brody, 1983, p. 320). This distinction recognizes the ambiguity of "ethics" in bioethics. There is not one framework of morality, nor one account of the nature or foundations

M.J. Cherry and A.S. Iltis (eds.), Pluralistic Casuistry, 109–130.

of morality. In particular, the ambiguity Brody notes opens the door to acknowledging that there is at best a cluster of moralities or ethics that may at best share family resemblances.

As Brody shows, halakhic requirements cannot determine the ethics ingredient in secular bioethics. Brody makes this point by contrasting what God demands of Jews with what God demands of Gentiles, implicitly placing both these requirements over against what is usually considered obligatory in the light of secular morality and bioethics. He does this *inter alia* by using the case of abortion, which offers one of the few instances in which halakhic requirements are more severe for Gentiles than for Jews.[2] The contrast depends in part on the circumstance that, unlike Jews who are bound by 613 obligations, the obligations for Gentiles in God's covenant with Noah are only seven. Brody summarized this point by quoting Talmud:[3]

> Seven precepts were the sons of Noah commanded: social laws; to refrain from blasphemy; idolatry; adultery; bloodshed; robbery; and eating flesh cut from a living animal. R. Hanania b. Gamaliel said: Also not to partake of the blood drawn from a living animal. R. Hidka added emasculation. R. Simeon added sorcery. ...R. Eleazar added the forbidden mixture [in plants and animals] (*Sanhedrin* 56^{a-b}).

The contrast depends as well on the circumstance that the halakhic interpretation of some obligations for bnai Noah are not compatible with the obligations for Jews. This essay draws on Brody's rich analyses and arguments to advance the position that a religious believer's obligations need not be reducible to the claims of secular morality. Moral norms must be recognized as falling within quite different frameworks, so that, as already noted, one confronts numerous ethics or moralities underlying a plurality of bioethics. In particular, there are the obligations of Jews, the obligations incumbent on gentiles because of God's covenant with Noah, and the obligations supported by various secular bioethics.

At the outset some disclaimers are in order: no attempt is made to reconstruct or present the full force of Brody's arguments in his article.[4] Instead, Brody's article is used to advance three points. The first is that secular morality does not exhaust the halakhic requirements for right behavior. Or to rephrase the claim, halakhic requirements cannot be reduced to secular morality. Second, there are conflicts between secular morality and halakhic requirements. Third, the first two points have force in part because Orthodox Jews, and for that matter Orthodox Christians, do not have a morality, a moral philosophy, or a theology, as these practices have come to be understood in Western European culture, especially after the first millennium. Or to put the matter more precisely, though Orthodox Judaism and Christianity have a morality in the sense of norms for behavior, and a theology in the sense of a recorded reflection on the experience of God, neither has a morality or theology as a practice independent of the religious life. Orthodox Judaism, as well as Orthodox Christianity, does not recognize the jurisdiction of an independent moral perspective that can critically bring into question the norms of behavior supported by a rightly-ordered religious life. In addition, there is no independent scholarly practice either as a moral philosophy or as an academic theology (e.g., moral theology) that can bring into question that which one knows religiously. Although the claims of a religiously informed morality can be at tension with those of a secular morality or with the requirements of moral philosophy and academic theology, the latter are not accepted as having the authority to reshape the former.

One might consider in contrast how moral philosophy as an intellectual, academic practice has recast both Western morality and much of Western Christian moral theology. This academically located practice has come to have a life and authority that is independent of religion, including the once religious culture within which it took shape. This practice has come to claim the prerogative to judge the significance and meaning of Scripture, established pieties, and traditional moral practices and commitments. Such an independent and authoritative academic practice does not exist for Orthodox Judaism (or for that matter for Orthodox Christianity). Instead, moral commitments are appreciated within an encounter and experience of God, which carries with it its own logic of exegesis. As a consequence, morality is understood in an encounter with God, not in terms of an independent secular morality or an authoritative rational moral perspective. As a consequence, a discursive and comprehensive, philosophically integrated account of theology and metaphysics is for Orthodox Jews and Orthodox Christians primarily of cultural but not of theological interest. To Plato's choice in the *Euthyphro* between a secular morality that should guide religious moral commitments and a set of religious moral commitments that would be in tension with secular morality, the Orthodox response is to stress that humans find themselves in relationship with a fully transcendent God Who establishes the canons of appropriate human behavior. The Orthodox engage a dialectic of suspicion against "ordinary morality" (i.e., secular morality) and in favor of the requirements of the religious life. In terms of the latter account, a morality prescinded from a rightly-ordered religious life is one that is radically distorted in not rightly recognizing the nature of the human condition.

Again in warning: though I draw on Brody's reflections regarding halakhic method, the points offered are nested within commitments sustained by an Orthodox Christian understanding of morality, theology, and metaphysics.[5] Although Orthodox Christian theology both dogmatic and moral, may be closer to that of Orthodox Judaism than, say, to Roman Catholicism in many of its theoretical commitments, the two Orthodoxies are clearly different in character.

II. THEOLOGY AS AN ACADEMIC PRACTICE: MORALITY AND THEOLOGY SEPARATING GOD FROM MAN

The claim that Orthodox Jews and Orthodox Christians lack a morality and a theology requires further qualification. First, the claim that Orthodox Jews and Orthodox Christians have neither a morality nor a theology amounts to the claim that Orthodox Jews and Orthodox Christians do not have a moral or theological perspective as an independent practice, as a standpoint outside of the religious life itself. Instead, they have a religious life that integrates their settled judgments about right conduct and about God that cannot be brought into question or revised by reference to an independent set of moral intuitions and judgments. There is no authoritative moral or theological perspective outside of the religious life itself. Rather, the perspective of the religious life of itself is normative and includes both its own content and the proper modes for its exegesis and application. For example, the law given to Moses, along with halakhic determinations, on the one hand, as well as the noetic experience of the Fathers, along with conciliar dogmatic statements, on the other hand, discloses the bounds and the character of both moral behavior and theological experience. The character of right conduct is grounded in a rightly-ordered relation to God, which depends on special revelation or religious experience.

Second, both Orthodox Jews and Orthodox Christians not only lack a moral or theological perspective outside of their religious life, they also do not recognize a philosophical or independent academic perspective with the authority and capacity to re-articulate and re-shape according to its demands the religious life and its commitments. Orthodox Jews and Orthodox Christians maintain their freedom from the jurisdiction of any would-be independent, critical, moral-philosophical and natural-theological perspective because they appreciate the radical cleft between the uncreated being of God and the being of creatures. There is a recognition that there is no analogy of being (i.e., no *analogia* entis) between the nature of God and the nature of man, on the basis of which one can build an independent critical standpoint able dialectically to reason behind, and thereby able to undermine through independent philosophical reflections, the Law and halakhic determinations.[6] So, too, for Orthodox Christians, the experience of God and conciliar articulations are not open to criticism by an independent, discursive, rational practice of moral philosophy or theology.[7] Very importantly, the Law of Moses for Orthodox Jews and the revelation of Christ for Orthodox Christians disclose transcendently anchored ways of life that are not defined by an independent morality, moral philosophy, or theological perspective.

The force of these claims becomes clearer when one considers the culturally dominant understandings of morality and theology in the Christian West, which in great measure were framed as the Roman Catholic church emerged after the 8*th* century and especially in the High Middle Ages. Roman Catholic theology, indeed Western theology in the mainline Western Christian churches in general, came to constitute an independent discursive rational practice that now has standing as a set of academic disciplines that can and does critically reshape the religious life. It brings both moral commitments and theological doctrines into question. This academic practice now claims the competence to reshape and develop ordinary religious practices and experience. Consider the following statement on Roman Catholic theology:

> Through the course of centuries, theology has progressively developed into a true and proper science. The theologian must therefore be attentive to the epistemological requirements of his discipline, to the demands of rigorous critical standards and thus to a rational verification of each stage of his research (Congregation, 1990, p. 120).

In this passage, theology is accepted as a discursive, academic practice with its own intellectual standards and procedures on the basis of which it can make judgments about, and thus revise, Roman Catholic dogmatic and moral-theological commitments.

Theology, both moral and dogmatic, has become for Roman Catholics an institutionalized academic practice embedded in the ethos of universities, which often have a robustly secular cast, even when they remain Roman Catholic in name. This theology has internalized many of the general procedural and content-full intellectual, moral, and social assumptions of the surrounding secular culture. As a consequence, the Western Christian practice of theology functions as (1) a critical standpoint set over against traditional Christian moral and theological commitments, (2) especially those that are at tension with metaphysical, moral, and social commitments current in the institutionalized academic practice of theological reflection, which understands itself as (3) empowered to bring those traditional, metaphysical, moral, and social commitments into question, so as (4) to make theological and religious life conform to the assumptions internalized by theology as an academic practice. This intellectual standpoint thus immanentizes and relativizes the transcendent in terms of the requirements of a particular account of discursive rationality and moral deportment. Theology so understood becomes an instrument for rendering theological reflection and morality post-traditional in the service of "advances," that is, new moral fashions in secular reflection and cultural understanding. The result is a dynamic dialectic between the substance of the Christian tradition and the critical requirements of academic theology. Orthodox Judaism has not embraced this dialectic and has resources to protect itself against its claims.

Cardinal to the Roman Catholic academic practice of theology, as opposed to the halakhic tradition of Orthodox Judaism, is the Roman Catholic reliance on philosophy as a secular intellectual project, which in turn determines the character of theological reflection. As Pope Leo XIII (A.D. 1810–1903) opines,

> [Theology's] solid foundations having been thus laid, a perpetual and varied service is further required of philosophy, in order that sacred theology may receive and assume the nature, form, and genius of a true science. For in this, the most noble of studies, it is of the greatest necessity to bind together, as it were, in one body the many and various parts of the heavenly doctrines, that, each being allotted to its own proper place and derived from its own proper principles, the whole may join together in a complete union; in order, in fine, that all and each part may be strengthened by its own and the others' invincible arguments (Leo XIII, 1879, § 6).

Unnoticed in all of this is that philosophy constitutes a perspective that can call theology as an experience of God into question and revise moral theology, as well as other elements of theology in terms of philosophy's own requirements and commitments. In the process, philosophy becomes a dialectical other to the religious life-world of Western Christianity, so that philosophical reason can reshape the way in which religion is lived and experienced. Since philosophy is now culturally the more vigorous of the two, its impact on theology is greater than theology's impact on philosophy. In addition, and crucially, the perspective of Western philosophy is at odds with that of the commitments of Christianity. This is by no means unanticipated, in that the presuppositions of Western philosophy are rooted in a pagan cultural past, not in a culture framed and directed by an encounter with the personal and radically transcendent God of Abraham.[8]

Pope John Paul II's (A.D. 1920–2005) call in *Fides et Ratio* for better philosophy and more committed philosophers, so as to meet the challenges of the contemporary world, illustrates Western Christian theology's dependence on philosophy. John Paul II appeals, for example, to philosophers

> to trust in the power of human reason and not to set themselves goals that are too modest in their philosophizing. ...I appeal now to philosophers to explore more comprehensively the dimensions of the true, the good and the beautiful to which the word of God gives access. ...The intimate bond between theological and philosophical wisdom is one of the Christian tradition's most distinctive treasures in the exploration of revealed truth. ...I appeal also to philosophers, and to all teachers of philosophy, asking them to have the courage to recover, in the flow of an enduringly valid philosophical tradition, the range of authentic wisdom and truth – metaphysical truth included – which is proper to philosophical enquiry (John Paul II, 1998, pp. 86 [§56], 148 [§104], 149 [§105], 151 [§106]).

John Paul II wished to reconstitute philosophy so that philosophy can appropriately guide and structure theology. He took for granted philosophy's centrality to theology. As a consequence of such views, Roman Catholic theology has come to hold that it can lay out the general character of moral obligations without reliance on either Scripture, tradition, or religious experience (Fuchs, 1970).

Consider, in contrast, the remarks of the Ecumenical Patriarch, Bartholomew I, in his reflections on how Orthodoxy and Western Christianity have

> become ontologically different. ... The Orthodox Christian does not live in a place of theoretical and conceptual conversations, but rather in a place of an essential and empirical lifestyle and reality as confirmed by grace in the heart [Heb. 13:9]. This grace cannot be put in doubt either by logic or science or other type of argument. Our conception of Holy Tradition moves upon the same track [unlike that of Western Christianity]. Holy Tradition for the Orthodox Christian is not just some collection of teachings, texts outside the Holy Scriptures and based on their moral tradition within the Church. It is this, but not only this. First and foremost, it is a living and essential imparting of life and grace, namely, it is an essential and tangible reality. (Bartholomew, 1997, pp. 1, 2).

Central to Bartholomew's warning is the recognition that the noetic empirical character of Orthodox theology is central and distinguishing.[9]

In summary, a crucial difference separating the moral and theological perspectives of Orthodox Judaism and Orthodox Christianity from Western religious thought is rooted in the circumstance that morality and even theology for Roman Catholics and for many other Western Christian religions function as perspectives outside of the religious life. The moral life is regarded as existing in a domain of intuitions, sentiments, convictions, and settled judgments able critically to engage theological and moral commitments. The domain of secular moral commitments is then considered to be fully expressed when appreciated within philosophy in general and moral philosophy in particular. The result is the recognition of the intellectual and moral authority of a self-sustaining, academic practice of discursive rational reflection able to revise key understandings of God, salvation, ecclesiology, worship, political theory, and morality, *inter alia*. Moral philosophy and theology so construed come to regard themselves as competent to re-assess and alter the character of traditional religious moral obligations.[10] Morality and theology become third things between the religious life and God. They claim the authority to mediate the relationship of the religious believer to God. Morality as a set of norms for deportment independent of the religious life and appreciated in terms of moral philosophy, when internalized by moral theology, can be used to recast Scripture, tradition, conciliar holdings, and papal statements in the service of bringing them into conformity with what are held to be moral and philosophical advances. In so doing, moral and dogmatic theology internalizes the conceits of the age, producing not just the development of doctrine, but an affirmation of such developments as a positive theological accomplishment. Moral and dogmatic theology become a dialectical other able to reshape the commitments of the religious life.

III. THE MORAL LIFE WITHOUT MORAL PHILOSOPHY OR MORAL THEOLOGY

Halakhic reflections are not grounded in an independent morality or governed by a university-based practice of critical philosophical reflection. In particular, there is no appeal to an independent set of canons of secular rational morality or an independent intellectual practice with authority over against halakhic reasoning. Halakhic understandings of appropriate and inappropriate deportment are lodged within a practice and appreciation of rectitude independent of secular canons of moral rationality, as well as the practice of secular morality philosophy as an academic discipline. As a consequence, Orthodox Jews do not have a moral theology, as do Western Christians. This is not to deny that even for Orthodox Jews there is not something that can be characterized in a restricted sense as the development of halakhic rulings and responsa. One might consider, for example, the passage in Baba Mezia, which can be taken as affirming that halakhic determinations are produced by (1) arguments based on Scripture, tradition, rulings of the Sanhedrin, and previous holdings and responsa, (2) all set within a tradition of precedential argument that excludes (3) further direct divine guidance.[11]

> On that day R. Eliezer brought forward every imaginable argument, but they did not accept them. Said he to them: "If the *halachah* agrees with me, let this carob-tree prove it!" Thereupon the carob-tree was torn a hundred cubits out of its place – others affirm, four hundred cubits. "No proof can be brought from a carob-tree," they retorted. Again he said to them: "If the *halachah* agrees with me, let the stream of water prove it!" Whereupon the stream of water flowed backwards. "No proof can be brought from a stream of water," they rejoined. Again he urged: "If the *halachah* agrees with me, let the walls of the schoolhouse prove it," whereupon the walls inclined to fall. But R. Joshua rebuked them, saying, "When scholars are engaged in a *halachic* dispute, what right have ye to interfere?" Hence they did not fall, in honour of R. Joshua, nor did they resume the upright, in honour of R. Eliezer; and they are still standing thus inclined. Again he said to them: "If the *halachah* agrees with me, let it be proved from

> Heaven!" Whereupon a Heavenly Voice cried out: "Why do ye dispute with R. Eliezer, seeing that in all matters the *halachah* agrees with him!" But R. Joshua arose and exclaimed: "*It is not in heaven.*" What did he mean by this? – Said R. Jeremiah: That the Torah had already been given at Mount Sinai; we pay no attention to a Heavenly Voice, because Thou hast long since written in the Torah at Mount Sinai, *After the majority must one incline.*
>
> R. Nathan met Elijah and asked him: What did the Holy One, Blessed be He, do in that hour? – He laughed [with joy], he replied, saying, "My sons have defeated Me, My sons have defeated Me" (*Babylonian Talmud*, Baba Mezia 59b [Soncino edition]).

Whatever one might wish to claim about the development of rabinnic discussions and reflections, they are not lodged within a practice rooted in and directed by secular norms. Instead, they are located in a covenant with God. Talmudic case-based reflection, nested within and shaped by previously established conclusions, norms of discourse, and traditional modes of argument, is precisely not nested in a practice (1) embedded in the general discursive rational assumptions of the surrounding culture and (2) aimed at authenticating and critically shaping a morality independent of relationships established by God.

The West attempted to marry Athens and Jerusalem: the offspring of the union were, among other things, a sense of morality, a moral philosophy, and a natural theology that were outside of the religious life and thus separated man from God. The West sought to elaborate an understanding of moral rationality and theological claims that could be appreciated as nested within a secular rational framework neutral to any Divine revelation. For Orthodox Jews, this did not occur. The Orthodox Jewish dialectical and logical exegetical tradition operates without reliance on an independent, discursive, rational theory, as this developed in Roman Catholic moral theology and more generally in the mainline church theologies of the West. To engage but nevertheless radically recast a Heideggerian idiom, both Orthodox Jews and Orthodox Christians not only reject the feasibility of the onto-theo-logy of Western metaphysics, but have not relied on it. Orthodox Jews and Orthodox Christians have not affirmed morality or theology as nested in a Being whose being can be grasped by a canonical, immanent rationality and for this reason can be articulated in a historically unconditioned fashion. But they have affirmed that morality as proper conduct and theology as the experience of God are not grounded in mere human *Gespräch*, dialogue, or conversation,[12] but instead in a reality whose being is beyond all being (i.e., beyond all created being) – the transcendent, personal God.

To some extent, the current cultural abandonment of onto-theo-logy, the classic framework of Western Christianity, in favor of a hyper-ecumenical, culturally syncretic Christianity, represents a reaction against some of the metaphysical assumptions that framed the Western Christian medieval synthesis.[13] The denial of many of its background framing moral assumptions may in part explain the collapse and disarray of mainline Christian religions.[14] "Anyone who has, through his own development, experienced theology, whether that of the Christian faith or that of philosophy, nowadays prefers to be silent about God so far as thinking is concerned. For the onto-theo-logical character of metaphysics has become questionable to thinking people, not because of some kind of atheism, but because of an experience of thinking in which the still unthought unity of the essence of metaphysics revealed itself in onto-theo-logy" (Heidegger, 1960, p. 51).[15] This reference does not imply that Heidegger (A.D. 1889–1976) in general drew the right conclusions from this partially correct diagnosis.[16] Instead, it underscores, with regard to Orthodox Jews and Orthodox Christians, that it is not only "onto-theo-logy" that is fundamentally denied, but also what one might term, through a similar neologism, "moral-onto-theo-logy."

Orthodox Jews and Orthodox Christians appreciate the impossibility of securing the moral and natural theological perspective promised by the conceptual synthesis at the roots of Western Christianity. Such a perspective cannot deliver a self-grounding, rational perspective that can critically bring religious tradition into question without in the end bringing itself into question. In part, this is the case because secular morality needs a veridical, canonical point of reference, so as to specify a particular and binding ordering of values and right-making conditions. Without a definitive foundation to secure such an ordering, all attempts to deliver a canonical moral perspective inevitably beg the question, argue in a circle, or involve an infinite regress (Engelhardt, 1996, chapters 1–3). Also, as Immanuel Kant (A.D. 1724–1804),[17] G.E.M. Anscombe (A.D. 1919–2001),[18] and others recognize, without a grounding of morality in God, one cannot guarantee that moral rationality should always trump prudential rationality. Moreover, only an encounter with or an experience of God can indicate how created being (e.g., humans) should relate to uncreated being (i.e., God). Both Orthodox Judaism and Orthodox Christianity from their encounter with God experience the Ground Who guarantees the coordination of the good and the right, a unique content for morality, and an accord between the justification of morality and the motivation to act morally.

In this light, one can better understand that to hold that Orthodox Jews and Christians lack a morality is to recognize that Orthodox Jews and Orthodox Christians do not have a morality outside of what they accept as the canons of deportment given by God, nor do they recognize an intellectual practice able critically to recast moral and theological claims. As a consequence, should the moral requirements of the religious life offend "ordinary," that is, secular, moral intuitions, sentiments, and settled judgments, then so much the worse for ordinary morality. If moral philosophical reflection attempts to fortify ordinary morality's claims against religious canons of deportment, this would supply for Orthodox Jews and Orthodox Christians further grounds for impeaching the underlying assumptions of such a moral philosophical approach. Philosophy as an independent moral perspective has not become integral to the moral reflection of Orthodox Judaism or Orthodox Christianity. For example, though Orthodox Christians have imported theological terminology from secular philosophies, and though discursive rational arguments have been employed in apologetics as well as in disputes with those outside the faith, Orthodox Christian theology and morality have been recognized as anchored in an enduring experience of God grounded in grace (the uncreated energies of God). It is this experience (which in Western terms would be characterized as mystical) that maintains a community of worship and belief over space and time. The result is that one can appreciate that theologians in the primary sense (i.e., those who immediately know God) need not be, and are quite frequently not, academics. They are rather those who experience God and are often without academic education or training.[19] Those who are merely academic theologians (the office of the author of this article) are so in a very restricted office. They serve as translators of the experience of the first genre of theologians into the language of the general culture, or as analyzers of the implications of the experience of theologians in the first sense. Theology in the second sense (i.e., mere discursive rational philosophical reflection regarding God and morality) thus cannot bring into question theology in the first sense (i.e., the experience of God and His commandments). Theology in the second sense cannot offer an independent moral-philosophical practice with critical reversionary authority, as it does for many Western Christianities, so as to bring theology in the first sense into question and then to revise it.

In short, there is a family resemblance between Orthodox Christianity and Orthodox Judaism, in that (1) both do not recognize morality as a set of moral obligations and/or intuitions able to be experienced and understood outside of religious life and experience,[20] and (2) they do not recognize the authority of intellectual practices such as moral philosophy or natural theology as able critically to revise and reshape religious life, along with the canons of appropriate deportment.

IV. THE LAW OF MOSES AND THE LAW FOR BNAI-NOAH: ABORTION ALLOWED AND FORBIDDEN

By showing that halakhic reflection is taken to allow abortion for Jews, while abortion constitutes a capital offense for Gentiles, Brody demonstrates that halakhic moral norms have a character at odds with secular moral reflection. Orthodox Jews with respect to abortion, as well as with respect to other moral norms, often take positions that run against the grain of the dominant assumptions not only of secular bioethics, but of contemporary Western morality and bioethics.[21] In arguing that abortion is permitted for Jews, while prohibited for Gentiles under threat of capital punishment, Brody sets halakhic moral concerns in a context of expectations radically other than that taken for granted by most, if not all, secular bioethical reflection, as well as by Western Christian moral theory grounded in natural law. Halakhic moral obligations for both Jews and Gentiles are often not just incompatible with the content, but also globally at odds with the framing context of Western academic moral reflection. In contrast with the assumptions of secular morality and bioethics, different groups of persons, given Brody's account, should consider themselves bound by different moral laws, which are neither grounded in, nor equivalent to (indeed, in many cases in tension with) general secular morality.[22] In contrast, a natural-law perspective would require harmonizing the law for Jews and the law for Gentiles, bringing both into harmony with the requirements of a secular view of right moral reasoning and conduct. The goal would be to bring both into line with secular requirements for a philosophically defensible morality neutral to religion and culture.

The contrast Brody develops with secular morality and bioethics is in this regard robust. The Talmud not only permits but indeed requires abortion for Jews in certain circumstances. "If a woman is in hard travail, one cuts up the child in her womb and brings it forth member by member, because her life comes before that of [the child]. But if the greater part has proceeded forth, one may not touch it, for one may not set aside one person's life for that of another" (*Oholoth* 7:6). Given this understanding, a Jewish hospital as well as Jewish physicians would be obliged not just to provide an abortion to a Jewish woman under such circumstances, but also to attempt to bring her to accept such an abortion. Such is not the case for a Gentile hospital or Gentile physicians (the issue remains as to whether the halakhic prohibition against the involvement of Gentiles in performing abortions would apply to a Gentile woman who submitted to an abortion at the hands of a Jewish physician, or what the obligations are of Jewish physicians regarding a Gentile woman). To make the point, Brody notes "On the authority of R. Ishmael it was said: [He is executed] even for the murder of an embryo. What is R. Ishmael's reason? –Because it is written, *Whoso sheddeth the blood of man within* [*another*] *man, shall his blood be shed.* What is a man within another man? –An embryo in his mother's womb" (*Sanhedrin* 57^b). Brody then quotes in further elaboration of this bifurcated (if not trifurcated) account of morality: "We have learnt that before the head emerges, one can dismember the embryo, limb by limb, and bring it out in order to save the mother, but such a procedure would be prohibited for a ben Noah since they are commanded against destroying embryos ... but perhaps it would be permissible even in the case of a ben Noah" (*Tosafot* 59^a x.v. leka, quoted in Brody, p. 327). The concluding hesitation does not undermine the general point:[23] halakhic reflection concerning abortion[24] is nested within a set of moral considerations that accords neither with the permissive character of the dominant secular bioethical view, nor with that of Western Christianity, especially as expressed in Roman Catholicism,[25] nor with that of the Christianity of the first millennium alive today, in Orthodox Christianity (Engelhardt, 2000, pp. 275–282).

Brody acknowledges the gulf between moralities through emphasizing "the theoretical pitfall of misusing halakhic material which does not apply to bnai Noah" (Brody, 1983, p. 328). Through the case of the halakhic permissiveness regarding abortion for Jews, in contrast with the prohibition of abortion by Gentiles under threat of capital punishment, Brody concludes that Talmudic requirements for Jews cannot directly guide secular bioethics. He shows as well by implication the gulf in general between requirements for bnai Noah and the commitments of secular morality and bioethics. From halakhic requirements for Gentiles or for Jews, one cannot advance considerations that will be able directly to bring secular moral reflections to similar conclusions. Given the distinction drawn between halakhic requirements for Jews versus halakhic requirements for Gentiles, as well as the implicit distinction of the latter from secular bioethics, the plurality of morality and bioethics is established.

It is not just that the integrity of Talmudic moral reflection defeats the aspirations of secular morality cum bioethics to establish a global, all-encompassing moral perspective, so that independent moral rationalities can and do exist. It offers a categorically different account of the character such a morality should have. In so doing, Talmudic reflection offers a global ethics and bioethics in two parts, one for Jews and the other for non-Jews. First, the global morality and bioethics grounded in Talmudic is lodged within a justificatory context different from the philosophical framework of secular ethics and bioethics. Second, the global morality and bioethics grounded in halakhic reflection have a content different from that of secular morality and bioethics. Third, the halakhic global morality and bioethics (i.e., the requirements for bnai-Noah) have a clearly God-directed character. In particular, in order for a Gentile to be a righteous Gentile, he must recognize the law as given by God, as Moses Maimonides (A.D. 1135–1204) and others[26] have held.

> Anyone who accepts upon himself the fulfillment of these seven mitzvoth and is precise in their observance is considered one of 'the pious among the gentiles' and will merit a share in the world to come. This applies only when he accepts them and fulfills them because the Holy One, blessed be He, commanded them in the Torah and informed us through Moses, our teacher, that Noah's descendants had been commanded to fulfill them previously. However, if he fulfills them out of intellectual conviction, he is not a resident alien, nor of 'the pious among the gentiles,' nor of their wise men. *Mishneh Torah*, Hilchot Melachim UMilchamotehem, viii, 11 (Maimonides, 2001, p. 582).[27]

The point to underscore is that, if one follows Moses Maimonides' contention that in order to be "one of the pious among the Gentiles," a person who will merit eternal life, one cannot simply live a good life, as Immanuel Kant would require or the natural law of Thomas Aquinas (A.D. 1225–1274) would mandate. The point here is that living a good life still falls radically short of living the required pious life.[28] In short, halakhic requirements for a global morality cum bioethics, the moral requirements for Gentiles, are incompatible with secular morality cum bioethics, not just because of differences in content and justification, but because all humans are required to take Divine Transcendence and commands seriously.

Brody nevertheless claims that halakhic reflections may prove heuristic for those considering issues of bioethics in non-halakhic contexts. There are grounds to doubt that this can be the case, save in the most adventitious or tangential sense. Halakhic determinations are nested within a context, a tradition, and a system of moral reflection and argument foundationally different from that of secular moral philosophical discourse. Talmudic examinations of the Law, along with the development of *responsa*, are set within an exegetical tradition whose basic premises, rules of evidence, and rules of inference are other than those that guide secular moral reflection. To engage a well-worked notion of paradigmatic differences,[29] Talmudic reflections on proper conduct, in contrast with those that underlie secular morality *cum* bioethics, presuppose quite different paradigms. They are separated by different understandings of the nature and possibility of metaphysics, moral epistemology, the sociology of experts, and what should count as exemplar cases of knowing and acting rightly. Even when there is agreement about particular obligations in particular circumstances, persons will nevertheless be divided by different moralities because they will have different understandings as to what is at stake in these particular obligations. The choices they confront are nested in foundationally incompatible life-worlds. These life-worlds are separated by a gulf that reflects the difference between immanent concerns and those that acknowledge obligations to a truly transcendent God Who commands.

V. RETHINKING THE EUTHYPHRO

Plato deploys Socrates in the Euthyphro in order to undermine a divine-command account of morality: the good, the right, and the virtuous cannot be (so Plato contends) good, right, and virtuous simply because God wills it so. Rather, God should will the good, the right, and the virtuous because they are good, right, and virtuous. Among the premises presupposed in such contentions is the affirmation that, were the good, the right, and the virtuous simply good, right, and virtuous because God wills this to be the case, there would then be a tension between ordinary moral intuitions, on the one hand, and divine commands, on the other. The result would be that at least in certain circumstances, according to secular moral criteria one would be pressed to judge some of God's commands to be bad, wrong, or vicious. The Euthyphro can be taken as integral to a much larger project, namely, the endeavor to ground morality, and by implication a moral theology, in an account of the secularly reasonable that has priority over any direct experience of God. The secular moral enterprise depends on the plausibility of this undertaking.

Such approaches at their roots rely on a number of key assumptions regarding moral and religious experience, as well as concerning the tie between discursive rational thought, on the one hand, and morality and theology, on the other. In particular, Plato's Socrates in the *Euthyphro* presupposes (1) that there is no radical gulf between created and uncreated being (i.e., God is not recognized as in His nature radically transcendent), (2) that God is not the origin and sustainer of all things, such that the ultimate meaning of created being (insofar as such can be appreciated in purely discursive rational reflection) is understandable apart from its Creator (in contrast, see Romans 1:18–32); (3) that the good, the right, and the virtuous can be understood independently of God's existence, that is, as having their standing independently of God's relation to the good, the right, and the virtuous, (4) that moral claims that cannot be justified in discursive rational terms are critically to be brought into question, and (5) that rational individuals will find any discordance between the requirements of ordinary morality (i.e., secular morality) and those of the religious life as impeaching religious claims, not the reverse.

Under the slogan "the unexamined life is not worth living" (Apology 38a), a dialectic of suspicion is engaged against moral claims grounded in claims of an encounter with or an experience of God. This dialectic of suspicion against religious and traditional moral claims is nested within an intellectual practice and discourse (1) in which supposedly nothing is taken for granted outside of the presuppositions of that discourse itself, with its own particular moral and philosophical rationality, and (2) in which it is presupposed that this particular presentation of rationality can deliver canonical moral content and guidance, even though (3) this project has failed to establish a foundation for its claims (i.e., foundationalism has failed). The difficulty is that the subsequent history of philosophy gives substantial grounds to doubt the truth of the claims made on behalf of the Socratic project: the project of rationally establishing a neutral moral perspective affirmable by all has not produced a resolution of substantive moral disputes through sound rational argument.[30] The Socratically examined life has fractured into numerous alternative narratives: the cacophony of post-modernity. It has not succeeded in delivering the promised canonical secular moral perspective.

In any case, over against Plato's assumptions, there is the Talmudic paradigm that recognizes the presence of a transcendent Creator God, His law, and the authority of a particular traditional exegesis of that law. The content of the bioethics grounded in the obligations of bnai-Noah (e.g., the absolute prohibition of performing abortions) can only be understood when nested within the recognition of a personal, transcendent God, Whose ways need not be our ways (Is 55:8) and Who intrudes into history with His demands. Because the Socratic project is confined within the bounds of created being, and because created being is radically different from uncreated Being (i.e., there is no *analogia entis*), discursive moral rationality is unable critically to judge claims regarding God's nature and His requirements for moral conduct. The Talmudic requirements express the conditions for coming into right relations with God, rather than with immanent goods. The conditions for being holy (i.e., in right relationship with God) are prior to and independent of the conditions for acting rightly or achieving the good.

Over against the positions of Orthodox Jews and Orthodox Christians, there is the secular moral framework held to be grounded in rational discourse and intuitions open to all and unconstrained by the requirements of an ultimately inscrutable God, or even for that matter by a recognition of His presence. It is a morality articulated within a concern for immanent goods and right-making conditions. The law of Moses and the law for Bnai-Noah in contrast are lodged within frameworks that acknowledge the presence of a transcendent Creator God. As a result, the frameworks of these laws possess a hierological rather than a rational discursive character. The bioethics for Jews and that for Bnai-Noah do not collapse into an understanding of bioethics that has its roots in the assumptions of Athens (i.e., that bioethics must have at its core an ethics whose content has been justified by sound rational argument without a recognition of the demands of God), because the cardinal focus is on the holy rather the good and the right, and because the holy exists and is understood independently of the good and the right. Or to put the matter slightly differently, since all goods are recognized as created goods, their full significance can only be appreciated only in right relationship with their Creator God. So, too, right-making conditions can only be understood rightly in terms of a right relation of the creature to the Creator.

In contrast, Western Christian religions have increasingly assimilated concerns with the holy to those with the good and the right, so that the holy was made compatible with the good and the right, not the good and the right compatible with the holy. This reduction of the holy has been the final outcome within these religions of the tension Plato sought to introduce between the requirements of God and the requirements of rational morality. This tension has served to affirm a dialectic that in the case of Western Christianity has recast the religious in terms of the requirements of a particular account of the discursively rational.[31] This dialectic troubles Western Christianity. Western Christian thought generally seeks both to affirm an independent rational moral account (e.g., as in natural law), while acknowledging the requirements of a transcendent personal God. This tension leads, for example, to the agony that Kierkegaard experiences in his reflections on the teleological suspension of the ethical in the face of the divine command that Abraham sacrifice Isaac (Gen 22:1–19). Other such tensions with secular morality can easily be located, such as with the invasion of Canaan regarding which Deuteronomy reports, "At that time we took all his [Sihon, king of Heshbon] towns and completely destroyed them – men, women and children. We left no survivors" (Deu 2:34). "We completely destroyed them [those of Og, king of Bashan], as we had done with Sihon king of Heshbon, destroying every city – men, women and children" (Deu 3:6). Similarly, Joshua slaughters all the inhabitants of Jericho. "They devoted the city to the Lord and destroyed with the sword every living thing in it – men and women, young and old, cattle, sheep and donkeys" (Joshua 6:21). For Western Christianity, there is a tension between what secular rational morality requires and what divine commands can impose.[32] For traditional Christianity, as with Orthodox Judaism, there is a recognition that God is the Lord of life and death so that He may properly order His creatures to bring about death.[33]

Similar tensions exist because of traditional Jewish and Christian condemnations of secularly morally accepted consensual acts between willing partners, ranging from homosexual liaisons to physician-assisted suicide. Orthodox Jews and Orthodox Christians recognize divine commands that appear unjustifiable within secular moral rationality and the intuitions it nurtures. In Western Christianity, the distance between what secular moral rationality can justify and what can be divinely required was once widely accepted. Roman Catholic moral theologians recognized the limited ability of secular moral rationality to account for or to warrant the full scope and content of Christian morality. One might consider one of the prominent contributors to the Second Scholastic, Francisco de Vitoria (A.D. 1485–1546), who acknowledged that natural-law reflections could not secure a categorical prohibition against lying, fornication, and usury.[34] The bottom line is that the requirements of God and the requirements of secular moral rationality cannot be shown to coincide.

VI. NOAH, MOSES, AND SOCRATES: MORALITY IN THE PLURAL

In summary, by affirming one bioethics for Jews and another for Bnai-Noah, neither of which is compatible with the dominant secular bioethics, Brody acknowledges the existence of incompatible moral understandings. Tensions between the requirements of God on the one hand and the requirements of secular morality on the other do not trouble Brody's exegesis in his article[35] (as this tension also would not trouble Orthodox Christian moral reflection), because the "rationalities" of the moralities do not touch each other. Indeed, Brody acknowledges at least three domains of moral discourse and reflection. First, there are the halakhic requirements for Jews. Second, there are the halakhic requirements for Gentiles, which offer the basis for a global bioethics for Gentiles. Third, there are the diverse requirements of moral conduct sustained within a plurality of secular moral and bioethical understandings (e.g., one might consider the differences between social-democratic versus libertarian approaches to bioethics in general and to health care resource allocation in particular). Halakhic requirements for Jews, halakhic requirements for Gentiles, and the requirements of secular morality are incompatible. These disparate categories of moral requirements frame incompatible moral life-worlds structured by disparate moral commitments. Neither the standpoint of Socratic dialogue nor the standpoint of the moral-philosophical (-theological) reflection that took shape in the second millennium in the West is able to claim governance over Talmudic argument. The disparate paradigms and their assumptions remain incommensurable. By laying out the integrity of halakhic reflection, Brody invites us to confront the circumstance that we face a heterogeneity of moralities, moral-philosophical accounts, and moral-theological understandings.[36]

NOTES

[1] In my article, I use bioethics as a generic term so as to place under the rubric bioethics much of what has been placed under medical ethics and medical moral theology, as well as all biomedical ethics and health care ethics.

[2] Another example of a more severe restriction on Gentiles than on Jews involves the absolute prohibition of eating a limb severed from an animal still showing muscular contractions (i.e., the animal was not completely dead). As long as the animal is ritually slaughtered, Jews do not have to notice whether the animal is still showing muscular contractions. See *Mishneh Torah*, Hilchot Melachim UMilchamotehem, ix, 13.

[3] According to Moses Maimonides, six of the seven requirements were imposed on Adam. They were the prohibitions against worship of false gods, cursing God, murder, incest and adultery, and theft, and the command to establish laws and courts of justice. The prohibition against eating flesh from a living animal was added for Noah. *Mishneh Torah*, Hilchot Melachim UMilchamotehem, ix, 1.

[4] The author of this article is by no means a Talmudic scholar (would that God had given me the grace), nor does he pretend to the learning needed to be such. What is offered is a set of non-Talmudic reflections on some Talmudic concerns in the service of better appreciating the limits of secular morality cum bioethics, as morality came to be understood in the shadow and light of Western European moral-philosophical commitments that were nurtured by the cultural and metaphysical synthesis of the early Western European 2*nd* millennium, which eventually issued in the secular bioethics that took shape at the end of the 20*th* century (Engelhardt, 2000).

[5] For a contemporary overview of Orthodox Christian morality, theology, and metaphysics, see Romanides, 2002, and Vlachos, 1998.

[6] While Orthodox Christianity recognizes that one can never know the nature of God, it recognizes as well that one can encounter and experience His uncreated energies. True theologians experience His energies so that they do not simply know about God, they know God. This relationship is achieved through rightly-directed prayer. Theologians in the primary sense are those who experience God, not those who reflect about God. "If you are a theologian, you will pray truly. And if you pray truly, you are a theologian." Evagrios the Solitary, "On Prayer," in Sts. Nikodimos and Makarios, 1988, vol. 1, p. 62. As a result, Orthodox Christianity's morality and theology are one with its experience of God.

"Therefore we do not engage in idle talk and discuss intellectual concepts which do not influence our lives. We discuss the essence of the Being Who truly is, to Whom we seek to become assimilated by the grace of God, and because of the inadequacy of human terms, we call this the image of the glory of the Lord. Based on this image, and in the likeness of this image, we become 'partakers of the divine nature' [2 Peter 1:4]. We are truly changed, although 'neither earth, nor voice, nor custom distinguish us from the rest of mankind. [To Diognetos 2, PG 2, 1173] This change, which is bestowed on us from the right hand of the Most High, remains hidden, secret and mystical to many. And thus, a life which is directed toward Him is called mystical. That which leads to divine grace are called mysteries. The entire change of both language and intellect is beyond comprehension and when directed by God leads to unspeakable mysteries" (Bartholomew, 1997, p. 3).

[7] From the very first centuries after Christ, Orthodox Christianity acknowledged the incomprehensibility and radical transcendence of God. See, for example, St. John Chrysostom, "On the Incomprehensible Nature of God," as well as the works of Dionysios the Areopagite. This view has been consistently affirmed by theologians. See, for instance, St. Symeon the New Theologian (A.D. 949–1022) and St. Gregory Palamas (A.D. 1296–1359).

[8] Western philosophy, as it developed in Greece during the 5th and 4th centuries before Christ, did so. within a view of reality that did not require considering what is involved in an experience of, and an encounter with, a radically transcendent, personal God, though concerns with the experience of a transcendent God mark neo-Platonism. As a consequence, there was little appreciation in the roots of Western philosophy of what would be involved in knowledge and experience of such a God. For Semitic Christian reflections on this theme, see the writings of St. Isaac the Syrian (A.D. 613–?).

[9] Orthodox Christian theology is empirical in the sense of holding that a non-sensuous [noetic] experience of God is not only possible but grounds its theology. "Therefore, experiencing the Dogma of the Church is not something that is taught through intellectual teachings, but it is learned through the example of Him Who, through Incarnation, joined Himself to us. To this point, dogma is life and life is the expression of dogma. However, a mere theoretical discussion on the meaning of life and dogma is unnecessary" (Bartholomew, 1997, p. 5).

[10] For an illustration of the Roman Catholic support of the doctrine of doctrinal development, namely, that the Roman church has changed its doctrinal commitments over time, see Cathleen Kaveny's study of John Noonan (Kaveny, 1992). Kaveny's and Noonan's positions reflect a position taken in Vatican II, namely, that "this tradition which comes from the apostles develops in the Church with the help of the Holy Spirit. ... For, as the centuries succeed one another, the Church constantly moves forward toward the fullness of divine truth until the words of God reach their complete fulfillment in her" ("Dogmatic Constitution of Divine Revelation [Dei Verbum]," chap. II, § 8, in Abbott, 1966, p. 116). The explicit Roman Catholic commitment to the notion of development of doctrine is indebted to John Henry Cardinal Newman's (A.D. 1801–1890) *Essay on the Development of Christian Doctrine* (1878). Newman understood full well that the only way to account for such novel Roman Catholic doctrines as papal infallibility was to claim that they had "developed."

[11] Recognizing the contrast between halakhic and Roman Catholic moral-theological reflection does not involve the claim that the hermeneutic procedures of dialectical and logical reasoning in halakhic reflection are free of any influence from the surrounding culture. It is rather to hold that rabbinic legislation, customs, precedents, and exegetical procedures are set within a practice that sustains halakhic reflection, and that this practice is not supported by an independent intellectual point of view supposedly grounded in a neutral moral rationality open to all without antecedent religious commitment. One might consider the contrast between the halakhic approach and those of the standard Roman Catholic textbook accounts of natural theology, which were produced until the early 1960s, and that presupposed that their reasoning was lodged in such a neutral intellectual practice. See Boedder, 1902; Gornall, 1962; Holloway, 1957; Joyce, 1923. These undertakings also included enterprises in rational cosmology (McWilliams, 1949) and psychology (Barrett, 1931).

[12] For an example of an attempt to construe Christianity as *Gespräch*, one might consider Vattimo, who claims "*interiore homines animat veritas*. This is exactly the way of realizing that being is *Gespräch*, is dialogue, because the dialogue takes place in political common life" (Vattimo and Rorty, 2005, p. 67). This leads Vattimo to identify post-modern nihilism as the truth of Christianity. "[F]rom the perspective

that I propose here, postmodern nihilism (the end of metanarratives) is the truth of Christianity" (Vattimo and Rorty, 2005, p. 51). The Socratic *Gespräch*, the dialogue of the Euthyphro, which presumes to judge the actions of the transcendent God, in the end asserts its independence over against that God and the deep roots of reality, so that Vattimo is able to assert that there are no facts, only interpretations.

[13] One might think of Santiago Zabala's attempt to sever Christianity and philosophy thought from God and instead to embed in history and conversation. "The task of the philosopher today seems to be a reversal of the Platonic program: the philosopher now summons humans back to their historicity rather than to what is eternal" (Zabala, 2005, p. 9).

[14] In reflecting on the decline if not the radical collapse in the late 20th century of many of the mainline Christian religions, as well as Marxism-Leninism, one might note how at the end of the 20th century both dialectical materialism and traditional Western Christian theology with their metaphysics fell into disarray. The metaphysics sustaining them no longer seemed plausible. It is important to observe that in the case of Western Christianity the radical collapse was not just due to the metaphysical scaffolding no longer being credible. More importantly, Western Christianity, in particular Roman Catholicism, was undermined when the pieties that sustained its life-world were brought into question and altered. These pieties were undermined by means of (1) a thoroughgoing reconstruction (not simply the translation into the vernacular) of its liturgy, (2) the general disarticulation of the life of the church and its liturgy from the mind of the Fathers, and (3) a loss of any sense of spiritual asceticism (e.g., the abandonment of the Friday fast and the near-complete abandonment of all Lenten asceticism). As a result of Vatican II, there occurred a disorienting rupture in the character of everyday Roman Catholic religious life. Its very life-world altered. Given the cultural prominence of Roman Catholicism, these changes had an impact on the life of other Western Christian sects. See, for example, Davies, 1977 and 1980, and Wathen, 1971. See as well Engelhardt, 2000, pp. 53–55. For an overview of some of the consequences of this collapse, see Jones, 2003.

[15] Another way to put matters, different from Heidegger's contentions with respect to atheism, is that Western philosophy from Anselm (A.D. 1033–1109) and Thomas Aquinas (A.D. 1225–1274) to Leibniz (A.D. 1646–1716) and Whitehead (A.D. 1861–1947) developed claims regarding the possibility of discursively demonstrating the existence of God that could in the end not be justified. Given the gulf between created and uncreated being, this failure was inevitable. As Michael Buckley observes, this failure lies at the roots of modern atheism (Buckley, 1987).

[16] Hans Jonas (1903–1993) correctly appreciates that, if one draws on Heidegger for theological guidance, there are significant difficulties. "My Christian friends – don't you see what you are dealing with? Don't you sense, if not see, the profoundly pagan character of Heidegger's thought? Rightly pagan, insofar as it is philosophy, though not every philosophy must be so devoid of objective norms; but more pagan than others from your point of view, not in spite but because of its, also, speaking of call and self-revealing and even of the shepherd. … Quite consistently do the gods appear again in Heidegger's philosophy. But where the gods are, God cannot be. That theology should admit this foe – no mean foe, and one from whom it could learn so much about the gulf that separates secular thinking and faith – into its inner sanctum, amazes me. Or, to express myself reverently: it passes my understanding." Jonas, 1966, pp. 248–249. Hans Jonas correctly notes philosophy's internalization of pagan attitudes and dispositions. Given the integration of Western philosophy into its theology, Western theology itself was similarly influenced. Western philosophical theology is marked by pagan notes.

[17] For Kant's account of the necessity of affirming the immortality of the soul and the existence of God as postulates of pure practical reason, see *Critique of Practical Reason*, Part I, Book II, Chapter II, "The Dialectic of Pure Reason in Defining the Concept of the Highest Good".

[18] Anscombe appreciates that, once the existence of God is no longer recognized, morality loses the force of its categorical claim on human action. As she puts it, "It is as if the notion 'criminal' were to remain when criminal law and criminal courts had been abolished and forgotten" (Anscombe, 1958, p. 6).

[19] For examples of Orthodox Christian theologians in the primary sense of those who experience God, see St. John of San Francisco (A.D. 1894–1966), Elder Joseph the Hesychast, the Cave-Dweller of the Holy Mountain (A.D. 1895–1959), Elder Paisios of Romania (†1993), Elder Paisios of Mt. Athos

(A.D. 1924–1994), Elder Porphyrios (A.D. 1906–1991), St. Silouan the Athonite (A.D. 1866–1938), and Archimandrite Sophrony (A.D. 1896–1993).

[20] The integral connection between morality and religious life, especially morality and right worship, is made by St. Paul in Romans 1:18–32, the dominance of perverted sexual passions was the result of perverted worship. It should be noted that St. John Chrysostom (A.D. 354–407) in his commentary on Romans 2:10–16 stresses that only the righteous Gentiles such as Melchizedek were able correctly to understand moral obligations. "But by Greeks he [St. Paul] here means not them that worshipped idols, but them that adored God, that obeyed the law of nature, that strictly kept all things, save the Jewish observances, which contribute to piety, such as were Melchizedek and his, such as was Job, such as were the Ninevites, such as was Cornelius" ("Homily V on Romans I.28, V.10," in Schaff, 1994, vol. 11, p. 363).

[21] On a number of important points, the moral commitments of Orthodox Jews collide with those of secular morality and secular bioethics. Consider, for example, the halakhic restrictions on removing life-sustaining treatment with their consequent prohibitions on the cessation of mechanical ventilation, even if the patient would have wanted such ventilation terminated. As a consequence, save in immediate proximity of death (i.e., when the patient is *goses*), life-sustaining treatment may not be discontinued (Jakovobits, 1959, pp. 119–125).

[22] Neither is the author's position relativistic, nor does it involve a moral metaphysical skepticism. The author acknowledges the unique truth of a particular moral, metaphysical, and theological perspective, which cannot be anchored or justified through discursive, philosophical argument, but rather requires noetic experience. See Engelhardt, 2000.

[23] The point about abortion being forbidden as a capital offense for Gentiles is reinforced by Moses Maimonides. "A gentile who slays any soul, even a fetus in its mother's womb, should be executed [in retribution] for its [death]. *Mishneh Torah*, Hilchot Melachim UMilchamotehem, ix, 4.

[24] The author of this essay acknowledges that on particular points of required behavior God may have given special indulgence to those under the Mosaic law (Matthew 19:7–8). Orthodox Christians may be able to understand the halakhic rulings permitting abortion to have reflected something of this sort – though the rulings fail to appreciate that which was originally required. The traditional Christian position has always been to recognize that all abortion is forbidden. See, for example, St. Basil the Great, Letter 188. For a further account of the difference between Orthodox Christian and Mosaic understandings of abortion, see Engelhardt, 2005.

[25] The history of Roman Catholic views regarding abortion is complex, multilayered, and contradictory. Moreover, it is governed by the evolution in the 19th century of an account of double-effect already implicit in Thomas Aquinas's reflection distinguishing between foreseen and intended consequences of actions. For an overview of the elements of the principle of double effect, see Kelly, 1958, pp. 12–14.

[26] Stephen Schwarzschild gives the following in support of the position taken by Maimonides. "*The Mishnah of R. Eliezer (The Midrash of Thirty-two Hermeneutic Rules)* also contains a passage which might be regarded as Maimonides' source: having discussed the term 'the pious of the nations of the world,' it finishes: 'But if they fulfill the seven commandments saying: "We have heard them from somebody or other," –or as a result of their own intellectual cogitations, –...then, even if they obey the entire Torah their reward will be limited to this world.'" Schwarzschild, 1962, p. 306.

[27] Moses Maimonides held that a righteous heathen will only have a portion in the world to come if he both observes the laws given to Noah and recognizes them as divinely revealed. In his commentary on *Mishneh Torah*, Hilchot Melachim UMilchamotehem, viii, 11, Rabbi Touger notes, "Thus, there are three levels in the gentiles' acceptance of their seven mitzvoth: a resident alien who makes a formal commitment in the presence of a Torah court; 'the pious among the gentiles,' individuals who accept the seven mitzvoth with the proper intent, but do not formalize their acceptance; and a gentile who fulfills the seven mitzvoth out of intellectual conviction" (Maimonides, 2001, p. 583). The last have no share in the world to come. For a further discussion of these issues, see Schwarzschild, 1962 and 1963.

[28] The point that living a moral life out of rational conviction but not out of obedience to God does not suffice was appreciated but radically rejected by Benedict Spinoza (1631–1677). "Maimonides ventures openly to make this assertion: 'Every man who takes to heart the seven precepts and diligently follows

them, is counted with the pious among the nations, and an heir of the world to come; that is to say, if he takes to heart and follows them because God ordained them in the law, and revealed them to us by Moses, because they were of aforetime precepts to the sons of Noah: but he who follows them as led thereto by reason, is not counted as a dweller among the pious, nor among the wise of the nations.' Such are the words of Maimonides, to which R. Joseph, the son of Shem Job, adds in his book which he calls 'Kebod Elohim, or God's Glory,' that although Aristotle (whom he considers to have written the best ethics and to be above everyone else) has not omitted anything that concerns true ethics, and which he has adopted in his own book, carefully following the lines laid down, yet this was not able to suffice for his salvation, inasmuch as he embraced his doctrines in accordance with the dictates of reason and not as Divine documents prophetically revealed." Spinoza, 1951, p. 80. Spinoza was not the only one who tried to think his way around the position enunciated by Moses Maimonides.

[29] I engage Thomas Kuhn's well-worn notion of paradigms and paradigm change in *The Structure of Scientific Revolutions* (1962) in order to underscore fundamental differences rooted in disparate metaphysical and epistemological commitments separating halakhic (and for that matter Orthodox Christian) morality from secular moralities. The notion of paradigm shift is useful as well for recognizing the multiple recastings of Western Christian theological perspectives that occurred as the result of changes in its basic philosophical categories. With changes in the surrounding secular culture, philosophical fashions changed. As different categories were imported into Western Christian theological reflections, given the central role of philosophy in Western Christian theology, Western theology itself was variously changed. As Hegel comments concerning cultural change, "All cultural change reduces itself to a difference of categories. All revolutions, whether in the sciences or world history, occur merely because spirit has changed its categories in order to understand and examine what belongs to it, in order to possess and grasp itself in a truer, deeper, more intimate and unified manner" (Hegel, 1970, § 246 Zusatz, vol. 1, p. 202).

[30] The impossibility of secular philosophy's securing basic moral premises and rules of evidence without begging the question, arguing in a circle, or engaging an infinite regress was already well recognized by the early third century and summarized within the *pente tropoi* of Agrippa (Diogenes Laertius, *Lives of Eminent Philosophers* IX.88, Sextus Empiricus, "Outlines of Pyrrhonism," I.164).

[31] The result of bringing the claims of God into harmony with the claims of philosophical rationality has been a reduction of moral theology to moral philosophy. This point is captured well by the claim of Charles Curran. "Obviously a personal acknowledgement of Jesus as Lord affects at least the consciousness of the individual and his thematic reflection on his consciousness, but the Christian and the explicitly non-Christian can and do arrive at the same ethical conclusions and can and do share the same general ethical attitudes, dispositions and goals" (Curran, 1976, p. 20).

[32] Moses Maimonides, for example, lays out the difference between the obligation imposed by God on Joshua to destroy all of the original inhabitants of Canaan, and the obligation in particular wars to kill only the males after their majority, but to spare women, children, and animals (Deut 20:14). For a discussion of these matters, see *Mishneh Torah*, Hilchot Melachim UMilchamotehem, vi, 4–5.

[33] Orthodox Jewish reflection takes a strong position with regard to the obligation to execute idolators. "A *yefat toar* [captive woman] who does not desire to abandon idol worship after twelve months should be executed. Similarly, a treaty cannot be made with a city which [desires to] accept a peaceful settlement until they deny idol worship, destroy their places of worship, and accept the seven universal laws commanded Noah's descendants. For every gentile who does not accept these commandments must be executed if he is under our [undisputed] authority." *Mishneh Torah*, Hilchot Melachim UMilchamotehem, viii, 9.

[34] See Francisco de Vitoria, "On Dietary Laws or Self-restraint," I.5, second conclusion.

[35] This is not to claim that Orthodox Jews cannot or have not recognized the tensions between religious and secular moral perspectives. It is rather to note that these perspectives are defined by two incommensurable concerns: an exegesis of God's requirements versus a commitment to an independent practice of philosophical reflection on moral obligations held to be able critically to reassess and revise religious moral commitments.

[36] This plurality of moralities gives no ground for endorsing a moral relativism. On the one hand, the different behavioral requirements for Jews versus bnai-Noah identify obligations under different

covenants with God. On the other hand, the aspirations of secular morality and bioethics from the perspective of Orthodox Judaism and Orthodox Christianity appear as radically one-sided and incomplete.

BIBLIOGRAPHY

Abbott, W.M. (ed.) (1966). *The Documents of Vatican II*. New York: Herder and Herder.

Anscombe, G.E.M. (1958). 'Modern moral philosophy,' *Philosophy*, 33, 1–19.

Barrett, J.F. (1931). *Elements of Psychology*. Milwaukee: Bruce Publishing.

Bartholomew (1997). "Joyful light", speech delivered on October 21 at Georgetown University, Washington, DC.

Boedder, B. (1902). *Natural Theology*. London: Longmans, Green.

Brody, B. (1983). 'The use of halakhic material in discussions of medical ethics,' *Journal of Medicine and Philosophy*, 8, 317–328.

Buckley, M. (1987). *At the Roots of Modern Atheism*. New Haven: Yale University Press.

Congregation for the Doctrine of the Faith (1990). 'Instruction on the ecclesial vocation of the theologian,' *Origins*, 20, 118–126.

Curran, C.E. (1976). *Catholic Moral Theology in Dialogue*. Notre Dame: University of Notre Dame Press.

Davies, M. (1980). *Pope Paul's New Mass*. Dickinson: Angelus Press.

Davies, M. (1977). *Pope John's Council*. Kansas City: Angelus Press.

Engelhardt, Jr., H.T. (2004–2005). 'Abortion and the culture wars,' *Pemptoussia*, 16, 49–59.

Engelhardt, Jr., H.T. (2000). *The Foundations of Christian Bioethics*. Salem, MA: M & M Scrivener Press.

Engelhardt, Jr., H.T. (1996). *The Foundations of Bioethics*, Second Edition. New York: Oxford University Press.

Fuchs, J. (1970). 'Gibt es eine spezifisch christliche moral?' *Stimmen der Zeit*, 185, 99–122.

Gornall, T. (1962). *A Philosophy of God*. New York: Sheed and Ward.

Hegel, G.W.F. (1970). *Hegel's Philosophy of Nature*, trans. and ed. M. J. Petry. London: George Allen and Unwin.

Heidegger, M. (1960). *Essays in Metaphysics: Identity and Difference*, trans. K.F. Leidecker. New York: Philosophical Library.

Holloway, M.R. (1959). *An Introduction to Natural Theology*. New York: Appleton-Century-Crofts.

Isaac the Syrian, St. (1984). *The Ascetical Homilies of Saint Isaac the Syrian*, trans. Holy Transfiguration Monastery. Boston: Holy Transfiguration Monastery.

Jakobovits, I. (1959). *Jewish Medical Ethics*. New York: Bloch Publishing.

John Paul II (1998). *Fides et Ratio*. Vatican City: Libreria Editrice Vaticana.

Jonas, H. (2003). *The Phenomenon of Life*. Chicago: University of Chicago Press.

Jones, K.C. (2003). *The Index of Leading Catholic Indicators*. Fort Collins: Roman Catholic Books.

Joyce, G.H. (1923). *Principles of Natural Theology*. London: Longmans.

Kaveny, M.C. (1992). 'Listening for the future in the voices of the past,' *Religious Studies Review*, 18, 112–117.

Kelly, G. (1958). *Medico-Moral Problems*. St. Louis: Catholic Hospital Association.

Kuhn, T. (1962, rev. 1970). *The Structure of Scientific Revolutions*. Chicago: Chicago University Press.

Leo XIII (1879). *Aeterni Patris*. [On-line] Available: http://www.vatican.va/holy_father/leo_xiii/encyclicals/documents/hf_l-xiii_enc_04081879_aeterni-patris_en.html.

Maimonides, M. (2001). *Mishneh Torah*, trans. E. Touger. New York: Moznaim Publishing.

Masterman, M. (1970). 'The nature of a paradigm,' in I. Lakatos & A. Musgrave (Eds.). *Criticism and the Growth of Knowledge* (pp. 59–89). London: Cambridge University Press.

McWilliams, J.A. (1949). *Cosmology*. New York: Macmillan.

Mendelssohn, M. (1983). *Jerusalem*, trans. A. Arkush. Hanover: Brandeis University Press.

Nikodimos and Makarios, Saints (1988). *The Philokalia*, trans. and ed. G.E.H. Palmer, P. Sherrard & K. Ware. Boston: Faber and Faber.

Oholoth (1989) *Babylonian Talmud*, ed. I. Epstein, trans. H. Bornstein. London: Soncino Press.
Romanides, J. (2002). *The Ancestral Sin*, trans. G.S. Gabriel. Ridgewood: Zephyr.
Sanhedrin (1994). *Babylonian Talmud*, ed. I. Epstein, trans. Jacob Shachter. London: Soncino Press.
Schaff, P. (ed.) (1994). *Nicene and Post-Nicene Fathers*, First Series. Peabody: Hendrickson Publishers.
Schwarzschild, S. (1962). 'Do Noachites have to believe in revelation?' *The Jewish Quarterly Review*, 52, 297–308.
Schwarzschild, S. (1963). 'Do Noachites have to believe in revelation?' *The Jewish Quarterly Review*, 53, 30–65.
Spinoza, Benedict de (1951). *A Theologico-Political Treatise*, trans. R. H. M. Elwes. New York: Dover.
Vattimo, G. & Rorty, R. (2005). *The Future of Religion*, ed. S. Zabala. New York: Columbia University Press.
Vlachos, H. (1998). *The Mind of the Orthodox Church*, trans. E. Williams. Levadia: Birth of the Theotokos Monastery.
Wathen, J.F. (1971). *The Great Sacrilege*. Rockford: Tan Books.
Zabala, Santiago (2005). 'A religion without theists or atheists,' in R. Rorty & G. Vattimo, *The Future of Religion*, ed. Santiago Zabala. New York: Columbia University Press.

SECTION III

BIOMEDICAL PUBLIC POLICY

CHAPTER 9

THE GOOD (PHILOSOPHY), THE BAD (PUBLIC POLICY) AND THE UGLINESS OF BLAMING FAMILIES FOR INEFFECTUAL TREATMENTS

ROBERT M. ARNOLD

Montefiore Universty Hospital, Division of General Internal Medicine, Pittsburgh, PA, U.S.A.

To do good academic work in clinical medical ethics requires a broad knowledge base. One must have a basic understanding of diseases and how they affect individuals, of how health care providers envision and carry out their work, and of the broader social and economic forces that influence the environment in which health care providers work. In addition, a scholar must be able to synthesize the above information to identify the salient ethical issues. Finally, the best clinical ethicists can draw on their knowledge of ethical theory to help patients, policy-makers or clinicians think more clearly about what they ought to do.

Baruch Brody's work is paradigmatic of good clinical medical ethics. His credentials in philosophy are incomparable. As a full professor in philosophy at Rice University, he is recognized for his work in political and social philosophy. As the Director of the Center for Medical Ethics and Health Policy at Baylor's College of Medicine, he has extensive practical experience in how medicine works. His work on the ethics of research on thrombolytics, for example, could be used to teach a course in either cardiology or clinical research methods. Health care providers see his work as relevant and useful to their daily work, in part because of his careful attention to the clinical facts. Finally, he has extensive knowledge of public policy, having worked for numerous federal agencies ranging from the Office of Technology Assessment (OTA) to the North Atlantic Treaty Organization (NATO).

Two examples highlight his ability to mesh clinical facts, public policy and philosophical analysis: his writings on death using neurological criteria and his work on futility. In the former, Brody attempts to determine if there is a "correct" answer to the question regarding the definition and criteria used to determine death, focusing in particular on the neurological criteria. Appealing to uncertainty theory, Brody argues that there is no single unitary definition. Instead, given that the definition is a mix of biological and social facts, he argues that the question of determining

M.J. Cherry and A.S. Iltis (eds.), Pluralistic Casuistry, 133–148.

the correct definition requires one to determine the goals the definition is being used to achieve and the consequences of false positives and negative determinations (Halevy and Brody, 1993; Brody, 2002). While his view has not changed public policy, it has been adopted by a wide range of philosophers and clinicians as the most intelligent solution to the problem of defining death (Truog, 1997).

In this paper, I will focus my attention on his arguments regarding the role of medical futility in clinical decision making. Again, I find Brody's theoretical arguments to be novel, tightly argued and generally correct. However, as his work was used by others to formulate hospital and state policy, careful attention to pluralistic values and process gave way to professional power. Moreover, I find the focus on futility as a way to decrease over treatment to be an ineffectual way to improve clinical practice.

I. THE GOOD: THE PHILOSOPHY OF FUTILITY

The early writings in bioethics and end-of-life care emphasized patient autonomy and surrogate decision making. The standard consensus was that doctors acted paternalistically in making end-of-life decisions. In reaction, philosophers emphasized (1) that when to forgo life sustaining treatment was a value decision; (2) that patients, not physicians, should be allowed to make treatment decisions based on their view of the therapy's benefit/burden ratio; and (3) that surrogates who are representing the patient's wishes should also be able to make treatment decisions. In addition to the procedural importance of autonomy, early bioethicists emphasized the "naturalness of death." The paradigmatic cases discussed in the law and ethics communities dealt with patients or families who wanted to stop treatment and allow the patient to die. The underlying assumption was that physicians valued the quantity of life over the quality of life and were "death-denying." Patients and families wanted to stop treatments that were no longer viewed as beneficial because they only prolonged dying. Medical ethicists tried to convince physicians that the previous distinctions they used to determine the appropriateness of treatment, such as ordinary versus extraordinary or stopping versus not starting, were unjustified. Instead, physicians should accede to the patient's view of a therapy's benefit/burden ratio and recognize that death was a natural phenomenon. Physicians' views that life should be prolonged at all costs were replaced by more nuanced views.

In the late 1980s and the 1990s, however, families, not doctors, insisted on the provision of high technology to keep their loved ones alive. In a paradigmatic case, the patient suffered irreversible cognitive impairment and was being kept alive in the ICU without hope of cognitive recovery. In another common scenario, families would insist on continued treatment in situations where the physicians thought treatment was unlikely to prolong the patient's life more than hours or days. Physicians argued that such treatment provided little benefit, and that they should be able to refuse to provide this treatment (Angell, 1991; Miles, 1991).

These cases pitted the emphasis on autonomy against the assumption that death is a natural phenomenon and that quality of life is important. Larry Schneiderman argued, in a now classic paper, that these cases represented futility and that in limited situations, based on their integrity, physicians could unilaterally refuse to provide requested interventions. This principle would avoid both physician-driven and surrogate-driven over treatment at the end of life and provide a corrective to the pre-eminence of patient autonomy over professional judgment (Schneiderman, Jecker, and Jonsen, 1990).

The opposing positions led to a heated debate within the bioethics literature (Rubin, 1998). Some argued that given the limits of empirical causality, one could never be sure that a treatment would not work. The decision over whether the probability of success was enough to try the treatment or whether the outcomes were "worth" the effort pointed out the value-laden aspects of these decisions. Appeals to futility were merely a mask for medical paternalism and thus no definition of futility was justified (Lantos et al., 1989; Youngner, 1990; Troug, Brett and Frader, 1992). Others suggested that futility be limited only to those cases in which the treatment could not achieve even its physiologic outcomes, claiming that these outcomes were least likely to involve important value considerations. Clinically, this argument basically invalidates futility as there rarely is a case in which the doctors can be certain that the treatment will not even achieve its physiologic purpose (e.g., that CPR will not allow the patient to regain circulatory function) (Council on Ethical and Judicial Affairs, 1999). Proponents of futility pointed out the difference between negative and positive rights. Futility reflected ethic's over-emphasis on autonomy. Advocates of futility argued that requests for these treatments go against core beliefs about what it is to be a doctor. Patients cannot demand non-end-of-life treatments when the physician believes that they will not work or the burdens outweigh the benefits, so why is end-of-life treatment different? In both cases, physicians should have the right to refuse to perform interventions which violate their professional conscience (Lantos et al., 1989; Callahan, 1991; Tomlinson and Brody, 1990; McCullough and Jones, 2001).

Into this debate came Halevy and Brody's ground-breaking paper in *The Journal of Medicine and Philosophy* (1995). Emphasizing that the role of futility judgments is to ground physician unilateral decision making, they argue that if the concept is going to accomplish this goal, it has to satisfy a number of conditions:

A. The Definition has to be Sufficiently Precise so that its Application to the Intervention in Question can be Determined

The precision requirement is to ensure that health care providers did not invoke futility in a capricious manner. There is some reason to be concerned given an empirical study showing wide variations in how doctors define futility (Curtis et al., 1995). If futility is merely whatever an individual doctor thinks it is, there is a great potential for bias and stereotyping.

B. The Definition must be Prospectively Applicable

Requiring that the concept can be defined prospectively is necessary if it is going to be useful in clinical practice.

C. The Definition must be Socially Acceptable

Brody and Halevy were trying to develop a policy that could withstand societal and legal scrutiny. The goal is to come up with a definition that the general public believes takes into account both family autonomy and physician discretion. One could argue with this position, asserting that this is not philosophy's role, but it shows their attempt to utilize philosophy to justify practical endeavors.

D. The Definition must Apply to a Significant Number of Cases

Brody and Halevy argue that from a practical perspective there is little point in spending the energy required for the development and implementation of futility policies if they can only be applied in very few cases. Moreover, if the number of cases is very small, then the potential for incorrect application of the policy may outweigh its usefulness.

E. The Definition must not Require Patient or Family Agreement

Brody and Halevy want to ensure that futility policies achieve their goal of resolving disagreements by allowing physicians unilaterally to override patient or family wishes.

Next Brody and Halevy try to characterize all proposed definitions of futility along two axes: the relationship between the intervention and the possible outcomes and the role of patient input into the definition. There are four possible outcomes by which one could categorize an intervention as futile. First, when the intervention cannot lead to its intended physiologic effect it is *physiologically futile*. Second, *imminent demise futility* is when an intervention—despite achieving its physiologic purpose—does not prevent the patient from dying in the very near future. The most common criteria used to evaluate this notion of futility are discharge from a hospital (although the intensive care literature often uses discharge from the intensive care unit). A more liberal criterion, *lethal condition* futility, argues that an intervention is futile if it does not alter the patient's underlying lethal condition and death results in the not too far future (weeks to months). Finally, an intervention can be defined as *qualitatively futile* if it fails to lead to an acceptable quality of life (in the futility literature this is typically defined by the health care provider rather than the patient). This definition is often invoked when it is claimed that an intervention is futile for PVS patients because even if they prolong life for an extended period of time, the quality is too low.

All of these criteria involve certain value judgments. For example, the physiologic criteria requires that we determine what the goal of an intervention is, and the imminent death criteria requires that we define how soon after discharge an individual must die to be defined as imminent. Brody and Halevy, therefore, argue that each of the above criteria for futile intervention can be further differentiated depending on the degree to which they allow the patient's values to influence the criteria. In *patient independent criteria*, there is a single professional-based answer to the value questions and patient/surrogate views have no weight; while in *patient dependent criteria*, patient values may influence when a health care provider believes a treatment has become futile. Thus, according to the latter criterion, if a patient believes death is imminent only when the patient has less than a week to live, this is the criterion the physician should use to determine if the treatment is futile according to imminent death criteria.

As a result of this distinction, Brody and Halevy came up with eight possible definitions for futility. They argue, however, that none of these definitions meet their requirements for a satisfactory definition. The physiologic criterion fails largely because it is difficult "to define a concept… that is precise, prospective, and applies to a significant amount of cases" (1995, p. 136). They note that the more confident one needs to be that CPR will not restore a heart beat, the fewer will be the number of cases to which it applies. They argue that the current data do not provide enough evidence to justify the claim that CPR will not work in a specific set of patients. The imminent demise criterion fails, they argue, because of problems with the definition of imminence. They argue that even if one accepts as the definition of imminent as hospital deaths (which they point out is an unusual way to define imminent, because it talks less about the time one lives, and focuses on where one lives), the numbers are not sufficiently precise prospectively to make a futility determination. Opponents of futility have relied on small cases series to make their argument, and the problem with this data is that the confidence intervals are large and, therefore, one can not be sure that "no one will survive." Brody and Halevy do not talk in any detail about how sure one must be that no one will survive, or alternatively the number of patients that can survive, and still have the intervention be called futile. Schneiderman argues that if there is a less than one percent chance of survival than the intervention can be called futile (Lantos et al., 1989). There are two problems with this number. First, based on subsequent studies, some done by Brody et al., using the most accurate prognostic scales, no hospital bed days were used by patients with greater than a 99% predicted hospital mortality (Sachdeva et al., 1996). So this number is unlikely to apply to many patients. Second, if one lowers the number, there will be false positives, e.g., people who survive (Halevy, Neal, and Brody, 1996; Atkinson et al., 1994). Third, it is unclear whether even this high a number would be socially acceptable. In studies of patients with incurable cancer, patients are much more willing to undergo toxic chemotherapy for even small increases in life expectancy than are health care providers. Moreover, as a society we are willing to expend large sums of money to rescue people from accidents when the chance of success seems much lower than one percent. Given this, it is unclear whether society would be willing to accept a one percent chance of surviving as futile.

When Brody and Halevy turned to analyzing whether the lethal condition criteria would pass muster they again rejected it, because of the difficulty of prospectively and precisely defining those conditions in which a physician can unilaterally override a patient's wishes. Again, the problem is that prognostic criteria (such as APACHE) are not precise enough to define even 90% one year mortality rates. One might argue that despite the lack of data for large groups of patients, physicians can determine when their individual patients are not going to survive. The data do not support this claim. The most common criteria used for prognostication in the ICU apply to groups of patients, not specific identifiable individual patients. Prognostic instruments for patients whom all agree are dying (e.g., patients in in-patient hospice programs) still lack accuracy (e.g., they can not predict survival within weeks) (Glare et al., 2003). To override patients or families wishes, one requires more than current science is able to provide.

Finally, Brody and Halevy turn to the qualitative definition of futility. They again point out the problem with prognostication. Currently, the data are unclear regarding how long one has to wait before determining that a patient with a closed head injury or stroke will not improve. There may even be questions about patients in persistent vegetative states. For example, there are studies from England that show that some patients who have been vegetative for up to a year may have some recovery (Andrews, 1993). If one waits longer, for example, three to five years for saying care is futile, the number of patients who meet these criteria will be relatively small. Thus, one wonders whether it is worth the effort to develop such a policy.

Brody and Halevy, therefore, conclude

> the advocates of the various conceptions of futility have not given enough thought to the conditions which must be satisfied by any adequate conception of futility that could be used as a basis for unilateral limiting of life prolonging interventions. As a result, they have put forward accounts and examples that are problematic just because there are no databases adequate to ensure that the relevant conditions are met (1995, p. 142).

In the intervening years we are unaware of any data that should change this conclusion.

This paper exemplifies the strengths of Baruch Brody's work. First, he chooses to write on a controversial topic that may have clear clinical implications. Second, while there had been hundreds of other articles about futility, many of which made the same points over and over again, he brought to the analysis a new way of looking at the picture. Therefore, rather than once again laying out the arguments for and against futility, he recognizes that the point of futility policies is to allow doctors unilaterally to override family decisions. To avoid "arbitrary, capricious or biased" use of the concept requires that the definition meet certain relatively non-controversial procedural and substantive criteria. Third, he integrates the empirical data into his philosophical argument. He is careful to point out that it may be possible in the future to come up with data to support one of the definitions of futility, but that the science is inadequate at this time. Finally, he followed his argument to its logical conclusion, even if that conclusion did not lead to an easy solution to the problem.

I am largely in agreement with Brody's analysis and feel that in general nothing has changed to dissuade us from his conclusion. (It is interesting that most proponents of futility policies have never argued against this analysis.) I would, however, add two further criteria that should be included when justifying a futility policy. First, one should ask whether there are less intrusive ways to deal with the number of cases in which physicians might have to override a patient's or family's wishes. I believe that whenever a futility policy is invoked, some harm has been done to the patient's/family's autonomy. One might be able to justify this harm, but nonetheless it would be a better decision if the patient or family was convinced that forgoing life-sustaining treatment was the right option. This criterion mirrors arguments made about involuntary hospitalization or violations of confidentiality. While these arguments recognize that patient autonomy is not the only value at stake, they do recognize that autonomy has value even when it is overridden by other values.

The adoption of this criterion would lead us to gather more empirical data in the futility disputes. One would want to better understand why families disagree with health care providers in cases where the health care providers have strong evidence that further treatment is unhelpful; that is, why families are making decisions that the health care providers believe are not rational. It might turn out that these disagreements are due to inadequate health care provider communication skills or hospital policies that systematically limit or lead to inadequate communication (Back and Arnold, 2005). If this were the case, before one could institute a futility policy, hospitals would need to take corrective action to minimize the need for futility policies.

Second, the consequences of instituting futility policies needs to be weighed against the policy's cost. Futility policies allow physicians to override patients or families in those cases in which they believe the continued treatment will do no good. The reason that physicians should be allowed to do so is related to physician autonomy and the importance of professional integrity. While this may be true in individual cases, it is also possible that errors will be made. Curtis' data regarding the very broad ways that doctors interpret a futility policy raises this concern (Curtis et al., 1995). In addition, determining whether the practice should be instituted as a hospital policy requires a broader look at the policy's effect on doctor-patient communication. One can imagine, for example, that the institution of futility policies leads to less communication between doctors and patients. Physicians, knowing that they can always override the surrogate if he or she does not agree with them, may spend less time or energy communicating and educating surrogates about the different options. This would lead to decreased informed consent in a large number of cases, not just those cases in which futility is at stake. One might conclude that even if a futility policy makes sense from the point of view of physicians' integrity in selected cases, the policy as a whole is not defensible.

I am not saying that this is the case. In fact, it might turn out that having a futility policy leads to better communication. The development of a futility policy may lead a hospital ethics committee to do education regarding how to decrease conflict. Moreover, before invoking futility, physicians might to want to be absolutely sure that there is nothing that they could have done to try to educate the surrogate. Empirical data will be needed to answer this question. I am merely pointing out that even if a futility policy is defensible in of itself, it may not be an ethically wise idea in a hospital in which every action causes a reaction.

II. THE BAD: HOSPITAL POLICY AND WILL WE KNOW FUTILITY WHEN WE SEE IT

One would have thought, given that Halevy and Brodys philosophical analysis, that futility was dead. However, this is not the case. In a series of articles they argue that "the basic problem is that the clinical reality of the uniqueness of patients and diseases results in judgments of futility that are not easily formulated into a general prospective definition" (1996, p. 572; 1998). They conclude that one needs to treat futility like pornography, "acknowledging while can not be defined, we certainly know it when we see it" (1996, p. 572). While never subjecting this view to their previous analysis, they convened an ad hoc group of representatives from most of the major Houston Hospitals, to see if they could come up with a justifiable futility policy.

Rather than defining futility the Houston group came up with a procedurally just policy, which gives "voice to all parties involved in the dispute." First, they argue they include all stakeholders in the debate. The ad hoc group represented most of the major hospitals in Houston, including teaching and non-teaching, private, public, secular and religious. The institutions accounted for the majority of the Greater Houston Hospitals bed capacity. Individuals who participated in the policy development included physicians, nurses, social workers, attorneys, chaplains, administrators, and ethicists. Moreover, public comment was solicited at a 1994 public conference hosted by the Houston Bioethics Network.

Second, the authors argued that the Houston process avoided problems that afflicted other policies: non-participation by the patient or surrogate; unilateral physician action without review; ignoring patient transfer options; and the potential for patient abandonment (Halevy and Brody, 1998, pp. 11–13). The policy states:

1. Disagreements between physicians and patient/surrogates about whether a treatment is medically inappropriate [medically inappropriate is synonymous with futile], should be resolved by conversation and negotiation.
2. If the surrogate decision-maker continues to insist on the treatment, the physician should discuss the option of patient transfer, to either another physician or another institution. Other health care providers should be involved in these conversations.
3. If disagreement continues, the physician must present the case for review to an institutional interdisciplinary body and provide the clinical and scientific information documenting that the intervention is medically inappropriate. The

physician must notify the patient or surrogate that this is happening. The patient or surrogate should be encouraged to come to the meeting.

4. If the institutional review body affirms the medically inappropriateness of the treatment then the patient should either be transferred to another physician or that treatment may forgone over the wishes of the patient or surrogate (1996; 1998).

Finally, Halevy and Brody argue that this policy "is firmly grounded in an accepted ethical principle, that of integrity" (1996; 1998, p. 12). Halevy and Brody argue that, in addition to individual professional integrity, institutions have principles and values. They point out that "care of actual patients is not accomplished solely by the physicians, rather it involves the dedicated efforts of many health care professionals and significant institutional resources" (1996; 1998, p. 12). Thus, their futility policy requires both the doctor and institutional committees to conclude that the disputed intervention is incompatible with their integrity. While no theory of integrity is provided, possible examples of integrity violations include treatment that causes "harm" or is "unseemly." Moreover, they draw a linkage between integrity and stewardship of resources. It is the health care providers' and hospital's duty to serve as responsible stewards of resources and, thus, when interventions are "wasteful," it can ground a claim that the treatment is futile.

There is certainly much to commend in this policy. It is clearly more humble than other futility policies, clearly specifying what cannot be proven in an honest way. Moreover, it attempts to develop a fair process for determining whether there is a professional consensus regarding the inappropriateness of a treatment. Finally, the emphasis on including the patient or patient surrogate in the process and ensuring non-abandonment are important concepts.

However, I remain unconvinced that the policy is procedurally fair either in its development or its enactment. First, the fact that this policy was developed by an interdisciplinary group from a variety of hospitals does not mean that it was open and fair. Would one conclude that an energy policy developed by different oil companies was open and fair? The process was developed by health care providers who came to the table because of their worry that futile care was being provided. Where was the diversity in the committee's ethnicity, religious or valued-based beliefs? Where were the nurses' aides, dieticians, respiratory therapists, and occupational therapists? There also is no evidence that people who might hold different views regarding futility were intentionally included in the process.

I also am concerned about the degree to which this policy underwent public scrutiny. Having it discussed at a conference on futility is unlikely to bring in a diverse group of stakeholders. One can compare this process to Denmark's attempt to develop a law on death using neurological criteria or Oregon's experience of developing a fair method of allocating health care resources (Rix, 2002). In these cases, there was a massive attempt to involve a wide cross-section of the public in debating the issue. There were television shows, newspaper articles and a well-financed public education campaign. Open discussions were held in demographically diverse areas to ensure that everyone had an opportunity to have their voice heard. One conference at The Center for Bioethics and Health Policy is unlikely to accomplish this goal.

Not only am I concerned about the procedural fairness in the policy's development, I am also concerned about how it will be operationalized. Negotiators and mediators understand that the most important part of the process is development of the rules. The process by which mediation occurs, if not carefully attended to, is likely to involve biases that lead to a pre-determined result.

The Houston policy as currently constituted is unlikely to lead to a fair and open discussion. When cases come to the institutional interdisciplinary body (futility board), the assumption is that the health care providers have done everything that they can to "convince" the family that the therapy is futile. The purpose of presenting to the futility board seems to be limited to presenting "clinical and scientific information pertinent to the determination that the intervention is medically inappropriate" (Halevy and Brody, 1996; 1998). My clinical experience and a growing amount of clinical data suggests that these irreconcilable differences occur because of poor communication. One wonders why there is no requirement for a mediator to meet with the family and health care providers before going to futility tribunal.

Second, one needs much more information about the futility board to believe that the review process will be fair. I am told nothing about the committee's composition, the review process or the manner of decision making. It is likely that the people who will volunteer to be on this committee will have a strong health care provider-centric view regarding futility. Committee members are likely to be health care employees who can take the time, often on very short notice, to attend the meetings.

I wonder how much community representation there will be on the committee. This is particularly important given the importance of financial issues in futility determinations. Hospitals work a very narrow profit margin. This profit margin is often tied to length of stay. Cases of futility often involve patients who have been in the hospital for very long periods of time. I, therefore, worry that the institution has a strong self-interest, not only because of their stewardship role, to view continued costly, non-reimbursed care as ineffectual. Involving the community in the decision making process is needed as a counterbalance.

The final problem with this policy is its emphasis on integrity. Halevy and Brody assume that there is a philosophically defensible notion of professional integrity. If such a concept exists, I am unaware of it. In areas as varied as abortion, physician-aid in dying, enhancement technologies, and even physicians' obligations to put themselves at risk to care for others, there is a diversity of professional opinion of what is the right thing to do. There is not even a consensus regarding the goals of medicine. Given this, how can the authors' appeal to integrity for these goals?

Assume, however, that Halevy and Brody are right and that there is a shared, coherent view of institutional integrity. Also assume that futile care violates institutional integrity because it causes harm without benefit or is unseemly. If this is the case, then the policy should not allow the patient to be transferred to another physician in the institution. The care will still be futile and violate physician integrity. Even transferring the patient to another institution seems wrong.

Finally, why should we assume that professional integrity trumps patient autonomy. The question is what society should do when surrogates' views violate professional integrity. Society may wish to say that health care providers must, in these rare cases, hold their nose and provide care (Gampel, 2006; Wicclair, 2000). I understand that society may not come to this conclusion. My point is only that such a view is not philosophically incoherent. And thus, I am unsure why health care institutions should be able to determine the next step rather than socially authorized bodies such as the courts.

In summary, I am concerned that in Brody and Halevy's move from philosophical analysis to public policy, they fail to pay sufficient attention to the basic rule of mediation—do not choose sides. The attempt to develop a process-based approach to medical futility fails because the process is health care-centric. Rather than developing a process that is independent of the two sides, the policy is written by and operationalized by one side—the health care providers. It emphasizes professional integrity rather than looking closely at the root causes of over-treatment and miscommunication.

III. PUBLIC POLICY AND FUTILITY DETERMINATIONS

Unfortunately, others have made things worse by legislating the Houston Hospital Policy into state law. The policy had no legal standing and thus, even when ethics committees agreed the treatment was futile, treating physicians were generally unwilling to withdraw life-sustaining treatment because of their fear of a lawsuit. The Texas Advance Directive Act of 1999 proved a legally sectioned extrajudicial process for resolving disputes about end-of-life decisions; this process included the creation of a safe harbor for physicians and institutions involved in end-of-life treatment disputes (Fine and Mayo, 2003; Fine, 2001).

The state law was modeled after the Houston hospital policy and as such, has the same problems. Despite more of an attempt to involve community representatives and those with a different point of view, the individual cases are not mediated in a neutral format. The ethics consultation committee is drawn from the same group that decided that the intervention was futile. While the law says that the physician cannot be on the ethics consultation committee, this does not preclude her colleagues, friends, and staff from the unit being on the committee. Moreover, given that most individuals on ethics committees are paid by the hospital and supposedly embody the institution's integrity, one wonders about their ability to be objective.

There are also problems because the state law confuses the ethics consultation committee's role with the Houston policy's "futility committee." In the state law, the ethics consultation committee consultants both mediate the process and serve as the "social arbitrators" of futility. One wonders the degree to which one can successfully arbitrate a process that, if unsuccessful, one has to serve as the final judge. This violates almost all principles of mediation.

Moreover, American Society of Bioethics and Humanities standards for ethics consultation clearly state that an ethics consultant's role is as a mediator, not an expert judge (Youngner, Arnold, and Aulisio, 1998). There are good reasons for this role differentiation. For example, we wonder regarding the ethics consultant's ability to determine what is "futile." Are ethics committees to serve as the community standard of futility, much like juries might determine if a film is pornographic? If so, one would want to make sure that an ethics committee is more representative of the community than is currently the case. If, on the other hand, their role is to make a scientific determination of futility, one wonders what their expertise in the evaluation of medical data is. If it is merely a scientific determination, one might ask for an external medical consultation rather than an ethics committee meeting. Moreover, given the lack of a substantive standard of futility by which to evaluate the data, we wonder how the committee will come to this decision. The problem is that this process assumes that ethics committees' values regarding futility are more correct than the health care providers, the patients, and the families. They are able to discern, when others disagree, the true meaning of futility. We believe that an argument needs to be made for why this is the case.

Little has been published regarding the experience of the state law, but what has been published makes us even more concerned. In Robert Fine's article summarizing the law's early experience, he presents a case of an African American minister with multiple strokes and recurrent ICU admissions for problems including aspiration pneumonia, gram negative sepsis, renal insufficiency, malnutrition, and decubitis ulcers (Fine and Mayo, 2003; Fine, 2001). The hospital team felt that further care was futile. The health care team tried to negotiate with the family who was divided in their views regarding what should be done. However, the primary decision maker, the patient's adult daughter, insisted that it was against their religious beliefs to stop any medical treatment. Fine reports, "the ethics committee agreed with the treatment team in that ongoing treatment other than comfort and care was *at least qualitatively futile*" (Fine and Mayo, 2003; Fine, 2001, pp. 68–71). Treatment was continued until the Texas law went into effect, at which time the family was told that life-sustaining treatment was going to be withdrawn over their objections. This was done and the patient died nine days later.[1]

This is a case that clearly does not meet quantitative notions of futility. From a vitalist point of view, it is clear that the treatment achieved the patient's goals. One wonders why the ethics committee's view of what is worthwhile should take precedence over the families'. One also wonders to what to degree there would be social unanimity about this case. If the Terri Schiavo case is any indication, there would be a sizeable minority of Americans, who believe that this care would not be futile.

IV. THE UGLY: BLAMING FAMILIES FOR COSTLY, INEFFECTUAL CARE

Texas is one of only two states that have these policies; although, after the Terri Schiavo case, a *New York Times* article reported that a number of other states were considering such policies. Given the rapidly rising health care costs, one might see decreasing futile care as a way to decrease ineffectual health care costs. The empirical data, however, suggest that cases of futility are relatively uncommon. Teno et al. analyzed data from the SUPPORT study; the largest study of critically ill dying patients ever conducted in America. They looked for patients who had a greater than a 99% chance of surviving for two months, based on APACHE, the most accurate prognostic scale available. Of the 4301 patients, 115 met these criteria (2.7%), and all but one died within 6 months. By forgoing or withdrawing life-sustaining treatment in accord with a futility policy, 199 of 1,688 hospital days (10.8%) would be forgone, with estimated savings of $1.2 million in hospital charges. Nearly 75% of the savings in hospital days would have resulted from stopping treatment for 12 patients, six of whom were under 51 years old, and one of whom lived 10 months (Teno et al., 1994).

Family demands for "futile" care thus do not seem to be a large driver of medically ineffectual care. On the other hand, in the U.S., the wide variations in health care intensity suggest that physicians have some responsibility for non-beneficial care. In a study of academic medical centers, Weinberg et al. found extensive variation in the percentage of deaths associated with a stay in the intensive care unit (8.4–36.8%). About a third of patients who were loyal to UCLA Medical Center or NYU Medical Center died in the hospital after an ICU stay compared to 20% at UCSF or Mount Sinai Hospital. Other studies have shown wide variations in other ICU practices ranging from DNR orders to time in the ICU prior to death (Weinberg et al., 2004). Most disturbingly, these variations do not seem to correlate with better health outcomes or patient preferences.

These findings fit my clinical experience. Demands for "futile" treatment are more likely to come from physicians than families. Some doctors are focused on biological life and will continue interventions to promote life even when the other health care providers feel that care is "futile." These doctors are not interested in talking to anyone about the patient's goals—they have decided that continued life-prolonging treatment is in the patient's interest.

If clinicians were really interested in policies that promote their integrity by minimizing "ineffectual or futile care," they might develop policies to limit variations in their medical practice. A hospital might require weekly review of all patients who met certain clinical criteria suggesting "futility." They might encourage clinicians who believe their colleagues are continuing ineffectual care to speak up. For outliers, a hospital could require data from the doctor about why he or she believes the therapy will work despite the lack of empirical data. We know of no hospital that has a policy for reporting or overriding clinicians who are providing care that their colleagues believe is ineffectual.

Current futility policies focus not on ineffectual care, but only on surrogates' ability to request such care. The issue is less about futility and more about decision-making power.

Moreover, if hospitals wanted to develop policies that decreased futility *and* attempted to respect surrogate decision making they would focus on improving communication rather than unilaterally overriding surrogates (Youngner, 1994). Empirical data suggests that when patients know the medical reality of their situation, they typically make the same choice that health care providers do. Similarly, studies show that when there are regular family meetings in the ICU, then families are less likely to demand ineffectual care (Lilly et al., 2000).

V. CONCLUSION

Dr. Brody accurately pointed out that futility is not the kind of problem that is resolvable by conceptual analysis. If bioethics wants to be helpful to the conflict between patients and doctors, we suggest stepping back from the debate. Like Baruch Brody did in his first article, we need to start again at the beginning. We should focus less on solving the problem and more on understanding it. Why have these cases arisen now, and what they tell us both about the goals of medicine and the society in which we live? What has led to the increased number of conflicts over end-of-life decision making and what does this tell us about informed consent and shared decision making? What are the ethical implications of choosing between the different interventions which may decrease ineffectual care? Are there ways to decrease the conflict by improving the environment in which doctors and surrogates work? What would a fair process for resolving debate in hospitals look like, and what are the institutional and economic barriers to instituting such as policy? What would a theory of professional integrity look like, and how can it be reconciled with a pluralistic society? The questions may not lead to easy answers. Still, they are the questions that need to be answered before we move forward. We can only hope that bioethics produces more scholars like Baruch Brody who are willing to tackle these difficult questions.

ACKNOWLEDGEMENT

Thanks to Debbie Seltzer for working with me to make this paper readable and coherent, to Stuart Youngner for talking through many of these ideas and to the editors for staying after me to finish this paper even when it seemed futile.

SUPPORT

Dr. Arnold was supported by the Project on Death in America Faculty Scholars Program, the Greenwall Foundation, Ladies Hospital Aid Society of Western Pennsylvania, the International Union Against Cancer (UICC), Yamagiwa-Yoshida Memorial International Cancer Study Grant Fellowship, and the LAS Trust Foundation.

NOTE

[1] For a summary of the Schiavo case, see Mayo, "Living and dying in a post-Schiavo world" (2005); Perry, Churchill, & Kirshner, "The Terri Schiavo case: legal, ethical, and medical perspectives" (2005).

BIBLIOGRAPHY

Andrews, K. (1993). 'Recovery of patients after four months or more in the persistent vegetative state,' *British Medical Journal,* 306, 1597–1600.

Angell, M. (1991). 'The Case of Helga Wanglie,' *New England Journal of Medicine*, 325, 511–512.

Arnold R.M., Youngner, S.J., & Aulisio, M. (1998). *For The SHHV/SBC Task Force On Standards For Bioethics Consultation. Core Competencies For Health Care Ethics Consultation.* Glenview: American Society for Bioethics and Humanities.

Atkinson, S., & Bihari, D., et al. (1994). 'Identification of futility in intensive care,' *Lancet*, 344, 1203–1206.

Back, A.L., & Arnold, R.M. (2005). 'Dealing with conflict in caring for the seriously ill: "It was mentioned more than a few times; we indicated that it was just out of the question."' *Journal of the American Medical Association*, 293, 1374–1381.

Brody, B.A. (2002). 'How much of the brain must be dead,' in S. Youngner, R.M. Arnold, & R. Schapiro (eds.), *The definition of death: Contemporary controversies* . Baltimore: Johns Hopkins University Press, pp. 71–82.

Brody, B., & Halevy, A. (1995). 'Is futility a futile concept,' *Journal of Medicine and Philosophy*, 20, 145–163.

Callahan, D. (1991). 'Medical futility, medical necessity. The-problem-without-a-name,' *Hastings Center Report*, 21(4), 30–35.

Council on Ethical and Judicial Affairs (1999). 'Medical futility in end-of-life care: report of the Council on Ethical and Judicial Affairs,' *Journal of the American Medical Association*, 281(10), 937–41.

Curtis, J. R., Park, D. R., Krone, M. R., & Pearlman, R.A. (1995). 'Use of the medical futility rationale in do-not-attempt-resuscitation orders,' *Journal of American Medical Association,* 273, 124–128.

Fine, R., & Mayo, T. (2003). 'Resolution of futility by due process: early experience with the Texas Advance Directives Act,' *Annals of Internal Medicine*, 138, 743–746.

Fine, R.L., (2001). 'The Texas Advance Directives Act of 1999: politics and reality,' *HEC Forum*, 13, 59–81.

Gampel, E. (2006). 'Does professional autonomy protect medical futility judgements,' *Bioethics*, 20, 92–104.

Glare, P., Virik, K., & Jones, M., et al. (2003). 'A systematic review of physicians' survival predictions in terminally ill cancer patients,' *British Medical Journal,* 327, 195.

Halevy, A. & Brody, B.A. (1998). 'The Houston process-based approach to medical futility,' *Bioethics Forum*, 14(2), 10–17.

Halevy, A., & Brody, B.A. (1996). 'A multi-institution collaborative policy on medical futility,' *Journal of the American Medical Association*, 276, 571–74.

Halevy A., & Brody B. (1993). 'Brain death: Reconciling definitions, criteria and tests,' *Annals of Internal Medicine,* 119, 519–525

Halevy, A., Neal, R.C., & Brody, B.A. (1996). 'The low frequency of futility in an adult intensive care unit setting,' *Archives of Internal Medicine*, 156, 100–104.

Lantos, J.D., et al. (1989) 'The illusion of futility in clinical practice,' *American Journal of Medicine*, 87, 81-84.

Lilly, C.M., De Meo, D.L., & Sonna, L.A., et al. (2000). 'An intensive communication intervention for the critically ill,' *American Journal of Medicine*, 109, 469–475.

Mayo, T.W. (2005). 'Living and dying in a post-Schiavo world,' *Journal of Health Law*, 38(4), 587–608.

McCullough, L.B., & Jones, J.W. (2001). 'Postoperative futility: a clinical algorithm for setting limits,' *British Journal of Surgery*, 88, 1153–1154.

Miles, S.H. (1991). 'Informed demand for 'non-beneficial' medical treatment,' *New England Journal of Medicine*, 325, 512–51.

Perry, J.E., Churchill, L.R. and Kirshner, H.S. (2005). 'The Terri Schiavo case: legal, ethical, and medical perspectives,' *Annals of Internal Medicine*, 43(10), 744–8.

Rix, B.A. (2002). 'Brain death, ethics and politics in Denmark,' in S. Youngner, R.M. Arnold and R. Schapiro (eds). *The Definition of Death. Contemporary Controversies*. Baltimore: Johns Hopkins University Press, pp. 71–82.

Rubin, S. (1998). *When Doctors Say No: The Battleground Of Medical Futility*. Bloomington: Indiana University Press.

Sachdeva, R.C., Jefferson, L.S., Coss-Bu, J., & Brody, B.A. (1996). 'Resource consumption and the extent of futile care among patients in a pediatric intensive care unit setting,' *Journal of Pediatrics*, 128, 742–747.

Schneiderman, L.J. (1994). 'The futility debate: effective versus beneficial intervention,' *Journal of the American Geriatric Society*, 42, 883–886.

Schneiderman, L.J., Jecker, N.S., & Jonsen, A.R. (1990) 'Medical futility: its meaning and ethical implications,' *Annual of Internal Medicine*, 112, 949–54.

Teno, J.M., Murphy, D., Lynn, J., Tosteson, A., & Desbiens, N., et al. (1994). 'Prognosis-based futility guidelines: does anyone win? SUPPORT investigators study to understand prognoses and preferences for outcomes and risks of treatment,' *Journal of American Geriatric Society*, 42, 1202–1207.

Tomlinson, T. & Brody, H. (1990). 'Futility and the ethics of resuscitation,' *Journal of the American Medical Association*, 264, 1276–1280.

Truog, R. (1997). 'Is it time to abandon brain death?' *Hastings Center Report*, 27(1), 29–37.

Troug, R.D., Brett, A.S. & Frader, J. (1992).'The problem with futility,' *New England Journal of Medicine*, 326, 1560–64.

Weinberg, J., Fisher, E.S., & Stuikel, T.A., et al. (2004). 'Use of hospitals, physician visits, and hospice care during last six months of life among cohorts loyal to highly respected hospitals in the United States,' *British Medical Journal*, 328, 607.

Wicclair, M.R. (2000). 'Conscientious objection in medicine,' *Bioethics*, 14, 205–227.

Youngner, S.J. (1994). 'Futility: Saying no is not enough,' *JAGS*, 42, 887–889.

Youngner, S.J. (1990). 'Futility in context,' *Journal of the American Medical Association*, 264, 1295–1296.

CHAPTER 10

A MATTER OF OBLIGATION: PHYSICIANS VERSUS CLINICAL INVESTIGATORS

E. HAAVI MORREIM

Health Science Center, The University of Tennessee, Memphis, Tennessee, U.S.A.

I. INTRODUCTION

In *Taking Issue: Pluralism and Casuistry in Bioethics* (2003) Baruch Brody offers a moral pluralism (p. 3) that invites us to acknowledge the complexities of our moral dilemmas, abjuring simplistic mandates and formulae. He urges us to appreciate the reality of "deep moral ambiguity" and to understand that solving moral problems cannot be reduced to some sort of metric. Rather, he provides "a case-dependent casuistric pluralism, because the resolution of which appeal takes precedence may vary from case to case" (p. 4). With this approach he takes the reader into three major areas of bioethics—research ethics, clinical ethics, and Jewish medical ethics—exploring a variety of the difficult dilemmas that pepper bioethics as a clinical field.

This essay will venture into the first of those areas, namely research ethics, focusing on one particular question. I will inquire what obligations investigators owe the people who enroll in their research studies, and will provide an analysis that follows the spirit of Brody's moral pluralism.

By way of overview, research differs in many ways from standard care—often adding uncertainties, for instance, and nontherapeutic interventions such as diagnostic tests whose only purpose is to measure the effects of the research intervention. Hence one may inquire whether a physician engaged in clinical research has the same obligations toward research subjects as those that ordinary physicians owe their patients, or whether they differ in any fundamental ways.

Perhaps the most common answer is that the relationship is the same. Physicians are fiduciaries who owe the highest sort of obligation to their patients. Physician-investigators, on this view, owe no less to the volunteers[1] who enroll in research trials. Each owes the best available medical care, which means that a physician can only justify enrolling a patient in research if the study meets the requirements of clinical equipoise—namely, that there is legitimate disagreement within the medical community as to whether the standard treatment or the investigational intervention is superior. In this way, he can be sure that he is not denying his patient the best available therapy.[2]

M.J. Cherry and A.S. Iltis (eds.), Pluralistic Casuistry, 149–165.

This essay argues that this view begins with the wrong question and ends with the wrong answer.[3] Physician-investigators are not, and cannot possibly be, fiduciaries of volunteers, for a variety of reasons to be discussed. This does not mean they do not have strong obligations to protect volunteers' welfare. Rather, their duties can best be understood as "side constraints" that strictly limit, but do not substantively direct, the ways in which research volunteers may be treated. A better understanding of these limits can, in turn, suggest better ways to protect the people who enroll human research.

As will become evident, my approach echoes Brody's rejection of simplistic mandates, in favor of making reasonable judgments and reaching nonmechanical conclusions (Morreim, 2005) that respond to the variable realities embodied in specific situations.

II. THE WRONG QUESTION

The "Common View," as I will call it, begins with an empirical claim. The most typical setting in which someone might be invited to enroll in a research trial, it is said, is one in which a patient approaches his physician, or a specialist to whom he has been referred, seeking treatment.[4] The key question then becomes "When may a physician offer enrolment [sic] in a clinical trial to her patient?" (Weijer and Miller, 2004, p. 571; see also Mann, 2002, p. 35; Kovack, 2002, p. 32; Weijer et al., 2000, p. 756; Glass, 2003, p. 5; Miller and Weijer, 2003, p. 93). Even in those less common instances where a physician is solely a clinical investigator, he is still a physician, a profession carrying certain fundamental obligations.

Given that this physician already has a strong duty to provide optimal medical care to his patient, the research must not compromise this fundamental obligation (National Placebo Working Committee, 2003, p. 37).[5] As emphasized by the Declaration of Helsinki and other international standards, the researcher must ensure that his patient receives the best available care (National Placebo Working Committee, 2003, p. 37; Rothman and Michels,1994, p. 394; Angell, 1997, p. 847). As that initial formulation evolved over time (Freedman, Weijer and Glass, 1996, p. 252), it has now essentially become the duty to ensure "clinical equipoise." Research can only be justified where there is legitimate disagreement within the medical community as to whether the standard or the research intervention is superior (Weijer, Shapiro, Glass and Enkin, 2000; Miller and Weijer, 2003), so that the patient will "not be agreeing to medical attention that is known to be inferior to current medical practice" (Freedman, Weijer and Glass, 1996, p. 253).

On this view, the investigator is a fiduciary of the volunteer (Miller and Weijer, 2003, p. 95; Saver, 1996, p. 221; Holder, 1982; National Placebo Working Committee, 2003). A fiduciary relationship exists where one party is far weaker than, and dependent upon, a stronger party who thereby has an obligation to promote the other's best interests. In the Common View, the investigator's duties to promote the volunteer's best interests are no less stringent than in the standard physician-patient fiduciary relationship. The volunteer's welfare must not be subordinated to the goals of research (Glass, 2003, p. 5).[6]

The first problem with the foregoing analysis is that it begins from the wrong starting point. Even if it happens to be empirically true that many or most volunteers are invited into research by their own physicians, this hardly entails that the physician-patient relationship is the best or only basis on which to define investigators' obligations toward volunteers. Many clinical studies, after all, take place completely outside the context of clinical care. Most drugs' phase I trials, for instance, are carried out on normal volunteers. These investigators usually are physicians, but their relationship with volunteers is completely independent of the treatment setting. Surely these relationships carry moral obligations, and if anything they would be the most clear instances of purely research-oriented duties.

More importantly, research in the treatment setting is fraught with mixed obligations. As discussed below, the treating physician begins with a strong set of duties to promote the best interests of his own patients. Even if occasionally he may have conflicting obligations, such as to warn the sexual partners of a patient with HIV, his paramount duties focus on the patient. In contrast, a researcher's fundamental obligation points in an altogether different direction, namely, to honor the requirements of the protocol. If he violates it every time that it might ill-serve an individual patient's best interests, he can thwart the study's scientific quality and thereby harm future patients. We cannot clearly sort out the duties of investigators unless we can distinguish them from the ordinary physician duties from which they may differ or even conflict.

III. A BETTER STARTING POINT

The fundamental flaw of the Common View is that it takes an empirically common situation (research in the clinical care setting), dubs it paradigmatic (even while acknowledging it to be deeply problematic), then derives ethical norms. Thereafter comes the doctrine of Clinical Equipoise in an attempt to reconcile the obligations of medical treatment with those of clinical research.

Far from presuming that it is paradigmatic for treating physicians to recruit their own patients into research, we must start with a cleaner question. We must look first at the research situation in its own right, and discern what obligations the investigator owes the volunteer. Only then can we then determine when (if ever) and under what restrictions a treating physician can undertake research on his own patients.

Prerequisite to that task, we must understand the basic differences between medical care and clinical research. A more detailed discussion is available elsewhere (Morreim, 2003), but for present purposes the most pivotal distinction is that medical care focuses on promoting the best interests of the individual patient, whereas clinical research focuses on creating generalizable knowledge. Scientific generalizability, in turn, typically requires that research be conducted according to a protocol that will control as many variables as possible, so that the resulting observed differences will be ascribable to the intervention (drug, device, procedure) and not something else. Accordingly, standard scientific controls such as randomization, blinding, and placebo controls are frequently used. More to the point, these techniques mean that volunteers will be treated, not according to what is individually best for them, but according to what the protocol says will happen to them. Care is not individualized; it is rigorously standardized.

Indeed, attention to individual participants' preferences and best interests is quite strictly limited. In the typical research trial there are only two avenues for attuning what goes on to suit the individual volunteer. First, the protocol may expressly permit certain areas of flexibility, such as to use certain adjuvant medications for symptom relief. Second, the investigator can remove from the study any volunteer who is unduly harmed or inconvenienced. Other than these two options, the investigator must do exactly what the protocol requires, even if it is suboptimal for a given individual. He must administer the dose the protocol dictates, even if a higher or lower dose might be better for a given volunteer; he must forbid any adjuvants that might help relieve symptoms unless they are expressly permitted; he must undertake a host of measurements and tests that have no benefit for the volunteer. As noted by Frank Miller and Howard Brody,

> [t]rials often include interventions such as blood draws, lumbar punctures, radiation imaging, or biopsies that measure trial outcomes but in no way benefit participants. RCTs [randomized controlled trials] often contain a drug 'washout' phase before randomization to avoid confounding the evaluation of the investigational treatment with the effects of medication that patients were receiving prior to the trial. These various features of research design promote scientific validity; they carry risks to participants without the prospect of compensating therapeutic benefit (Miller and Brody, 2003, p. 22; see also Miller and Rosenstein, 2003, p. 1383).

IV. INVESTIGATORS ARE NOT, AND CANNOT BE, FIDUCIARIES

To explore the respective obligations of physicians versus investigators, we must first understand that in studies where patient-volunteers receive both medical care and investigational interventions, those aspects of care that are standard medical treatment must still be regarded as clinical care, and that those who provide it are still bound by traditional physician obligations. Thus, if the study features a surgical innovation, it must still be performed with standard techniques of antisepsis. Any failure to do so would be a breach of the surgeon's standard obligations qua physician. Similarly, if someone enrolled in a trial of a new arthritis drug develops an infection, that problem must be treated according to medical standards, and traditional physician obligations remain intact (assuming such infections are not specifically addressed in the protocol). Thus, I do not propose that all physician duties are suspended during a research trial. Rather, in a trial featuring both, one must distinguish familiar physician functions from the specific functions of research, and inquire what duties adhere to the investigational elements of the activity. Each role must be considered in its own right, not on the basis of any happenstance overlap because one person may be playing both roles. Of note, several courts agree that this line must be drawn.[7]

However, the Common View declines to draw such a distinction, holding that MD-investigators still act fully as physicians, even with their investigator "hat" on. Accordingly, its proponents also insist that investigators are fiduciaries of research volunteers in the same way that physicians are fiduciaries of their patients. To understand why clinical investigators are not, and cannot be, fiduciaries, we must first understand what a fiduciary is.

A. The Nature and Purpose of Fiduciary Relationships

Admittedly, "[f]iduciary obligation is one of the most elusive concepts in Anglo-American law" (DeMott, 1988, p. 879). It arose in the law of trust, in which a trustee holds property for the benefit of another party. Over time it expanded to include agency, in which one person represents another—again to promote the interests of the latter. Its subsequent development encompassed additional relationships, such as partners, directors and officers, executors, attorneys, priests, and others.[8]

While one would be hard-pressed to define any single clear, pervasive concept of the fiduciary relationship, Shepherd explains that "[t]here are three traditional classifications of fiduciaries. From the law of trusts, we have a class which we will call here property-holders; that is, those who hold or manage property on behalf of another. The second class is that of representatives, stemming, obviously, from the law of agency. Finally, based largely in the law of undue influence, we have a third class which we will call advisers" (Shepherd, 1981, p. 21). The trustee, as Shepherd explains, has control of property, whether monetary or real, that he must use to benefit the person named as beneficiary. The agent stands as the representative or surrogate of the principal, acting on his behalf for the sake of the latter's betterment. The advisor, in turn, can become a fiduciary by providing advice and expertise on which the advisee must rely. Within this third group, the bare fact that one person decides to trust another's judgment does not, of itself, create a fiduciary relationship; rather, certain other conditions must also be met. Although it is debatable whether physicians are fiduciaries in the strictest sense (Rodwin, 1995), their relationship with patients would best fit under this third category, as advisors (Shepherd, 1981, p. 29).

Across these three classifications of fiduciary, a fairly distinctive set of features characterizes virtually all fiduciaries. First, fiduciaries invariably have discretion and power. To manage property, or to represent someone, or to advise, the fiduciary must have sufficient leeway to make the kinds of judgments that can, under a wide diversity of circumstances, affirmatively promote the best interests of the other—variously called "beneficiary," "principal," "fiducie," or "entrustor" (the term used in this essay) (Davis, 1986, pp. 4, 6). "If the relationship, as the parties structure it, does not confer discretion on the 'fiduciary,' then his actions are not subject to the fiduciary constraint. Even a designated 'trustee' may not be a fiduciary if he entirely lacks authority and thus has no discretionary power" (DeMott, 1988, p. 901). "The United States Supreme Court has noted that the central purpose of fiduciary law is to govern the exercise of discretion in making decisions that are not, and cannot be, controlled in advance by legal means" (Jacobson and Cahill, 2000, p. 160).[9]

Because the fiduciary necessarily has this power, and because it is usually difficult if not impossible for the entrustor to monitor the fiduciary's performance,[10] the fiduciary has a broad opportunity to exploit the entrustor's dependency and vulnerability for his own gain or for other purposes of his choosing. That is, the very power and discretion that enables him to do the job also enables him to harm the very one whose benefit he must promote. For this reason, the law imposes the strongest duty of loyalty. "A trustee is held to something stricter than the morals of the market place. Not honesty alone, but the punctilio of an honor the most sensitive, is then the standard of behavior" (Meinhard v. Salmon, 164 N.E. 545 [N.Y. 1928]; see also Davis, 1986, p. 3). "The law defines a fiduciary as a person entrusted with power or property to be used for the benefit of another and legally held to the highest standard of conduct" (Rodwin, 1995, p. 243).

A second feature of fiduciary relations, then, is a set of specific obligations, led by the duty of loyalty. The fiduciary's discretion and authority are to be exercised for the benefit of the entrustor. "The *Restatement (Second) of Agency* defines agency as a 'fiduciary relationship' in which the agent [has] a duty 'to act solely for the benefit of the principal in all matters connected with his agency.' Similarly, the *Restatement (Second) of Trusts* defines a trust as a 'fiduciary relationship with respect to property,' with the trustee being under a duty 'to administer the trust solely in the interest of the beneficiary'" (Davis, 1986, p. 25). As similarly noted by Shepherd, "[a] fiduciary relationship exists whenever any person acquires a power of any type on condition that he also receive with it a duty to utilize that power in the best interests of another" (1981, pp. 35, 93).[11]

Other commentators echo the theme. "A fiduciary relationship involves a duty on the part of the fiduciary to act for the benefit of the other party" (Scott, 1967, p. 39); to "act in the entrustor's best interests" (Frankel, 1983, p. 823); to "act in the interest of another person" (Scott, 1949, p. 540); to "give priority to the beneficiary's interests" (DeMott, 1988, p. 906; Davis, 1986, p. 19; Shepherd, 1981, p. 35; Finn, 1977, p. 3). "Anything that compromises the fiduciary's loyalty to the fiducie or the fiduciary's exercise of independent judgment on the fiducie's behalf creates a conflict of interest" (Rodwin, 1995, p. 244).

The loyalty duty has at least two important aspects. Most obviously, the fiduciary must not exploit the entrustor to promote his own gain. He must not enter into avoidable conflicts of interest that would pit his own welfare against the entrustor's or, when unavoidable he must disclose such conflicts of interest and permit the entrustor to decide whether he may handle this transaction, or indeed continue as fiduciary.[12]

Somewhat less obviously, the fiduciary also should not compromise the entrustor's welfare for the benefit of third parties. As noted by Finn, the fiduciary's duty in the "service of his beneficiaries' interests" includes "a duty not to act for his own beneift, or for the benefit of any third person" (1977, p. 15). Of course sometimes it may be impossible to avoid such conflicts of obligation. Nevertheless, the fiduciary's presumption is that the entrustor's interests are generally to take priority over third parties' interests as well as the fiduciary's benefit.

Fiduciary relationships can have other characteristics less pertinent here.[13] For present purposes, the task now is to show that investigators are not, and cannot possibly be, fiduciaries of their research volunteers.

B. Investigators are not, and Cannot be, Fiduciaries

To begin with, investigators do not fit into any recognized classification of fiduciaries. Obviously they are not trustees of property, nor are they agents who represent the interests of the volunteer in business transactions other specified forums.

Neither are they advisors in the way that physicians are. Most research studies are highly protocolized, dictating precisely what will happen at every turn, so there are few opportunities for "advising," Perhaps the investigator might advise the volunteer about the appropriate use of whatever flexibility the protocol permits; but this is a more physician-like than investigator-like activity. Or he might advise the prospective volunteer whether he should enroll in or drop out of the study. However, arguably the investigator should refrain from advising on this matter. After all, recruitment and retention present a thoroughgoing conflict of interest. Even if the investigator does not directly stand to earn money by enrolling people in his study, he will gain something of value for completing the research, whether prestige, promotion, enhanced grant-gaining eligibility, or other nonfinancial benefits.[14] Hence, investigators should abjure the only real areas available for substantive advising.

If it is fairly obvious that investigators do not fit any of the three classes of fiduciaries, it is even more important to understand that their activities do not fit the key elements of the fiduciary role discussed above.

A fiduciary's first key characteristic, as noted, is the need to exercise discretion. A treating physician must often exercise considerable discretion, as he determines which diagnostic and therapeutic interventions would be most appropriate for a given patient's constellation of signs and symptoms, and as he determines when to modify his initial plan. An investigator, in contrast, has very little discretion. Once the protocol is set he is not free to modify or deviate from it. He is not free to determine which arm the volunteer enters, if randomization is required. If a certain test must be performed at week #4, then he must perform that test, not some other test—at week 4, not 3 or 5. In the typical trial the only discretion he has is to decide whether to remove someone from the study, and how to use the protocol's available flexibility to suit the needs of a given volunteer.[15]

In the second characteristic from above, a fiduciary's primary focus must be upon the entrustor, not on his own benefit or even that of a third party. In a routine treatment relationship this means that the physician's primary goal must be to benefit his patient rather than, e.g., to build his own bank account. In contrast, the research volunteer is not, and cannot be, the investigator's primary focus. His first allegiance is already pegged on something else, namely, the research protocol and on the "third parties" (future patients) who will be helped by it.

Indeed, it cannot even be said that the investigator is benefiting the volunteer at all by placing him in a research study. Research, by definition, is not designed to benefit any specific individual, because its objective is to gain generalizable knowledge via tightly protocolized interventions (Morreim, 2003, pp. 16–17). A volunteer may of course benefit, but that is by good fortune and not by design. He can just as well fail to benefit or even be harmed. He may be assigned to placebo, or the research intervention may turn out to do more harm than good, or it may do less good than standard care would have done. More properly, we should say it is the volunteer who is contributing the benefit, as future patients may be helped by the study's results.

Admittedly, any fiduciary can find himself in a situation of mixed allegiance, as where a physician may need to breach his usual obligation of confidentiality in order to warn the sexual partner of an HIV patient. However, the investigator's conflicts are not this sort of occasional, often avoidable event. His is a systematic, thoroughgoing requirement that he follow the protocol, even where it may be suboptimal for the volunteer. Undue harm or risk may be reason to remove the volunteer entirely from the protocol, and of course the investigator may use protocol-permitted flexibility to suit the individual's needs and wishes. But otherwise, the volunteer's interests must generally be subordinated to the protocol. If he has been randomized to an arm that might be less desirable than another arm, or if he is receiving a dose that seems too small or too large, or if he would be better off with otherwise-prohibited adjuvant medications—so be it, unless the problems are so great that he should cease participation altogether. If the investigator deviates from the protocol every time it might suit the volunteer, he will destroy its scientific validity. In this way, the investigator cannot possibly ascribe primacy to the volunteer's welfare—the very thing that any fiduciary would be required to do.

It is worth noting that supporters of the Common View do not address this problem adequately. Their focus is on a physician's initial decision whether to invite his patient into a study, with justification accepted if the conditions of clinical equipoise are met. Yet this answer only speaks to the question whether the patient should be exposed to the investigational intervention or its alternative. In fact, volunteers are inevitably subjected to a number of interventions that not only do not benefit them, but may be seriously suboptimal even if they do not actively harm them. A volunteer may be randomized to an arm that clearly does not suit him, or may have to forego adjuvant medications for symptom relief. Any time such a thing happens, a physician-fiduciary under the Common View would have a clear obligation to adjust the dose to help the patient, or switch him to a different arm, or do whatever is optimal for that individual—the very thing that an investigator must not do.

To this challenge, Charles Weijer simply suggests that such risks must be minimized and must be reasonable in relation to the knowledge to be gained (2002; 2000). Clearly, however, this is hardly the "optimal medical care" the Common View mandates (Steinberg, 2002, p. 27). If the physician-investigator nevertheless pretends that he is still honoring the patient-volunteer's best interests above all else, he fosters the 'therapeutic misconception' in himself as well as the volunteer (Appelbaum, 2002; Appelbaum, Lidz and Grisso, 2004; Miller and Brody, 2003; Miller and Rosenstein, 2003, p. 1384; Miller, 2004, p. 111).

One more argument that investigators are fiduciaries deserves debunking. Angela Holder argues that because investigators must treat enrolled subjects with great care, deference, and loyalty, then the relationship must be fiduciary. She further suggests that, since courts are unlikely to excuse injury and unfairness to research participants on the ground that the investigator is not a fiduciary, this too must be reason to conclude that the investigator is a fiduciary (Holder, 1982, p. 6; Morin, 1998, p. 216; see also Vodopest v. MacGregor, 913 P.2d 779, 788 [Wash. 1996]).

Such reasoning runs backwards. Whereas the logical chain of reasoning says that a fiduciary relationship must exist before fiduciary duties can be imposed, Holder reasons in the opposite direction, moving from the fact that investigators have duties to subjects, to infer that the relationship must therefore be fiduciary. This argument commits a classic logical fallacy called "affirming the consequent." One begins with the premise "if A is true, then B is true." If one then conversely reasons that because B is true, then A must also be true, he is committing the fallacy. Example: it is clearly true that "if (A) it is raining heavily, then (B) the sky is cloudy." However, the reverse is not necessarily true: "if (B) the sky is cloudy, then (A) it must be raining heavily."[16]

V. RESOLVING THE PROBLEM

A. An Initial Attempt

The Common View presumes that if investigators are not fiduciaries, then there is no room for them to have strong obligations to protect research volunteers. Essentially, they commit a version of Holder's fallacy by assuming that since investigators must bear strong duties toward volunteers, they must be fiduciaries. "Neither moral theory nor legal principle permits making patients' interests secondary to answering a research question. Physicians have been held to the same legal standard of care in providing services whether or not they are investigators"(Glass, 2003, p. 5). "If a physician treats a patient with either an accepted or an experimental therapy, a physician-patient relationship is established that obliges the physician to care for that patient. It would be a violation of our common notion of responsibility if clinical investigators could administer an experimental therapy and then be absolved of the responsibility to manage the consequences of their intervention" (Steinberg, 2002, p. 27).

Surely this view is mistaken. Simple logic can tell us that moral duties need not be specifically fiduciary in order to be strong and binding. And indeed some scholars have argued that investigators have powerful duties toward volunteers, even if they are different in content and basis from those of physicians toward their patients. Miller, Brody, Emanuel and Rosenstein (Miller and Brody, 2002; 2003; Miller and Rosenstein, 2003; Miller, 2004; Emanuel and Miller, 2001), for instance, collectively suggest that research is a fundamentally different kind of activity from clinical care, and accordingly carries its own distinctive ethical requisites.

First, they rightly point out that the investigator's primary duty is not to benefit the individual volunteer, but rather to gather generalizable knowledge that will benefit future patients. At the same time, they understand that investigators have strong duties toward volunteers, but propose that these are mainly negative duties of protection rather than affirmative duties to promote welfare. Hence, rather than fiduciary obligations, this view suggests that investigators' major duty, given their systematic conflicts of interest and of obligation, is to refrain from exploiting volunteers. This, in turn, translates mainly into ensuring that volunteers are not exposed to undue risk and that they are sufficiently informed to provide genuine consent.

B. A Richer Approach

Although non-exploitation is surely a crucial duty toward volunteers, the Common View criticism may perhaps strike a chord. Many research volunteers are ill and vulnerable, and it may seem anemic simply to tell investigators to avoid undue risk and to inform volunteers sufficiently to achieve genuine consent. While those duties may ultimately turn out to be investigators' key responsibilities, it would be good to situate them, and perhaps other duties, within a broader framework. Since that framework cannot be a fiduciary relationship, some alternative should be identified.

That framework, I propose, is to be found in Robert Nozick's concept of a moral side constraint (1974, pp. 29–33). Nozick distinguishes between honoring a right as the moral goal of one's conduct, versus honoring a right as a side constraint on conduct whose goal is something else. In other words, in some cases the direct goal of the activity is to promote someone's right; e.g., as when an attorney defends his client's innocence or when a physician diagnoses and treats a patient who has sought his help. In many other activities, however, promoting someone's rights is not the goal of the activity. Rather, honoring those rights constitutes strong limits on the means by which the goal can be pursued. If I wish to build a road, I cannot simply build on land belonging to another person; that person's ownership poses a side constraint on pursuing my goal. This does not make me a fiduciary of the landowner; it simply places a strong limit on my activities. As Nozick puts it: "Side constraints express the inviolability of others ... The moral side constraints upon what we may do, I claim, reflect the fact of our separate existences ... that no moral balancing act can take place among us …" (1974, pp. 32–33).

This analysis fits the current challenge well. Medicine's goal is directly to promote the well-being of the patient. In contrast, clinical research has a very different goal, namely, to pursue generalizable knowledge for the benefit of future patients—a process requiring them to use human volunteers. However worthy that goal, investigators are not free to use volunteers in whatever ways will best satisfy science. They must observe strict limits to protect the underlying moral inviolability of those human beings. They must, that is, honor a host of side constraints.

Those side constraints emerge at several levels, going beyond just the obligation to refrain from exploitation.[17] The research design must pursue a medically worthy project, in scientifically credible ways. Otherwise, volunteers will be exposed to pointless inconveniences and perhaps also needless risks. And of course the design must minimize risk.

During the research, investigators must first ensure that each volunteer actually meets eligibility requirements, so that they neither face undue risk nor taint the results and thereby diminish or negate the contributions of everyone in the research. They or their representatives must also, of course, be fully apprised of the purpose, risks, potential benefits, and all the other elements required for genuinely informed consent.

As the project progresses, investigators should use the protocol's available flexibility to secure each volunteer's greatest comfort and risk minimization. Volunteers must also be apprised of any developments, either via information gained during the study or via their own adverse reactions, that may prompt them to reconsider continuing their participation. Under certain circumstances an investigator may also be obligated to encourage or require a volunteer's withdrawal or even to halt the study entirely.

C. Additional Protection

This side constraint approach is hardly anemic. It offers aggressive protections, but with a recognition that the goal of research is not to promote the best interests of any individual volunteer. It thereby also avoids deceiving volunteers by inviting them (or investigators) into the therapeutic misconception that somehow the project is designed to seek each volunteer's personal best interests.

Several further protections might also be considered. The Common View is right about one thing. Much clinical research takes place in a treatment context, in which a patient seeking care is invited into a study. The situation is fraught with hazard, as a physician considers whether to invite his own patient into research. The patient's expectations may be compromised, and he may not fully appreciate that the goals of his care have changed profoundly. For this reason, it may be best for treating physicians simply to avoid inviting their own patients into their research projects, but rather to convey this task to others (Levine, 2002). Often this will be quite feasible. For outpatient research, many sites are now devoted exclusively to clinical studies, and a patient interested in participating can be referred to one of them. Likewise, many inpatient trials are conducted at specialized sites. If patients are clearly informed in advance that this is not simply a referral to a specialist, but rather to a dedicated research site, the separation can help greatly to reinforce the difference between treatment and research.

Sometimes a complete separation between treatment site and research site may not be possible. For people with relatively rare diseases, such as slceroderma, it may be impossible to participate in research by traveling to another site. Rheumatology is not a common specialty, and the disease has no fully effective treatment. Many of the people with this disease are eager to participate in research, and cannot do so except through their own physicians. Analogously, in smaller communities with relatively few physicians it may not be possible to establish separate research sites, leaving one's own physician as the only opportunity.

Even in these instances, however, extra protections can be incorporated. To emphasize the difference between treatment and research, investigator-physicians can make greater efforts to emphasize that the goals of research are different, and that their participation in the study raises special conflicts of interest of obligation. With care and effort, these distinctions can be effectively taught to most patients (Appelbaum et al., 1987; Appelbaum, Roth and Lidz, 1982; Appelbaum, 2002; Appelbaum, Lidz and Grisso, 2004). Additionally, physician-investigators can create physical distinctions to demarcate research from treatment, as by wearing a different lab coat in the research setting, and perhaps even reserving separate exam rooms for that purpose.

Additionally, it has been suggested that volunteers be routinely paid for their participation in research, to emphasize that they are making a contribution and that research is not standard treatment (Grunberg and Cefalu, 2003, p. 1388; Miller and Rosenstein, 2003, p. 1385).

VI. CONCLUSION

One of the most important questions in research ethics asks what obligations investigators have toward the people who volunteer for clinical trials. Although the Common View rightly emphasizes strong duties to protect, it wrongly proposes that investigators must affirmatively promote volunteers' interests in the same ways that physician-fiduciaries do. This view is potentially pernicious, as it invites investigators to embrace, and to bring their patients into, the therapeutic misconception. It is also simply incorrect, because investigators do not satisfy the major criteria by which fiduciaries are identified.

In the final analysis, we need not falsely claim a fiduciary relationship in order to ensure that investigators bear powerful obligations to protect volunteers. These duties are best found in the concept of a side constraint—a concept that embodies the inviolability of the human person even while recognizing that the research does not aim to promote any particular volunteer's personal best interests.

ACKNOWLEDGEMENT

The author acknowledges with gratitude the very helpful comments provided on earlier drafts by David Resnick, JD, PhD, Ken Kipnis, PhD, Carl Coleman, JD, Jessica Berg, JD, Frank Miller, PhD, and Howard Brody, MD, PhD.

NOTES

[1] A variety of terms can designate those who enroll in research trials, and each has its problems. "Subject" may seem mechanistic and cold; "patient" too easily loses the distinction between treatment and research; "participant" can equally refer to investigators and others who take one or another role in the project. This essay will use "volunteer" to emphasize that a research trial is not part of routine treatment, and that it is a means by which someone can choose to make a contribution, even where he may also hope for benefit to himself.

[2] Prof. Brody provides an important critique of the doctrine of clinical equipoise in the context of placebo-controlled trials. See Brody, B.A. (1995). Ethical Issues in Drug Testing, Approval, and Pricing: The Clot-Dissolving Drugs, pp.112-131.

[3] This essay is based on Morreim, E.H. "The clinical investigator as fiduciary: Discarding a misguided idea" (2005).

[4] "Patients are not typically recruited for clinical trials by unrelated third parties but are recruited within the context of an established patient-physician relationship" (Mann, 2002, p. 36). See also Mann and Djulbegovic, 2003, p. 4; Miller and Weijer, 2003, p. 93.

[5] National Placebo Working Committee. National Placebo Initiative, Draft Report, October 2003; found at : http://www.cihr-irsc.gc.ca/e/services/pdf_19320.htm, at 37.

[6] As proposed by Steinberg, "[t]he notion that clinical investigators do not have a duty to provide optimal medical care is nonsense" (2002, p. 27).

[7] Several courts have discussed cases featuring interventions that were purely research, with no physician-patient relationship involved.

In Payette v. Rockefeller University, for instance, a student volunteered for three studies about diet. After taking iodine injections and oral potassium iodide the student developed enlarged thyroid and hypothyroidism. The court found that because this study was not medical practice, it could not be medical malpractice. Therefore, the applicable statute of limitations was the three-year limit of ordinary negligence, not the two and a half year term applicable to medical malpractice. To emphasize the difference, the court noted that the student was not seeking medical assistance by enrolling in the study, her relationship with the investigators was not physician-patient, and the interventions she received were not undertaken to diagnose and treat her. Hence, the court found, even though doctors were involved and performed medical procedures requiring professional skill, this was not medical practice, hence could not be medical malpractice. See Payette v. Rockefeller University, 643 N.Y.S.2d 79 (App Div 1 Dept. 1996).

In Craft v. Vanderbilt Univ., 18 F.Supp.2d 786 (M.D. Tenn 1998), a mid-1940s study gave pregnant women a "vitamin drink" that, unbeknownst to them, contained radioactive iron isotopes. Although the objective was to learn more about how iron is absorbed during pregnancy, it was allegedly known at the time that radiation posed health risks. As with the radiation cases above, a central focus in litigation was on whether the statute of limitations had expired. A federal district court found first that the Tennessee statute of repose for medical malpractice "does not apply to conduct that does not involve medical care" (at 796) which it defined as "actions ... taken in an effort to benefit or cure the patient" (id.). Since this study clearly did not aim to benefit the pregnant women, the rules of ordinary negligence applied rather than the (shorter) statute of repose for medical malpractice.

Vodopest v. MacGregor, 913 P.2d 779 (Wash. 1996), combined a sporting activity with medical research, as investigators attempted to implement specialized breathing techniques to cope with high altitude during a mountain-climbing expedition. The plaintiff developed increasing signs of altitude sickness that were not recognized by the nurse-investigator, who encouraged the plaintiff to continue using the special breathing technique. The plaintiff developed cerebral edema leading to brain damage. The court found that, although the project's exculpatory clause could apply legitimately to the purely sporting aspects of the expedition, it could not apply to the study of the breathing technique, which constituted medical research.

Another study was determined to be quality review, not research. In Ancheff v. Hartford Hospital, 799 A2d 1067 (Conn. 2002), a hospital had instituted a protocol for aggressively treating a particularly difficult bone infection. The plaintiff who developed hearing and balance problems from antibiotics claimed that the hospital had engaged in research without appropriate informed consent, on the ground that a protocol was used, that it was 'radical,' that the hospital gathered data on how patients did, and shared information with other health care providers. The hospital responded that this was simply a systematic implementation of the best available information, that there was no attempt to use control groups or to compare two research 'arms' to see which approach might be better, nor other key features of research. The plaintiff tried repeatedly to enter the Belmont Report in its entirety into the record, a move the judge consistently denied as prejudicial because of its references to Nazi research horrors and other abuses. By the time the plaintiff finally sought only to admit the Report's definition of "research" the court had had enough and permitted no further discussion of the matter. The state supreme court found no obvious error in this response.

[8] Frankel, T. "Fiduciary law" (1983, p. 795); Scott, A.W. "The fiduciary principle" (1949, p. 541); Cooter, R., Freedman, B.J. "The fiduciary relationship: its economic character and legal consequences" (1991, p. 1045).

See, e.g., Scott v. Dime Sav. Bank of New York, FSB, 886 F.Supp. 1073, 1078 (S.D.N.Y 1995), affd w/o opin. 101 F.3d 107 (2d Cir. 1996), cert. denied 520 U.S. 1122 (1997).

See also Doe v. Roe, 681 NE2d 640, 646 (Ill.App.1 Dist. 1997); Calhoun v. Rane, 599 N.E.2d 1318 (Ill.App. 1 Dist. 1992); Coughlin v SeRine, 507 NE2d 505 (Ill.App.1Dist. 1987); Moore v. Regents of the Univ. Of Cal., 793 P.2d 479, 483 (Cal. 1990); Lockett v. Goodill 430 P.2d 589, 591 (Wash. 1967); Miller v. Kennedy, 522 P.2d 852, 860–61 (Wash. Ct. App. 1974), aff'd, 530 P.2d 334 (Wash. 1975); Shadrick v Coker, 963 SW2d 726, 735 (Tenn 1998); Collins v Nugent, 443 NE2d 277 (Ill. App. 1 Dist. 1982); Lownsbury v. VanBuren, 762 N.E.2d 354 (Ohio 2002).

[9] As noted by Rodwin, "[f]iduciaries advise and represent others and manage their affairs. Usually they have specialized knowledge or expertise. Their work requires judgment and discretion" (1995, p. 243). See also Cooter and Freedman, 1991, p. 1046; Frankel, 1983, p. 810.

[10] "Often the party that the fiduciary serves cannot effectively monitor the fiduciary's performance. The fiduciary relationship is based on dependence, reliance, and trust" (Rodwin, 1995, p. 243).

[11] Other commentators echo the theme. "Anything that compromises the fiduciary's loyalty to the fiducie or the fiduciary's exercise of independent judgment on the fiducie's behalf creates a conflict of interest" (Rodwin, 1995, p. 244). "A fiduciary relationship involves a duty on the part of the fiduciary to act for the benefit of the other party;" to "act in the entrustor's best interests;" to "act in the interest of another person;" to "give priority to the beneficiary's interests" (Scott, 1967, p. 39); see also Frankel, 1983, p. 823; Scott, 1949, p. 540; DeMott, 1988, p. 906. See also Davis, 1986, p. 19; Shepherd, 1981, p. 35; Finn, 1977, p. 3.

[12] "Fiduciary law creates a cluster of presumptive rules of conduct compendiously described as the duty of loyalty. The obligations comprising this duty restrict the permissible scope of a fiduciary's behavior whenever possible conflicts of interest arise between the [1054] principal and the fiduciary" (Cooter and Freedman, 1991, p. 1053). "Other rules of fiduc conduct include, for example, the rule against conflicts of duty, the rule against self-interested transactions, the rule against bribes and secret commissions, the rule against purchasting trust property, and the rule regarding fiduciary opportunities" (Cooter and Freedman, 1991, p. 1053 n. 19). See also Rodwin, 1995, p. 244; Frankel, 1983, p. 824; Shepherd, 1981, p. 41.

[13] They are voluntary undertakings, for instance. Fiduciary relations may not be imposed on anyone unwilling to assume them even if, once assumed, the duties binding the fiduciary are not optional. See, e.g., Frankel, 1983, p. 820; Scott, 1949, p. 540.

[14] Moreover, the law looks with a rather dim view on the advice that is given by a fiduciary laboring under a conflict of interest. In the face of such a conflict the law actually presumes "that the fiduciary has misused his advice-giving powers to the detriment of his beneficiary" (Shepherd, 1981, p. 39). See also Cooter and Freedman, 1991.

[15] Interestingly, this point was expressly noted in the proceedings surrounding the case of Hyman v. Jewish Chronic Disease Hospital, in which live cancer cells were injected subcutaneously into elderly, debilitated nursing home patients. As noted by the Board of Regents, University of the State of New York, " the investigators' claim that they had acted as compassionate physicians was irrelevant. The experiment had not been part of 'the usual doctor-patient relationship,' so there was 'no basis for the

exercise of their usual professional judgment.' Thus, even though Sougham and Mandel wore white coats, they were researchers, not clinicians" (Lerner, 2004, p. 629).

[16] It is interesting to note that, even the leading case emphasizing investigators' duties toward subjects does not deem the relationship fiduciary, but rather a "special relationship" Grimes v. Kennedy Krieger, 782 A.2d 834, 843 (Md. 2001).

[17] Many of these points are highlighted in Emanuel, E.J., Wendler, D., Grady, C. "What makes clinical research ethical?" (2000). [Q1]TS: Please set the notes part as endnotes.

BIBLIOGRAPHY

Angell, M. (1997). 'The ethics of clinical research in the Third World,' *The New England Journal of Medicine*, 337, 847–49.

Ancheff v. Hartford Hospital, 799 A2d 1067 (Conn. 2002).

Appelbaum, P.S. (2002). 'Clarifying the ethics of clinical research: a path toward avoiding the therapeutic misconception,' *The American Journal Of Bioethics*, 2(2), 22–23.

Appelbaum, P.S., Lidz, C.W., & Grisso, T. (2004). 'Therapeutic misconception in clinical research: frequency and risk factors,' *IRB*, 26(2), 1–8.

Appelbaum, P.S., Roth, L.H., & Lidz, C. (1982). 'The therapeutic misconception: informed consent in psychiatric research,' *International Journal of Law & Psychiatry*, 5(3-4), 319–329.

Appelbaum, P.S., Roth, L.H., Lidz, C.W., Benson, P., & Winslade, W. (1987). 'False hopes and best data: consent to research and the therapeutic misconception,' *Hastings Center Report*, 17(2), 20–24.

Brody, B.A. (1995). *Ethical Issues in Drug Testing, Approval, and Pricing: The Clot-Dissolving Drugs* (pp. 112–131). New York: Oxford.

Brody, B. (2003). *Taking Issue: Pluralism and Casuistry in Bioethics*. Washington, D.C.: Georgetown University Press.

Calhoun v. Rane, 599 N.E.2d 1318 (Ill.App 1 Dist. 1992).

Collins v Nugent, 443 NE2d 277 (Ill. App. 1 Dist. 1982).

Cooter, R., & Freedman, B.J. (1991). 'The fiduciary relationship: its economic character and legal consequences,' *New York University Law Review*, 66, 1045–1075.

Coughlin v. SeRine, 507NE2d 505 (Ill.App. Dist. 1987).

Craft v. Vanderbilt Univ., 18 F.Supp.2d 786 (M.D. Tenn 1998).

Davis, K.B., Jr. (1986). Judicial review of fiduciary decisionmaking—some theoretical perspectives,' *Northwestern University Law Review*, 80, 1–99.

DeMott, D.A. (1988). 'Beyond metaphor: an analysis of fiduciary obligation,' *Duke Law Journal*, 819, 879-924.

Doe v. Roe, 681 NE2d 640, 646 (Ill.App.1 Dist. 1997).

Emanuel, E.J., Wendler, D., & Grady, C. (2000). What makes clinical research ethical? *Journal of the American Medical Association*, 283, 2701–2711.

Emanuel, E.J., & Miller, F.G. (2001). 'The ethics of placebo controlled trials—a middle ground,' *The New England Journal of Medicine*, 345(12), 915–19.

Finn, P.D. (1977). *Fiduciary Obligations*. Sydney: The Law Book Company Limited.

Frankel, T. (1983). 'Fiduciary law,' *California Law Review*, 71, 795–836.

Freedman, B., Weijer, C., & Glass, K.C. (1996). 'Placebo orthodoxy in clinical research II: ethical, legal, and regulatory myths,' *The Journal of Law, Medicine, & Ethics*, 24, 252–59.

Glass, K.C. (2003). 'Clinical equipoise and the therapeutic misconception,' *Hastings Center Report*, 33(5), 5-6.

Grunberg, S.M., & Cefalu, W.T. (2003). 'The integral role of clinical research in clinical care,' *The New England Journal of Medicine*, 348, 1386–88.

Holder, A.R. (1982). 'Do researchers and subjects have a fiduciary relationship?' *IRB*, 4(1), 6-7.

Jacobson, P.D. & Cahill, M.T. (2000). 'Applying fiduciary responsibilities in the managed care context,' *American Journal of Law & Medicine*, 26, 155–173.

Kovach, K. 'Distinguishing dilemmas in the ethics of placebo-controlled trials,' *The American Journal of Bioethics*, 2(2), 32–33.

Lerner, B.H. (2004). 'Sins of omission–cancer research without informed consent,' *The New England Journal of Medicine*, 351, 628–30.

Levine, R.J. (2002). 'Placebo controls in clinical trials of new therapies for conditions for which there are known effective treatments,' in H.A. Guess, A. Lkeinman, J.W. Kusek, L.W. Engel (eds.). *The Science of the Placebo: Toward an Interdisciplinary Research Agenda*. London: BMJ Books, pp. 264–280.

Lockett v. Goodill 430 P.2d 589, 591 (Wash. 1967)

Lownsbury v. VanBuren, 762 N.E.2d 354 (Ohio 2002).

Mann, H. (2002). 'Therapeutic beneficence and Patient recruitment in randomized controlled clinical trials,' *The American Journal of Bioethics*, 2(2), 35–36.

Mann, H., & Djulbegovic, B. (2003). 'Clinical equipoise and the therapeutic misconception,' *Hastings Center Report*, 33(5), 4–5.

Meinhard v. Salmon, 164 N.E. 545 (N.Y. 1928).

Miller v. Kennedy, 522 P.2d 852, 860-61 (Wash. Ct. App. 1974), aff'd, 530 P.2d 334 (Wash. 1975).

Miller, F.G. (2004). 'Research ethics and misguided moral intuition,' *Journal of Law, Medicine & Ethics*, 32, 111–116.

Miller, F.G., & Brody, H. (2002). 'What makes placebo-controlled trials unethical?' *The American Journal of Bioethics*, 2(2), 3–9.

Miller, F.G., & Brody, H. (2003). 'A critique of clinical equipoise: therapeutic misconception in the ethics of clinical trials,' *Hastings Center Report*, 33(3), 19–28.

Miller, F.G., & Rosenstein, D.L. (2003). 'The therapeutic orientation to clinical trials,' *The New England Journal of Medicine*, 348, 1383–86.

Miller, P.B., & Weijer, C. (2003). 'Rehabilitating equipoise,' *Kennedy Institute of Ethics Journal*, 12, 94–118.

Moore v. Regents of the Univ. of Cal., 793 P.2d 479, 483 (Cal. 1990).

Morin, K. (1998). 'The standard of disclosure in human subject experimentation,' *The Journal of Legal Medicine*, 19, 157-221.

Morreim, E.H. (2003). 'Medical research litigation and malpractice tort doctrines: courts on a learning curve,' *Houston Journal of Health Law and Policy*, 4(1), 1–86.

Morreim, E.H. (2005). 'The clinical investigator as fiduciary: Discarding a misguided idea,' *Journal of Law, Medicine & Ethics*, 33(3), 586–598.

National Placebo Working Committee (2003). *National Placebo Initiative, Draft Report*. October 2003 [On-line] Available: http://www.cihr-irsc.gc.ca/e/services/pdf_19320.htm, at 37.

Nozick, R. (1974). *Anarchy, State, and Utopia*. New York: Basic Books, Inc.

Payette v. Rockefeller Univ., 643 N.Y.S.2d 79 (App Div 1 Dept. 1996).

Rodwin, M.A. (1995). 'Strains in the fiduciary metaphor: divided physician loyalties and obligations in a changing health care system,' *American Journal of Law & Medicine*, 21, 241–257.

Rothman, K.J., & Michels, K.B. (1994). 'The continuing unethical use of placebo controls,' *The New England Journal of Medicine*, 331, 394–98.

Saver, R.S. (1996). 'Critical care research and informed consent,' *North Carolina Law Review*, 75, 205–271.

Scott, A.W. (1949). 'The fiduciary principle,' *California Law Review*, 4, 539–55.

Scott, A.W. (1967). *The Law of Trusts*, Third edition (pp. 34–49). Boston: Little, Brown and Company.

Scott v. Dime Sav. Bank of New York, FSB, 886 F.Supp. 1073, 1078 (S.D.N.Y. 1995), affd w/o opin. 101 F.3d 107 (2d Cir. 1996), cert. denied 520 U.S. 1122 (1997)

Shadrick v Coker, 963 SW2d 726, 735 (Tenn 1998).

Shepherd, J.C. (1981). *The Law of Fiduciaries*. Toronto: The Carswell Company Limited.

Steinberg, D. (2002). 'Clinical research should not be permitted to escape the ethical orbit of clinical care,' *The American Journal of Bioethics*, 2(2), 27–28.

Vodopest v. MacGregor, 913 P.2d 779 (Wash. 1996).

Weijer, C. (2000). 'The ethical analysis of risk,' *The Journal of Law, Medicine & Ethics : A Journal of the American Society of Law, Medicine & Ethics*, 28, 344–61.

Weijer, C. (2002). 'When argument fails,' *American Journal of Bioethics*, 2(2), 10–11.
Weijer, C. & Miller, P.B. (2004). 'When are research risks reasonable in relation to anticipated benefits?' *Nature Medicine*, 10, 570–73.
Weijer, C., Shapiro, S.H., Glass, K.C., & Enkin, M. (2000). 'For and against: clinical equipoise and not the uncertalinty principle is the moral underpinning of the randomised controlled trial,' *British Medical Journal*, 321, 756–58.

CHAPTER 11

IS WITHHOLDING ARTIFICIAL NUTRITION AND HYDRATION FROM PVS PATIENTS ACTIVE EUTHANASIA?

LORETTA M. KOPELMAN
Department of Medical Humanities, Brody School of Medicine, East Carolina University, Greenville, NC, U.S.A.

> It seems to me that there is another route available to those who want to justify withholding food and fluids from PVS patients. It is to accept the licitness of killing PVS patients if that had been their expressed wish in the past, and to view withholding nutrition as a psychologically easier way to engage in such acts of euthanasia (Brody 2003, p. 180).

I. INTRODUCTION

It is both an honor and joy for me to contribute to the celebration of the life's work of my good friend, Baruch Brody. He is a wonderful colleague and friend, and has been for so many years. I cannot recall when we first met, but each year brings me new amazements about his gifts, accomplishments, and wisdom. He is a philosopher of the highest quality who has brought his insights to bioethics. He is also a Rabbi who understands the importance of religious traditions. He is so familiar with the medical traditions, language, and literature, that physicians believe he is a doctor of medicine. He is gifted with clinical case consultation and family counseling. He is also an avid reader of novels and an expert on all movies, old and new. His leadership in the field of bioethics has helped shape a fledgling field of study into a thriving discipline. During the last year, I reread with great pleasure many of his essays. Being philosophers, I honor him by picking a point upon which we disagree to help advance the debate on an important issue, namely our obligations to patients in a persistent vegetative state (PVS).

The debates over how to provide good end-of-life care are ancient and momentous. They raise complex questions about the meaning of life and duties to our families, our communities, and ourselves. The most contentious arguments center around the legal, religious, or moral permissibility of *suicide* (voluntarily and intentionally taking one's own life), *assisted suicide* (when someone helps another person, by an act or omission, to end his or her life voluntarily and intentionally),

M.J. Cherry and A.S. Iltis (eds.), Pluralistic Casuistry, 167–178.

active euthanasia (mercy killing when someone terminates the life of another person for compassionate reasons) and *letting die* (allowing someone to die for compassionate reasons, sometimes called "passive euthanasia"). The boundaries between these activities are also subjects in these ancient debates. Is it active euthanasia or letting die when death occurs following administration of sufficient medication to deaden the pain, which also has the foreseen but unintended consequences of suppressing respiration?

A contemporary twist on these debates concerns the consequences of new technologies. While they can rescue patients from the edge of death and return them to flourishing lives, they sometimes tragically leave them in a persistent vegetative state. Such patients are permanently unconsciousness but can exist for years with dedicated medical attention. Many argue that this maintains the life of a moral non-person, which they view as an existence worse than death, so families should be able to decline medical interventions and communities may set low funding priorities for their care. Others argue that the family and community must find the emotional and financial resources to support them because they are not dying. Thus, a new problem in an old debate is whether withholding or withdrawing of life-saving medical treatments (LSMT) from non-dying PVS patients is active euthanasia or letting die. There are correlative problems concerning whether it is permissible for people who lapse into a PVS to decline LSMT in their advance directives, whether guardians can refuse LSMT on their behalf, and whether communities can make care for such patients a low funding priority. The debate over how to treat PVS patients has recently centered on the permissibility of withholding artificial nutrition and hydration (ANH) from PVS patients, such as tube feeding into patients' stomachs.

In what follows, I consider three influential policy options regarding whether it is permissible for guardians to refuse ANH for non-dying individuals in a PVS. Brody recommends the second and third.

> Policy 1: The delivery of ANH to PVS patients is a medical intervention that can be declined by people in their advance directives should they fall into a PVS, or refused by their guardians, with the advise and approval of doctors taking care of their PVS relative; withholding or withdrawing ANH from PVS patients is not active euthanasia. (See Gert et al., 1998; Kopelman, 2001; Kopelman and De Ville, 2001; American Academy of Neurology, 1989; AMA, 1990.)
>
> Policy 2: The delivery of ANH to PVS patients is basic care like keeping someone warm and dry, and it is obligatory to provide it; withholding or withdrawing this basic care from non-dying patients is active euthanasia. (See John Paul II, 2004; Brody, 2003.)
>
> Policy 3: "Because PVS patients are not persons, in any plausible account of personhood, society should give their life-prolonging care a very low priority as it develops priorities for the allocation of health-care resources, and individual providers should give them a very low priority in clinical triaging decisions" (Brody, 2003, p. 174).

Defenders of these three policy options generally agree that active euthanasia is not permissible for PVS patients. Some have principled objections to active euthanasia, believing it to be always wrong. Others have non-principled objections to it, believing that there are less contentious and more effective policies or that it is not a genuine option given current policies and attitudes (Gert et al., 1998; Kopelman, 2001). Active euthanasia is impermissible in most jurisdictions (Kopelman and De Ville, 2001) and has a high disapproval rate among doctors and nurses (Payne et al., 1996). Their second shared assumption is that basic care must be provided to all patients, and the third is that it is sometimes permissible to withdraw or withhold LSMT from extremely debilitated patients. While I focus on the above three policy options, there are, of course, other positions currently defended, such as those holding that active euthanasia is permissible for PVS patients or that it is never permissible to withhold or withdraw LSMT.

A. Policy 1

Policy 1 is an option regarding delivery of ANH to PVS patients as being a medical intervention that can be declined by people in their advance directives should they fall into a PVS, or refused by their guardians, with the advise and approval of doctors taking care of their PVS relative. Defenders argue they may authorize the removal of life-sustaining medical interventions, including ANH, so that these patients may be allowed to die (Gert et al., 1998; Kopelman, 2001, Kopelman and De Ville, 2001). The influential American Academy of Neurology states:

> The artificial provision of nutrition and hydration is a form of medical treatment and may be discontinued in accordance with the principles and practices governing the withholding and withdrawal of other forms of medical treatment (1989, p. 125).

The same notion that ANH are medical treatments and *may be* withheld is also found in the quotation from the Council on Ethical and Judicial Affairs American Medical Association (AMA):

> Even if death is not imminent but a patient is beyond doubt permanently unconscious, and there are adequate safeguards to confirm the accuracy of the diagnosis, it is not unethical to discontinue all means of life-prolonging medical treatment. Life-prolonging medical treatment includes medication and artificially or technologically supplied respiration, nutrition or hydration (1990, p. 429).

This policy is consistent with a considerable body of moral, legal and social support for the view that, unless there is an emergency, doctors must obtain consent from patients or their guardians for medical interventions (*Miller v. HCA*, 2003).

The moral or legal right to consent for medical interventions is meaningless without a right to refuse them. The legal right to refuse medical interventions is grounded in a right to privacy, which may be asserted on behalf of the patients by his or her guardian (*Vacco vs. Quill*, 1997; *Washington vs. Glucksberg*, 1997; Kopelman and De Ville, 2001). Moral and legal literature support the view that doctors and nurses must honor patients' or guardians' competent refusals unless they believe the judgments are so incompetent, uninformed, or non-voluntary that they have a duty to seek a court order to override the decision. Once their refusals or requests are found to be sound, their request should be honored (*Miller v. HCA*, 2003).

Policy 1 has also been sanctioned by professional groups such as the American Academy of Neurology (1989) and the American Medical Association (1990) as noted above. Some courts have also explicitly endorsed this view and some state laws, such as North Carolina's, also stipulate that ANH are medical procedures and that guardians have the authority to withhold or withdraw them when their wards are in a PVS. Hospitals around the country routinely offer options for patients and families to refuse ANH if they lapse into a PVS. Most of those who defend this position do so because they believe ANH are forms of medical treatment for which one must obtain consent from patients or their guardians, unless there is an emergency.

Allowing patients to write advance directives or their guardians the right to refuse medical interventions including ANH, does not mean that they can refuse basic care such as keeping patients warm and dry; but the legal and moral literature indicate that they can refuse ventilator support, antibiotics, and artificial nutrition and hydration (Gert, Culver and Clouser, 1998). Gert, Culver and Clouser (1998) found that one of the obstacles to adopting this policy was the misconception that death from lack of ANH would be painful, and that many legislatures accepted a form of the first policy once it was clear that this is generally untrue for sentient patients and almost certainly untrue for PVS patients.

Some critics such as Brody (2003) and the late Pope John Paul II (2004) have argued that permitting ANH to be withheld from individuals in a PVS is wrong because it kills non-dying patients and crosses the line over to active euthanasia. They defend another policy option, which is the second of the policies considered.

B. Policy 2

Policy 2 regards the removal of ANH from individuals in a PVS to be active euthanasia because defenders view the delivery of food and water, even by artificial means, as basic care, like feedings by mouth or keeping patients warm and dry.

Brody defends this second policy in his article, "Special Ethical Issues in the Management of Persistent Vegetative State (PVS) Patients" (2003). Allowing guardians, with the approval of the PVS patients' doctors, to withdraw ANH, he argues, is more like killing them than letting them die because the PVS patient dies of starvation. He argues this would be an inappropriate extension of what is currently permitted and is "a psychologically easier way to engage in such acts of euthanasia" (Brody, 2003, p. 180). Brody argues that it would, however, be permissible to honor advance directives of people requesting removal of ANH in the event they lapse into a PVS. He also defends Policy 3 that allows communities to set low funding priorities for maintaining PVS patients because they are moral non-persons.

Pope John Paul II also defended this second position but took a somewhat different view from Brody. He agreed that withholding or withdrawing ANH from non-dying PVS patients constitutes active euthanasia. In contrast to Brody, Pope John Paul II argued that it is impermissible, even if it is burdensome for families and communities, to deny them life-saving treatments and that families and communities have duties to find the emotional and financial resources to provide life-saving care to PVS patients. His message that food and water, whether delivered by mouth or by artificial means, is basic care sparked a debate over whether or not his views departed from the traditions of the Roman Catholic Church. Repenshek and Slosar argue that his message was consistent with the Roman Catholic Church's teaching of allowing extraordinary procedures to be withheld or withdrawn. They write, "The address holds quite simply that medically assisted nutrition and hydration for PVS patients cannot always and everywhere be considered either proportionate or disproportionate; instead its status depends upon the circumstance of the case" (2004, p. 16). Others, such as Shannon and Walter (2004), saw the pope's message as a clear departure. The religious debates over duties of the faithful in a faith-based tradition differ from what secular policy can be justified in a diverse culture. I will focus upon the secular objections raised by Baruch Brody over whether this is a permissible social policy.

As noted, Baruch Brody considers and rejects the policy option (that I call Policy 1) for dealing with PVS patients that allow guardians, with the approval of the PVS patients' doctors, to withdraw ANH. He frames the issue in terms of whether to *extend* to PVS patients the "relatively clear consensus about...decision making at the end of life" (Brody, 2003, p. 174).

> The main outlines of that familiar consensus are that life sustaining care can be withheld from patients providing that the appropriate decisional processes are undertaken, and that these processes are ways of finding out patient wishes either by contemporaneous decision, advance directives, or surrogate decision making based on substitute judgment or surrogate assessment of best interest (Brody, 2003, p. 173).

Brody states that the central issues about the view I have called "Policy 1" is whether:

> ...the standard consensus should be extended to the withholding of nutrition and hydration (often described as the withholding of artificial nutrition) because they are forms of medical care (Brody, 2003, p. 178).

First, he says this position needs to be clarified because while PVS patients are "normally fed through a gastrostomy tube or some other device such patients can sometimes (the percentage is, I believe, not known) be fed by hand." He concludes from this that the position under consideration needs to be restated and "...to be understood as saying that they will die shortly only if both artificial nutrition and attempts at normal feeding (in such cases where that is possible) are withheld from PVS patients." His restatement is:

> Some PVS patients will only die in a reasonable amount of time if all forms of nutrition and hydration are withheld from them, so the regular consensus must be extended to cover these modalities of therapy as well (Brody, 2003, p. 178).

Brody's restatement, however, changes the meaning of the routinely defended view I have called "Policy 1." PVS patients may die from many causes, such as infections, strokes and so on. Moreover, one can (and many would) be willing to defend the position as articulated by the American Neurological Association or the AMA, and yet, hold that if a patient can be fed by hand, they should be. Feeding by mouth seems more like basic care than a medical treatment. Thus, the restatement offered by Brody is problematic. It changes the meaning of the original policy option as typically defended, which permits guardians to refuse, with the advice and approval of the patient's doctors, the removal of *artificial* nutrition and hydration from PVS patients. This restatement is problematic because it transforms Policy 1 into a less plausible position than that defended in the literature. Henceforth, I will assume agreement that PVS or other patients who can be safely fed by mouth should be, and limit my discussion to those PVS patients unable to be fed by mouth and who would choke or aspirate on food or fluid if this were attempted.

In what follows, I will argue that the burden of proof falls on those, such as Brody and Pope John Paul II, who deny guardians have, with the advice and approval of the PVS patients' doctors, the authority to withhold or withdraw artificial nutrition or hydration from non-dying PVS patients because 1) ANH are more like a technical medical intervention than basic care; 2) this is *not an extension* of current policy, but a clear application of a widely held and defended policy that guardians may sometimes refuse medical interventions on behalf of PVS patients and that incompetent patients with advance directives that do not want such interventions should have their wishes honored; and 3) it is a less contentious policy than denying them this authority since this would be a break with carefully considered legal and moral practices giving individuals such rights and their relatives in consultations with the doctors, rights to individualized and compassionate end-of-life decisions.

II. BASIC CARE OR A MEDICAL INTERVENTION?

In this section, I will argue that the burden of proof is on those claiming that ANH are basic care. This is not a definitive reason since what is or is not basic care may change, but it generally is a stable notion. Basic care is non-optional and always required, if it is possible to provide it, and includes keeping individuals comfortable, warm, and dry. Many caretakers lacking any medical or nursing degrees perform such basic services. In contrast, providing ANH involves intravenous fluids, cut-downs, keeping lines clear, nasogastric tubes, or surgical placements of ports, which are clearly medical procedures requiring trained and licensed professionals. Doctors, nurses, or technicians must fulfill a variety of state and federal regulations to provide and monitor their delivery. Moreover, professionals must judge when the benefits outweigh the burdens of such aggressive interventions. ANH are medical procedures with potential benefits and harms. ANH require the services of licensed, highly skilled, and specialized professionals acting in institutions regulated by the state. The comparison to giving food and water by mouth are understandable but misleading. The best evidence suggests that PVS patients do not get thirsty or hungry and, I have agreed, those that can be fed by mouth should be. In contrast, there is never a question about the need to provide basic care.

Second, another mark of basic care is that it must be provided irrespective of the patients' or the clinicians' therapeutic goal. It would always be wrong let caretakers, whether they are relatives in the home or clinicians in the hospital, let patients get cold, wet, hungry, or uncomfortable, if it could be prevented. People are not even permitted to allow their animals to suffer in such ways, if it can be avoided. In contrast, professionals must make a technical decision about when benefits outweigh burdens of ANH to PVS or other patients. The selection of medical treatment plans should be evaluated in terms of the procedure's meaning in the life of the person being treated. For example, antibiotics might not be used to treat individuals when it would only prolong dying. Aside from basic care, the family or guardian of the incompetent individual is authorized to make medical choices about which medical interventions are best, including medications and ANH, unless there is an emergency or their choices are not in the patients' best interests.

III. EXTENDING OR USING CURRENT POLICIES?

Brody claims that permitting what I have called "Policy 1" would be an extension of current policy. I will argue that this claim is implausible. This is not a definitive reason to adopt Policy 1 because policies are sometimes bad or may change; yet it is a carefully considered view and so the burden of proof is on those rejecting Policy 1 in favor of Policy 2.

Brody's claim that what I have called Policy 1 is an extension of the current consensus is puzzling because he also acknowledges that such policies have been widely endorsed by the courts. States and hospitals around the country routinely offer patients and guardians options to refuse ANH along with such medical interventions as ventilators. Denying them these options would be a break from carefully considered legal and moral practices giving individuals the right to consent for or to refuse medical interventions, and their relatives, in consultations with the doctors, to do so on their behalf. This gives them control over end-of-life choices that reflect individualized rankings of how to understand duties to prolong life and to avoid excessive burdens. As noted, people have well-established rights to refuse medical interventions and guardians have this right to consent or refuse on behalf of their wards; unless a case can be made that there is an emergency or that the guardians are endangering their relatives and the courts intervene to take custody and authorize treatments, doctors must excuse themselves accept the decisions of the guardians (*Miller v. HCA*, 2003).

If ANH are basic care, however, they should be provided even if patients' advance directives instruct doctors not to provide them and even if patients who are not in a PVS do not want them. That is, if a patient or family member wanted a PVS patient to be kept cold and wet and never turned to avoid bedsores, the hospitals and clinicians would be justified in refusing. It would not only compromise their patients' care but their values. Regarding ANH as basic care that cannot be refused, therefore would be highly contentious and controversial.

IV. MORE OR LESS CONTROVERSIAL?

Third, the burden of proof should be on those forbidding removal of ANH from PVS patients, because forbidding this option would be a more contentious policy than allowing it (Kopelman and De Ville, 2001). This reason, of course, is not definitive since some justifiable policies are controversial and some unjustifiable policies are not. Yet one goal of ethics is to minimize strife and promote harmony and opportunities in society, thus we should seek less contentious policies, if possible. These end-of-life policies, however, have received considerable attention, and forbidding established options that are regarded as a matter of personal choice would be likely to be widely regarded as an invasion of privacy and an unjustifiable state intrusion into personal choices (Kopelman, 1997). This seems especially important when our nation is so deeply divided over how to respond to patients in PVS, as the Terri Schiavo case illustrates. Despite efforts by the executive and legislative branches of government, both the Florida and Federal Courts supported her husband's decision to remove her feeding tube after many years of existing in a PVS. The U.S. Supreme Court refused to set aside lower court rulings (Associated Press, CNN, 2005). A Gallup Poll conducted after she died showed 80% of the American public believed "that it was appropriate that Terri Schiavo's feeding tube was removed and she was allowed to die" (Gallup News Service, 2005, p. 1).

Policy 1 is a less contentious policy than Policy 2, or regarding ANH to be so basic they could not be withdrawn or withheld based upon a patient's advance directive or a guardian's informed and competent choice. Consequently, the burden of proof should be on those forbidding withdrawing or withholding ANH from PVS patients.

Gert, Culver, and Clouser (1998) argue that support for competent refusal by patients or guardians of medical interventions already exists in laws and policies. They argue we must distinguish between valid refusals and valid requests, and ought to honor legitimate refusals of medications, food and water, as well as other medical interventions. This need not be restricted to terminal patients, since guardians have rights to make decisions for their relatives if it is not a breach of their duties. Moreover, honoring the competent refusals of medical interventions including ANH allows patients and their families individualized choices about what they consider best; it also may undercut most of the calls for assisted suicide or euthanasia (Kopelman and DeVille, 2001).

Moreover, Brody's view, as captured in Policy 2, is more contentious for another reason. If ANH are basic care, guardians should not be able to refuse them and they should also be provided whether or not people leave advance directives refusing them and whether or not families and communities have the resources to provide them. As discussed, if keeping someone warm, dry and clean is basic care then hospitals should not honor an advanced directive that they be left cold and dirty. Hospitals have to follow regulations and accrediting agencies would not be persuaded they should leave their patients dirty or cold just because someone left an advance directive to that effect. If something is basic care, moreover, it should be provided by families or communities who may have to take resources from other needs to provide it. As we shall discuss,

Brody's defense of Policy 3, giving PVS patients a low priority rating in terms of funding, seems to fly in the face of his argument to regard ANH as basic care.

V. POLICY 3

Policy 3 is another option that Brody proposes, namely giving PVS patients a low priority in terms of community funding. Brody's solution depends upon being able to allocate resources away from such PVS patients because of their condition. Brody's solution, however, would be far less acceptable socially and politically than allowing patients and guardians to control the decision to remove artificial nutrition and hydration, along with other medical procedures from their relatives who are in a PVS, or so I well argue.

First, even if we assume he is correct that PVS patients lack personhood on any plausible moral account, they are still legally persons, and this will convince many that their guardians should control decisions for them. The proposals that Brody offers would have society make the allocation decisions for non-dying PVS patients unless families can pay for it. This would match what is done in many countries where resources are scarce. The current policy in this country, however, tries to equalize choices for families at the end-of-life. People who could not pay would lose control, and we know from the Terri Schiavo case we are strongly divided. Many agree with Pope John Paul II. (See Bopp, 1990; Doerflinger,1989; Koop, 1989; Reagan, 1986.) In addition, some doctors would not want to withhold or withdraw ANH, perhaps for religious reasons, such as those of Pope John Paul II. Brody says that doctors will have to understand that they have conflicting loyalties and that "conflicting loyalties has always been part of the patient-physician relation. Physicians have, in addition to their obligation to any individual patient, obligations to their other patients, to their families, to themselves, to their society, to their religion, etc." (2003, p. 192). But in a deeply divided country, this solution is likely to be contentious and the monetary savings uncertain.

Second, Brody's solution of permitting resources to be withdrawn from PVS patients does not seem to avoid the criticism he raised with respect to the position allowing families, with the approval of doctors, to withhold or withdraw ANH from PVS patients. If the one is euthanasia, then the other is as well. Brody seems to concede the problem when he says, "Life-prolonging therapy may certainly be withheld/withdrawn from them, although there are difficult questions surrounding nutrition and hydration and active euthanasia…it may nevertheless be withheld/withdrawn in at least some cases where economic considerations and considerations of provider integrity take precedence" (2003, p. 192). If he believes that it is euthanasia to withhold or withdraw ANH from PVS patients when families and doctors make these decisions for individualized and compassionate reasons, then it should also be euthanasia when it is done for economic reasons.

Brody's policy that would allow the community or state to decide based upon allocation choices seems more contentious in light of the fact of the deep division of opinion about whether PVS patients must have all medical support including ANH, antibiotics, ventilators, and so forth, but not basic care. Good decisions should take into account what reasonable and informed persons of good will would consider best for the PVS patient and still make individualized decisions. They should consider the facts of the case, the available options, the consequences of those options, the weightiest values of the providers and families, the previous wishes of the patient, if any, as well as what reasonable people would want for themselves and the constraints of social and legal policy. Many support individualized decision making for incompetent individuals and this is reflected in a variety of policies. (See American Academy of Pediatrics, 1994; 1995; 1996; Buchanan and Brock, 1989; Byock et al., 2001; Faden, R. et al., 1986; Lantos,1996, Paris et al., 1990; 2001; NHO, 1990; Stinson et al., 1983.) [1]

I have argued elsewhere that families must justify decisions for their incompetent relations in terms of three necessary and jointly sufficient conditions of the Best-Interests Standard and have offered the following analysis:

> ...the 'negative version of the Best-Interests Standard' which applies to incompetent individuals of all ages (1) instructs decision makers to decide what act(s) are in the incompetent individual's immediate and long-term interests and maximize his or her net benefits and minimize net burdens and setting that act(s) as a prima facie duty; (2) presupposes a consensus among reasonable and informed persons of good will about what choices for the incompetent individual are, all things considered, not unacceptable; and (3) determines the scope of the Best-Interests Standard in terms of the scope of established moral or legal duties to incompetent individuals (Kopelman, 2005b).[2]

Reasonable people can disagree and more than one course of action can be tolerated. Policy 1 allows individualized choices that reflect the values of the PVS patient, if known, or of the person's guardian. All things considered, persons should be able to leave advance directives indicating refusal of medical interventions, including ANH, and guardians should be able to refuse it on their behalf. The "all things considered" clause indicates that there may be other restraints. For example, a family may decide that they want some form of maximal life support for their PVS relative but neither they nor the community can provide it.

VI. CONCLUSION

The burden of proof is on those who wish to forbid guardians of PVS patients, with the approval and consent of their doctors, from removing or withholding ANH. First, guardians have the authority to give consent for or to decline medical interventions, and ANH should be viewed as medical intervention. While the analogy to taking nutrition by mouth is understandable, ANH for PVS patients are medical interventions and not basic care. They are highly regulated, technical, and burdensome procedures that cannot be set up by unlicensed or untrained individuals. Consequently, the burden of proof should be on those claiming they are basic care.

Second, if ANH are not technical, burdensome and regulated procedures but basic care, competent people could not refuse them for themselves should they lapse into a PVS and guardians could not refuse them on their relative's behalf in such circumstances. A right to consent is meaningless without a correlative right to refuse. This would be a break with carefully considered legal and moral practices giving individuals and their relatives, in consultations with the doctors, control over medical interventions as long as they fulfill duties to act in the patient's best interests. This includes non-terminal as well as end-of-life procedures where they can use their own values to assess reasonable hopes for benefits with duties to prolong life and avoid excessive burdens.

Third, this is a less contentious alternative than the policies advocated by Brody. Brody recommends leaving it up to the state to make priority funding decisions and giving support for PVS patients low priority ratings. Such allocation choices would be highly controversial because a sizeable number of people believe we have duties to find the resources to support them. In addition, the financial benefits of such an arrangement are uncertain, especially if the relatives choose to go to the courts to challenge such decisions, as the Terri Schiavo case shows. On Brody's view, PVS patients are owed basic care and ANH are basic care; consequently, it is euthanasia to remove them. But if their removal is done for economic reasons by the community as he recommends, it seems more like killing. On the view I have defended, they are not basic care but medical treatments, so persons should be able to leave advance directives refusing medical interventions, including ANH, guardians should be able to refuse it on their ward's behalf, and communities should be able to decide how best to allocate medical resources other than basic care.

NOTES

[1] One notable exception is the "Baby Doe" rules which I have argued elsewhere should be abandoned (Kopelman, 2005(a) and (b)).

[2] A more recent and fuller analysis of the Best Interests Standard than I used here may be found in Kopelman (2007).

BIBLIOGRAPHY

American Academy of Neurology (1989). 'Certain aspects of the care and management of the persistent vegetative state patient,' *Neurology,* 39(1), 125–126.

American Academy of Pediatrics, Committee on Bioethics (1994). 'Guidelines on foregoing life-sustaining medical treatment,' *Pediatrics*, 93, 532–536.

American Academy of Pediatrics, Committee on Bioethics (1996). 'Ethics and the care of critically ill infants and children,' *Pediatrics*, 98(1), 149–153.

American Academy of Pediatrics, Committee on Fetus and Newborn (1995). 'The initiation or withdrawal of treatment for high-risk newborns,' *Pediatrics*, 96, 362–364.

AMA Council on the Scientific Affairs and Council on Ethical and Judicial Affairs (1990). 'Persistent vegetative state in the decision to withdraw or withhold life support,' *Journal of the American Medical Association*, 263, 426–430.

Associated Press (2005). *Supreme Court Denies Schiavo Parents' Appeal* [On-line]. Available: http://transcripts.cnn.com/TRANSCRIPTS/0503/24/lol.03.html

Bopp, J. (1990). 'Choosing death for Nancy Cruzan,' *Hastings Center Report*, 20, 42–44.

Brody, B. (2003). 'Special ethical issues in the management of the persistent vegetative state (PVS) patient,' *Taking Issue: Pluralism and Casuistry in Bioethics* (pp. 173–195). Washington, W.C.: Georgetown University Press.

Buchanan, A. E., & Brock, D. W. (1989). *Deciding for Others: The Ethics of Surrogate Decision-making*. Cambridge: Cambridge University Press.

Byock, I. R., Caplan, A., & Snyder, L. (2001). 'Beyond symptom managements: Physician roles and responsibilities in palliative care,' in L. Snyder & T. E. Quill (Eds.), *Physician's Guide to End-of-life Care*. Philadelphia: American College of Physicians, American Society of Internal Medicine.

Doerflinger, R. (1989). 'Assisted suicide: Pro-choice or anti-life?' *Hastings Center Report*, 19,16–19.

Faden, R., Beauchamp, T., & King, N. (1986). *History and Theory of Informed Consent*. New York: Oxford University Press.

Gert, B., Culver, C., & Clouser, K. (1998) 'An alternative to physician-assisted suicide: A conceptual and moral analysis,' in M. P. Battin, R. Rhodes, & A. Silvers (Eds.), *Physician Assisted Suicide: Expanding the Debate* (pp. 182–202). New York: Routledge.

John Paul II (2004). *Life-Sustaining treatments and vegetative state; Scientific advances and medical dilemmas*. [On-line] Available: http://www.vatican.va/holy_father/john_paul_ii/speeches/2004/march/documents/hf_jp-ii_spe_20040320_congress-fiama_en.html.

Koop, E. C. (1989). 'The challenge of definition,' *Hastings Center Report*, 19, 2–3.

Kopelman, L. M. (2007). 'The best interests standard' for incompetent or incapacitated persons of all ages,' *Journal, of Law, Medicine, and Ethics*, 35(1), 187–196.

Kopelman, L. M. (2005a). 'Are the 21-year-old baby Doe rules misunderstood or mistaken,' *Pediatrics*, 115, 797-802.

Kopelman, L. M. (2005b). 'Rejecting the baby Doe rules and defending a "negative" analysis of the best interests standard,' *Journal of Medicine and Philosophy*, (30)4, 331–352.

Kopelman, L. M. (2001). 'Does physician assisted suicide promote liberty and compassion?' in L.M. Kopelman & K.A. De Ville (Eds.) *Physician Assisted Suicide* (pp. 87–102). Dordrecht: Kluwer Academic Publishers.

Kopelman, L.M. & De Ville, K. A. (2001). 'The contemporary debate over physician assisted suicide,' in L.M. Kopelman & K. A. De Ville (Eds.) *Physician Assisted Suicide* (pp. 1–25). Dordrecht: Kluwer Academic Publishers.

Kopelman, L. M. (1997). 'The best-interests standard as threshold, ideal, and standard of reasonableness,' *The Journal of Medicine and Philosophy*, 22(3), 271–289.

Lantos, J. (1996). 'Seeking justice for Priscilla,' *Cambridge Quarterly of Healthcare Ethics*, 5(4), 485–492.

Miller v. HCA, 47 Tex. Sup. J. 12, 118S.W.3d 758 (2003).

Newport, F. (2005). 'The Terri Schiavo case in review, support for her being allowed to die consistent,' *Gallup News Service*. [On-line] Available: http://www.gallup.com/poll/content/

National Hospice Organization (NHO) (1990). *Standards of a Hospice Program of Care*. Arlington, Virginia.

Paris, J. J., Ferranti, J., & Reardon, F. (2001). 'From the Johns Hopkins baby to baby Miller: What have we learned from four decades of reflection on neonatal cases?' *The Journal of Clinical Ethics*, 12(3), 207–214.

Paris, J. J., Crone, R. K., & Reardon, F. (1990). 'Physician's refusal of requested treatment: The case of baby L.' *New England Journal of Medicine*, 322, 1012–1015.

Payne, K., Taylor, R .M., Stocking, C., & Sachs, G. A. (1996). 'Physicians' attitudes about the care of patients in the persistent vegetative state: a national survey,' *Annals of Internal Medicine*, 125(2), 104–110.

Rachels, J. (1975). 'Active and passive euthanasia,' *New England Journal of Medicine*, 292(2), 78–80.

Reagan, R. (1986). 'Abortion and the conscience of the nation,' in J. D. Butler & D. F. Walbert (Eds.) *Abortion, Medicine and the Law, 3rd Edition* (pp. 352–358). New York: Facts on File.

Repenshek, M. & Slosar, J. P. (2004). 'Medically assisted nutrition and hydration: A contribution to the dialogue.' *Hastings Center Report*, 34(6), 113-116.

Shannon, T. A. & Walter, J.J. (2004). 'Implications of the papal allocution on feeding tubes,' *Hastings Center Report*, 34(4), 18–20.

Stinson, R. & Stinson, P. (1983) *The Long Dying of Baby Andrew*. Boston: Little Brown.

Vacco vs. Quill, 117 S. Ct. 229 3(1997)

Washington vs. Glucksberg, 117 S. Ct. 225 8 (1997).

SECTION IV

CRITICAL APPLICATION AND ANALYSIS

CHAPTER 12

THE VIRTUE OF INTEGRITY IN BARUCH BRODY'S MORAL FRAMEWORK

J. CLINT PARKER

Department of Medical Humanities, Brody School of Medicine East Carolina University, Greenville, NC, U.S.A.

I. INTRODUCTION

Baruch Brody's contributions to philosophy in general and bioethics in particular have been broad in scope and substantive in nature. His work combines philosophical depth, soundness and rigor with a pragmatic approach to real world problems. It is a model for those who wish to engage in careful and deliberative thinking in the pursuit of useful answers to the complicated philosophical dilemmas facing modernity. One example of his gift for offering practical and insightful analysis of a perplexing and intransigent problem is his work on the issue of futility. As befitting a paper in tribute to Brody, let me begin with a case. An attending physician in our institution recounted the following case in one of our monthly Internal Medicine ethics conferences.

Oncologists told an elderly female patient with a metastatic gynecological cancer that there was nothing more that they could do for her cancer. Some time later, however, she was admitted to an Internal Medicine service with mental status changes and was subsequently found to be in acute renal failure. The family wanted the patient to be started on dialysis. The attending physician felt that starting the patient on dialysis would only prolong her suffering until she eventually succumbed to the cancer. The nephrology service was consulted, and the nephrology attending also did not believe that starting dialysis in this patient was appropriate, yet the family was insistent as they thought that the patient with treatment for her kidney failure might wake up one last time to talk to them.[1]

When this case was discussed, the attending medical physician objected that dialysis in such a patient was medically inappropriate. The physician admitted, however, that despite the patient's poor prognosis for meaningful recovery, dialysis treatment was consistent with the goals of the family, which were simply to prolong the patient's life and attempt to reverse her acute mental status changes in an attempt to talk to her before she died. The wishes of the patient were unknown. When pressed as to why he did not want to pursue dialysis in this patient, the attending physician

M.J. Cherry and A.S. Iltis (eds.), Pluralistic Casuistry, 181–190.

expressed that he simply did not want to be involved in providing aggressive treatment because he believed that he was prolonging a life full of suffering and that dialysis would ultimately produce more harm than benefit.

These types of cases are not uncommon in modern medicine and they represent a dilemma for physicians. Physicians have recognized a strong duty to respect the autonomy of their patients and surrogates. However, sometimes they are asked to pursue treatments that they believe are morally wrong or that conflict with their own core values—values that animate and motivate their practice of medicine. In such situations, physicians may judge that they are being asked to sacrifice something morally important and may resist such sacrifices by invoking seemingly value neutral language such as medical inappropriateness. Such descriptions, however, do not seem adequately to capture the deep philosophical and moral issues that underlie physician's judgments that morally something is lost by acceding to these requests.

In his nationally recognized work on futility policy, Baruch Brody has noticed these concerns and has argued that many such claims are best conceptualized not as violations of an objective, value neutral concept, but as violations of a highly value-laden notion of integrity, and that these conflicts are best solved through a process that attempts to align patients and surrogates with physicians that share their values (Brody, 2003, pp. 198–199, 204; Fine, 2001). Brody's work helped guide the formulation of a futility policy within Houston's Texas Medical Center, which in turn served as an important model for the later legislative efforts in Texas that resulted in a process-based, statewide futility policy. These efforts have had ripple effects and are being studied as a model in other states (Kopelman et al., 2005).[2]

In this paper, I will explore Brody's conception of integrity. I will begin by briefly describing the moral framework within which Brody places appeals to integrity. I will then offer analysis of Brody's conception of integrity and address some important objections to this conception. In the end, I will conclude by offering a way of understanding Brody's conception of integrity that avoids these objections but still captures the moral concerns of health care workers such as illustrated in the opening case.

II. A BRIEF DESCRIPTION OF BRODY'S MORAL SYSTEM

In *Life and Death Decision Making*, Brody lays out an ethical framework for evaluating whether a given action is morally permitted, prohibited, or required. On this account of morality, there is a plurality of moral appeals, each of which may play an independent role in the moral evaluation of a given action. Some of the moral appeals that Brody identifies are familiar components of prominent monistic systems (Brody, 1988, p. 9).[3] In these systems, these moral appeals play a central organizing role and other moral appeals are conceptualized as derivative from and subordinate to these central appeals. Brody notes that within the history of moral philosophy these monistic systems have often been pitted against one another when analyzing particular cases (Brody, 1988). The soundness of monistic systems is often judged by whether they can correctly predict and explain our commonsense moral judgments regarding cases. However, oftentimes, one system seems to give the right answer in one case while another system seems to give the right answer in a different case. Brody notes that none of the monistic systems seems able adequately to predict and explain enough of our commonsense moral judgments to make it easy to set aside the claims of the competing systems.

One could look at these monistic frameworks as a series of failed attempts to organize and explain our moral intuitions. Brody, however, chooses to view these monistic systems as "a series of attempts to articulate different moral appeals, all of which will have to be combined to frame an adequate moral theory for helping us deal with difficult cases" (Brody, 1988, p. 10). He insists:

> we should not attempt to take one moral appeal from one theory and use it as the basis for all of morality, using other moral appeals derivatively and only to the extent that they can be based on the fundamental moral appeal. Rather we should regard each of these theories as correct in advocating a moral appeal whose use is certainly legitimate but limited. Thus, we need to incorporate all of these moral appeals, each treated as an independent appeal whose validity is not based on its derivability from some other moral appeal, into a pluralistic moral theory (Brody, 1988, p. 10–11).

Brody describes five major types of independent ethical appeals. These appeals include consequences of actions, rights, respect for persons, cost effectiveness, justice, and finally, virtues.[4] Within Brody's ethical framework each appeal is identified, its relative importance is judged on the basis of a separate account that delineates the importance of an appeal in a given case, and then an holistic judgment about whether the action in question is morally permitted, prohibited, or required is made.

III. A DESCRIPTION OF INTEGRITY IN BRODY'S SYSTEM

In Brody's system, integrity is listed under the appeal to virtues. In appealing to virtue, it is important to note that Brody is not making a moral assessment of an individual; rather, he is using virtue as part of a moral assessment of actions (Brody, 1988, p. 35). Brody remains neutral as to whether a virtuous action requires as a necessary condition a concomitant right motive, or whether an action that would have been done by a virtuous person for the right motive would count as a virtuous action regardless of what motive the actual individual possessed (Brody, 1988, p. 35). On the latter account, an appeal to virtue would mean that in deciding whether to act in a certain way, the fact that the act exemplifies a virtue (i.e., it is the act that a virtuous person would have done in those circumstances) counts in favor of doing the act; while on the former account, in addition to the act exemplifying virtue, the agent must also have a right motive in order for the act to be fully virtuous, and thus both conditions must be met in order for virtue to be a reason for doing the act.

Virtue is one of many moral appeals that may come into play in a given case. It may be that a given act does not exemplify certain virtues; however, if the act does fulfill other moral appeals, it still may be the act that is, all things considered, morally required or permitted. That is, each moral appeal in Brody's system plays some role in the final outcome regarding the moral status of the act in question; however, the importance of this role may vary from case to case. In what follows, I will focus on the individual moral appeal to the virtue of integrity, keeping in mind that this appeal is not necessarily determinative in every case.

What does Brody mean when he appeals to the virtue of integrity? The word *integrity* can be used in different senses. Let me begin the analysis by looking at senses of integrity upon which Brody does *not* focus. First, he does not focus on the idea of integrity as steadfastness in one's own moral convictions regardless of possible personal sacrifice.

> Sometimes, when we talk about people of integrity, we are referring to those who do what is morally required even if that leads them to sacrifice personal benefits. The moral obligations to which these people are faithful arise out of objective obligation-generating values such as producing good consequences and respecting individual rights. Although this is an important aspect of the virtue of integrity, I am also interested in another, more neglected aspect of that virtue (Brody, 2003, p. 15).

Second, he does not focus on the idea of "the integrity of the individual who is concerned not merely with seeing that wrong is not done but particularly with seeing that he or she does not do what is wrong" (Brody 1998, p. 36). As Brody notes, this understanding of integrity features prominently in Bernard Williams's critique of utilitarianism. As Williams puts it,

> A feature of utilitarianism is that it cuts out a kind of consideration which for some others makes a difference to what they feel about such cases: a consideration involving the idea, as we might first and very simply put it, that each of us is specially responsible for what he does, rather than for what other people do. This is an idea closely connected with the value of integrity (Smart and Williams, 1973 p. 99).

Brody emphasizes a different sense of integrity—*fidelity to values*. In particular, Brody is interested in fidelity to *personal values*.

> ... I am also interested in another, more neglected aspect of that virtue [integrity]. This is the virtue displayed by those who are faithful to their own personal values even when doing so is costly to them. My analysis connects the virtue of integrity with being faithful to values, but it differentiates two types of values to which one may be faithful: objective values and personal values. Integrity in both of these senses is not, of course, a virtue specific to physicians, but physicians also need to be mindful of that virtue in their behavior. They should not be "hired guns," ready to do anything within their technical capacities that is requested by a patient or those who speak on behalf of the patient (Brody, 2003, p. 13).

With what types of personal values is Brody concerned? He does not give us a full description of the distinction between objective values and personal values, nor does he seem to place many limits on what can count as a personal value in regards to the virtue of integrity. He does, however, give us examples of the types of values to which it is virtuous for physicians to be faithful.

> The [futility] policy we developed in Houston turned instead to a process by which a set of providers (both physicians and other professionals working in the hospital, such as nurses and therapists) could decide in a fair and transparent process whether the provision of the requested intervention was compatible with their personal values. If they decided it was not, they would be entitled as an act of integrity to refuse to provide the requested interventions, leaving the patient/surrogate free to find, if possible, other providers who were comfortable providing the requested care. For this reason, the policy our task force developed is best described as an integrity-based procedural approach to questions of medical inappropriateness or futility. What are the values that might lead to such judgments by professionals? Three have been mentioned in the various articles presenting such policies: Prolonging the patient's life is producing too much suffering without enough compensating benefit or is producing a life that is otherwise of little value. Here, the importance of listening to the patient/surrogate in a fair and transparent process is particularly important, because the providers may have missed a reason why there is compensating value in the life of the patient and may become comfortable with providing the care if they understand these compensating values. Providing the care is unseemly. We have described cases in which, for example, multiple limbs were amputated from a dying child because the parents could not accept the fact that the child was dying. Providing the care is an inappropriate stewardship of scarce resources (Brody, 2003, p. 17).

As mentioned earlier, on Brody's account once a moral appeal is identified its significance must be assessed. This assessment involves a separate account of when a given moral appeal is more or less important. With regards to integrity, Brody identifies two factors that can be used to judge its relative importance when compared to other moral appeals within a particular case. The first factor that he cites is the "centrality of the values with which a particular decision is in accord or in conflict" (Brody, 2003, p. 89). He writes:

> Not all of the values and goals that a person has adopted are of equal importance in the life of that particular individual. Some values and goals are central to our lives because we have made a great commitment of time, energy, and resources to them. Others are far less central because we have made far less of a commitment to them. A particular decision's integrity is of great significance if it consists of a consonance between that decision and a value that is central in the life of the decision maker. A particular decision's integrity is of lesser significance if it consists of a consonance between the decision and a value that is less central to the life of the decision maker (Brody, 2003, p. 90).

The second factor that Brody identifies is "the relationship between the values and the life circumstances of the person" (Brody, 2003, p. 89). He writes:

> Regarding the second factor, consider the patient who is contemplating risky surgery in the hope of restoring his or her health so that he or she can engage in various strenuous physical activities. Suppose, moreover, that engaging in those strenuous activities has been important in the life of that patient until now. The decision for surgery would then be a display of integrity. But how important is that? .... if the person's physical health changes engaging in these strenuous activities may retreat in centrality, and engaging in other activities may take their place without any change of mind about the value of the activities. These values may be more changeable because of changes in circumstances. We need to take this into account before we assess the significance of the integrity of a decision (Brody, 2003, p. 90).

This latter factor seems to involve the idea of stability of a value over time and may be seen as a special case of the first factor. If a value maintains its centrality over time, or perhaps is the type of value that has more of a propensity to maintain its centrality over time despite other changing circumstances, then fidelity to this value is more important.

IV. AN ANALYSIS OF BRODY'S ACCOUNT OF INTEGRITY

Brody's account of integrity emphasizes fidelity to personal values. In the simplest sense, to value something is simply to have some sort of positive or pro-attitude (desire, appreciation, and so forth) toward the thing in question. In this sense, a "value" is simply an object of an agent's pro-attitude, and the fact that there is an agent that is the subject of such an attitude makes the value *personal*. A second sense in which one might be said to value something, however, involves making a judgment that something actually possesses a quality that makes it *fitting* to have a positive attitude toward the thing, independently of whether one *actually* has such a positive attitude toward the thing (Frankena, 1967). The property that makes such positive attitudes fitting (some form of goodness) may be either intrinsic or extrinsic to the thing itself. The property may be extrinsic either in the sense that it leads to something else that intrinsically has the property or in the sense that it is part of some whole that intrinsically has the property (Frankena, 1967). A value in this second sense would be the thing toward which such a judgment is directed. Moreover, if the judgment were true, then such a value might be described as objective in the sense that it is the type of thing toward which a positive attitude truly is fitting. Certainly, there can be overlap in the two senses. In addition to being the type of thing toward which a positive attitude is fitting, the same thing can also be the actual object of such an attitude.

Brody, in his explanation of integrity, mentions objective values in contrast to personal values. However, I do not think the distinction between personal and objective values that Brody makes maps neatly on the distinction that I have just drawn. Brody writes of "objective obligation-generating values such as producing good consequences and respecting individual rights," which he contrasts with personal values. However, objective values that generate obligations seem to be only a subset of objective values in the sense just discussed. For example, a walk on the beach at sunrise may have qualities toward which a positive attitude would truly be fitting. However, this does not seem to be the type of thing that generates specific obligations. The world seems full of things toward which positive attitudes would truly be fitting but which generate no normative implications regarding *actions* directed at those things. Thus, it may be best to construe "personal values" simply as "non-objective obligation-generating values," which may include both objects toward which positive attitudes would truly be fitting and those toward which positive attitudes would not be fitting. The latter category would also include objects toward which a negative attitude would be fitting.

V. OBJECTIONS TO BRODY"S ACCOUNT

In this section, I will analyze two important objections to Brody's account of integrity. Brody argues that fidelity to core personal values is virtuous. Consonance between a course of action and one's core personal values counts as a right-making property of that act and dissonance with one's core personal values counts as a wrong-making property of the act. Suppose, however, that one values something extraordinarily bad. Consider the following case:

A. The Sadist Torturer

> Suppose there is a sadist torturer who enjoys, appreciates, delights—in short, personally values—causing exquisite pain in other persons and has made a career out of doing so for numerous years. This torturer is extremely good at what he does, and in addition to his regular duties of torturing people, he also gives lectures to others about torturing and has been featured in prominent torturing publications. He has even been elected president of a torturing society and has spent much of his life advocating for torture. One day, however, the torturer is tortured himself and comes to the stark realization that inflicting pain on others is bad and that he should radically change his way of life.

Does the fact that a person has valued torture and been committed to torture throughout his life provide *any* moral reason to continue to torture? Would a radical conversion to a different way of life in *any* way be considered non-virtuous simply because of infidelity to things that are not fit to be valued and which in some cases are intrinsically bad? I think the answer to both of these questions is no.[5]

Does the case of the sadist torturer then show us that Brody's account of integrity is wrong? I think not. However, the case of the sadist torturer does seem to tell us about the type of values fidelity to which could be virtuous. Earlier I examined the distinction between personal and objective values. I noted that a personal value might be simply that which is valued by a person. That which is valued by a person may be objectively good, bad or neutral. It seems clear that fidelity to things that are objectively bad cannot be virtuous. On the contrary, conversion is what is virtuous in such cases.

Consider another case that may pose problems for Brody's account of integrity.

B. The Scholar and the Orphans

> Suppose that there is a scholar who has spent his life committed to learning. Everyday he awakens and engages his studies with energy and passion. He loves learning. He uses the knowledge that he gains to teach others around him and to write books so that future generations will be enlightened. One day while reading, he comes across a story about an orphanage in a third world country. The orphans there are in desperate straights and the scholar feels convicted to go and spend time and energy helping the orphans. He knows, however, that his intellectual pursuits will be compromised while he is away. He will have to engage in counseling, teaching, and helping to supply the material needs of the children. He would only be able to spend a fraction of the time that he had in the past studying. But he desires to go and help and decides to forego the life of the scholar in order to help the orphans.

In going to help the children, would the scholar be compromising his virtue? Is the fact that he has been committed to doing something objectively good in the past a reason for him not to pursue this other objective good? I think not. One might think that there is no moral obligation for him to go off to help the children, but certainly it would not be an act of vice if he decided to go and help the children even though this would compromise a personal value that he has held in the past.

Part of the explanation of this case involves the fact that the scholar has chosen another objectively good thing to value. That is, in compromising his personal value of learning, which is good, he is acting in consonance with another objectively good value—helping the orphans. However, there also seems to be another component in the scholar's decision that is important in determining whether his act is a violation of integrity. Consider a variation in the case.

C. The Scholar and the Orphans (2)

> In this case, rather than reading about the orphans, a group of concerned citizens from the town come to the scholar and ask him to go and help the orphans. The scholar, while wanting to do his part, does not feel passionately about leaving his life of studies in order to help the orphans. He offers some financial assistance but the citizens are insistent. They show him graphic pictures of starving children and tell him that more children will die if he does not sacrifice his life of learning in order to go and help the children. Finally, driven by guilt the scholar gives up his scholarly life in order to go and help the children.

In this case, there seems to be something missing. In sacrificing his own personal values to go and help the children, the scholar is giving up an important part of himself. By allowing outside factors to tear him away from those things that he values which are also objectively good, he does seem to be compromising his virtue.[6]

Why is this the case? I think that part of the moral concern regarding compromise of integrity derives from a more general moral concern: it is good for moral agents to actively organize their lives around what is good. Fidelity to personal values that have some degree of goodness involves both a commitment to a good—the good that is valued—and also a commitment to *pursuing the good.* The latter idea captures what I think is at stake in the second scholar case. There is something good in an individual organizing and sustaining his life around values that are objectively good. It is not just the outcome of this activity that is good but also the activity itself. When an individual allows outside forces to undermine this process of individually pursuing what is good then his virtue is compromised. This compromise takes place even if what displaces the personal values of the individual is itself something worthy of being valued.

VI. CONCLUSION

In this section, I offer some concluding remarks regarding Brody's conception of integrity. First, while on Brody's account integrity is fidelity to personal values,

I think that the nature of that which is valued must also be taken into account. Certainly, it seems that fidelity to something that is objectively bad and not fit to be valued to begin with should not be considered a virtue. The case of the torturer brings this point out. Rather, integrity is better understood as fidelity to personal values that are objectively good or at least neutral. Here it is also important to note that objectively good values may or may not generate obligations and even if they do not fidelity to these values can be considered virtuous.

Second, it seems strange to think that fidelity to one personal value that is good should take precedence over a second personal value that is good, even if one had the first value longer and it has been central to one's life. The strangeness of this, I think, comes from the fact that both are *personal* values—that is, values that originate from within the person who chooses to act on one rather than another. In the first case of the scholar, it is because the value of helping orphans originated from him as a part of his own pursuit of the good that I do not think that his choice to forego the scholarly life involved a compromise of integrity. We have here both a freedom and constraint in regards to our integrity. We are free to pursue and maintain any number of values that are good, but we should not commit ourselves to fidelity to values that are bad.

Finally, if our actions involve values that are foreign to us and which compromise our own objectively good values then even if these actions involve pursuit of another objective good, I think that our own integrity is compromised. I think that this is what the second case of the scholar shows. Our integrity is linked to our *own* pursuit of those things that are good. Given the reality of our finitude we must make choices as to what we will value with our acts. This act of choosing and maintaining values that subsequently go on to organize our life is a virtuous activity when the values we choose are objectively good. Compromises of integrity involve performing actions that represent an abandonment of this autonomous pursuit of the good because of an imposition of foreign values that may or may not be good.

Brody's appeal to integrity helps physicians conceptualize the moral concerns that arise in cases such as the one mentioned at the beginning of this paper. In asking the attending to be involved in prolonging the suffering of the patient, the family was asking him to compromise a core personal value that was also objectively good. It is not clear that the alternative of prolonging a suffering life simply to let a family talk to a person was even morally neutral; however, even if one assumes that this would also exemplify a value that is good, it is not the physician's own value. In pursuing it, the physician would be forced to compromise his own pursuit of good and in so doing compromise his integrity.

NOTES

[1] Thanks to Dr. Chirag Patel and Dr. Eric Anderson for this case.

[2] See Kopelman et al., "The Benefits of a North Carolina Policy for Determining Inappropriate or Futile Medical Care" (2005). In this article the authors argue that legislation similar that in Texas should be adopted in North Carolina. In addition, the North Carolina Medical Society has adopted a resolution to study the Texas legislation as a possible model for similar legislation in North Carolina.

[3] For example, two important monistic systems from which Brody draws moral appeals are Kantianism and Utilitarianism. In *Life and Death Decision Making* Brody writes, "The field of moral theory is dominated by the conflicts among a wide variety of competing views. Utilitarianism in its various forms, deontological theories (such as Kant's theory of the categorical imperative), natural rights theories (such as Locke's theory of the rights of individuals), virtue theories (Aristotelian and otherwise), and social contractarian theories from Hobbes to Rawls) all have their adherents....I think we need a new way of looking at the moral theories that have been advocated in the past. We need to recognize that each has emphasized a particular moral appeal whose legitimacy is unquestionable. We also need to recognize that each has failed because it has recognized only one of the many legitimate moral appeals" (1988, p. 9).

[4] These appeals play a different role in Brody's system than in monistic systems. In Brody's system no appeal has an *a priori* position of dominance. Each appeal's importance is determined by the contingencies of the individual case.

[5] Certainly, the torturer can show steadfastness to his values and may be honestly mistaken in his judgment regarding the goodness or badness of these values, but in so far as integrity is understood as a virtue, I think that the object of such steadfastness must be at least morally neutral.

[6] Certainly there are degrees here. If the scholar was asked only to forego an afternoon of reading in order to collect funds for a charity to help the orphans then this does not seem to rise to the level of a violation of his integrity. Thus, in addition to the two factors mentioned earlier, the degree to which one's personal value is compromised by the act in question seems to operate as a third factor that can be used to judge the relative importance of the moral appeal to integrity when compared to other moral appeals within a particular case. Thanks to Loretta Kopelman for this point.

BIBLIOGRAPHY

Brody, B. (1988). *Life and Death Decision Making*. New York: Oxford University Press.

Brody, B. (2003). *Taking Issue: Pluralism and Casuistry in Bioethics*. Washington D.C.: Georgetown University Press.

Fine, A. (2001). 'The Texas advance directives act of 1999: politics and reality,' *HEC Forum*, 13, 59–81.

Frankena, W.K. (1967). 'Value and valuation,' in P. Edwards (ed.), *The Encyclopedia of Philosophy*, vol. 8. New York: Macmillan Publishing Co., Inc. & The Free Press, pp. 229–232.

Halevy, A. and Brody, B. (1996). 'A Multi-institutional community policy on medical futility,' *Journal of the American Medical Association*, 276, 571–574.

Hurka, T. (2001). *Virtue, Vice, and Value*. New York: Oxford University Press.

Kopelman, A.E. et al. (2005). 'The benefits of a North Carolina policy for determining inappropriate or futile medical care.' *North Carolina Medical Journal*, 66 (5), 392–4.

Smart, J.J.C. and Williams, B. (1973). *Utilitarianism: For and Against*. Cambridge: Cambridge University Press.

CHAPTER 13

PARADIGMS, PRACTICES AND POLITICS: ETHICS AND THE LANGUAGE OF HUMAN EMBRYO TRANSFER/DONATION/RESCUE/ADOPTION[1]

SARAH-VAUGHAN BRAKMAN

James Madison Program in American Ideals and Institutions, Department of Politics, Princeton University, Princeton, N.J., U.S.A. (2006–2007) and Department of Philosophy, Villanova University, Villanova, PA, U.S.A.

I. INTRODUCTION

Nearly 400,000 frozen embryos created as a result of in vitro fertilization (IVF) are in storage in the United States (Hoffman et al., 2003; Eydoux et al., 2004). Approximately 88% of these are in use by the couples who commissioned them. The remaining 12%, or 48,000 embryos, are considered "spare" (Hoffman et al., 2003; embryoadoption.org). Many countries in addition to the U.S. are facing ethical as well as practical issues regarding the ultimate fate of these stored embryos. In France, for example, the number of frozen embryos in 2001 was estimated at over 100,000 with an expectation of a continued increase by 20,000 each year (Eydoux et al., 2004). This essay concerns one option for the disposition of these spare embryos: permitting couples to give them to other individuals or couples to gestate. If a pregnancy results, the woman will give birth to a child whom she (and perhaps her spouse) will raise.[2] This practice has been referred to as heterologous human embryo *transfer* (HET), embryo *donation*, embryo *rescue* and embryo *adoption*. Until I specifically address the issue of terminology in section IV of this chapter, I will use HET.

Over ten years ago, John Robertson speculated:

> *If donor embryos were readily available* and *the practice was known more widely*, it is likely that more couples and individuals would, because of the low expense of the procedure, seek embryo donation. As demand grows, embryo donation is likely to fill an important niche in the array of assisted reproductive techniques available to infertile couples (Robertson, 1995, p. 886, emphasis added).

Currently, though, HET does not have any appreciable niche in the fertility industry compared to other assisted reproductive technologies (ART) offered. One could

M.J. Cherry and A.S. Iltis (eds.), Pluralistic Casuistry, 191–210.

argue that this is the case in large measure because the premises of Robertson's above prediction have gone unmet. Let us consider each of these premises in turn to better appreciate the issue.

If donor embryos were readily available: While a significant quantity of frozen "spare" embryos exists (the 88,000 mentioned above), the supply for couples who wish to gestate them is quite low. One program claimed to have 98 people on a 3 year waiting list (Kovas, Breheny and Dear, 2003). Interestingly, a majority of couples indicate a willingness to consider donation of excess embryos to other couples prior to undergoing fertility treatment, yet apparently after couples are finished with their reproductive projects, they usually do not chose this option (Lacey, 2005). I will explore the reasons for this phenomenon in this chapter, showing in particular how an appreciation of the lived experience of the couples who must decide about disposition demonstrates the ill-fit of current practices in reproductive medicine surrounding HET. I will then show why I think HET is distinct as a practice from other forms of ART and thus not best served by the conceptual framework or paradigm operative in reproductive medicine: one that views embryos as mere genetic property in the same way as gametes are considered. I argue here that as a practice, HET has more similarities to what goes on with traditional infant adoption (at least in the US) than it does to other types of ART and the policies governing these. Following this, I conclude that the adoption paradigm provides the proper lens through which to analyze the moral issues raised by HET (e.g., decision-making and disclosure) and from which we may provide future guidance regarding policy and practices.

HET poses many ethical challenges including the nature and extent of the rights of genetic parents and those of the potential social parents. Also at stake here are the obligations of clinicians, and most especially, justice concerns for any children born of the procedure. Establishing the appropriate paradigm for HET will contribute substantially to the ethical evaluation of the challenges the practice engenders for all concerned—practioners, those with "spare" embryos, those wishing to gestate them and most especially for the children born of such an arrangement.

Returning to Robertson's second premise: if *the practice were known more widely*—this has only recently come to pass but primarily for reasons of a political nature. In 2001, six years after Robertson's article, HET began to receive national attention when Congress passed an appropriations bill for the Department of Health and Human Services (HHS), including yearly one million dollars in grant money for the creation of "Public Awareness Campaigns on Embryo *Adoption*" [emphasis added].[3] In 2003, The Snowflakes Program of the Nightlight Christian Adoptions agency received the first such grant, over half of that money ($506,000), for their embryo *adoption* program (http://www.nightlight.org/snowflakeadoption.htm).

In addition to political concerns over governmental promotion of one type of fertility procedure over any other (http://www.news-star.com/stories/082102/New_15.shtml), what appears at issue in the media and in medicine is the political motivation behind the creation of the grants and the use of the terminology of embryo *adoption* for a practice known almost exclusively in reproductive medicine as embryo *donation*. For example, the president of the National Abortion and Reproductive Rights Action League has been quoted as saying, "Using the term 'adoption' rather than 'donation' makes it appear that the program views embryos as children" (Meckler, 2002).

What I have found curious is that this terminology debate is given cursory treatment by mainstream bioethicists. Among those in the field, there is an almost uniform rejection of "adoption" by those who have considered it and this without any inquiry (sustained or otherwise) into whether "adoption" is in any way an accurate descriptor of the practice. Language choice has political consequences, but it is in part reflective of reality and constructive of reality. In the words of Austin, "a sharpened awareness of words" will "sharpen our perception of, though not as the final arbiter of, the phenomena" (Austin, 1961, 3rd edition 1979, p. 130).[4]

My work connects two aspects of the HET discussion—the analysis and experience of the practice itself and the issue of the language of adoption—in the service of clarifying what is truly at stake in the ethical evaluation of HET. I show first that the similarities between HET and adoption as a practice are greater than the similarities between HET and most other types of ART. From this it follows that the paradigm of adoption and the procedures and policies that flow from it are at present a better fit for understanding HET than those of ART. I also give a general indication of the type of practical implications using the adoption paradigm will have for HET. Second, I examine the various terms used to describe HET when it is used primarily by couples who wish to raise a child born of this procedure. I show why all the proposed alternatives are unhelpful and inaccurate as descriptors of the practice. I consider the critiques of use of the word "adoption" in reference to HET and show how these are mistaken. Finally, while the descriptor *embryo adoption* is the most accurate among the options currently being used, I suggest an alternative terminology compatible with the adoption paradigm, but which focuses on the individuals who are making embryo disposition decisions and which speaks to the best interests of the children to be born.

Several caveats before beginning: For purposes of my argument in this chapter, I will neither assume 1) any claim about the moral status of the embryo nor 2) any claim about the moral legitimacy of pursuing parenthood through the use of HET or even through donor gametes. The debate about HET and other ARTs has stalled thus far in large measure because participants view their opponents' positions as arising from indefensible first principles/positions/intuitions. In this essay, I wish to show that viewing HET *as adoption* may be supported regardless of one's views on the moral status of the human embryo.

Second, this is an essay written in honor of my first bioethicist mentor and teacher, Baruch A. Brody. Though the ethics of ART is not one of his research interests, the overall concerns of this paper echo his early work on abortion, and they certainly fall within a general interest of his concerning beginning/end of life issues. More pointedly, though, I have written about this topic in a style that is in keeping with Brody's work on clinical bioethics. Hopefully those familiar with his signature blend of attention to clinical detail coupled with clear philosophical analysis and rigor will see my attempt in this work. Finally, though not precisely that of Brody's pluralistic casuistry, my inquiry here is casuistical in nature and grounded on argumentation and in the validity of moral intuitionism: it considers the fit of HET within two different practices, governed by two different paradigms or worldviews, in an attempt to situate it in light of issues where there are settled moral intuitions and/or philosophical agreement.[5]

II. WHAT THE DATA TELL US: ART AND HET

Though rarely done until recently, HET has been in existence since 1983 (Eisenberg and Schenker, 1998). As seen in chart 1, HET is of clinical interest to those couples with gametic insufficiency in both partners but who wish to experience pregnancy and birth, those with gametic insufficiency in one partner but the couple prefers not to have only one person genetically related, those couples with hereditary disorders in both partners and/or those who find IVF too expensive (Eisenberg and Schenker, 1998, p. 52; Lee and Yap, 2003, p. 992; Robertson, 1995, pp. 885–886). HET is also of interest to those who view embryos as early human life deserving protection from destruction and stasis.[6]

Lastly, couples pursue HET as an alternative to traditional infant adoption, which some may find too costly or who are given low priority by adoption agencies (Robertson, 1995, p. 886; Lee and Yap, 2003). HET in fact appears on average to be significantly less costly (estimates for HET at $3,600 to $4,000, see below) than domestic infant adoption which can cost anywhere from $4,000 to $35,000 (usual range $15,000 to $25,000), with international adoption costs ranging between

CHART 1. Comparison of HET with Other ARTs

The Similarities	The Differences
Couples with infertility, especially female and male gamete deficiencies experience pregnancy and birth.	HET significantly less expensive than IVF-ET under any conditions.
Medical interventions and hormone treatments for gestational mother, similar to what happens for surrogate mothers.	HET medically safer for female than IVF because ovarian stimulation is not involved for gestational woman. HET different than surrogacy because gestational woman becomes social mother.
Implantation and clinical pregnancy rates comparable to ART alternatives (Lee and Yap 2003).	Unlike most other treatments (with the exception of the use of donor sperm and oocytes), neither parent will be genetically related to the offspring.
As with ART that uses donor sperm, the manner of conception and genetic heritage may be easily withheld from others (including the child) because of achievement of pregnancy.	Child may have full genetic siblings that are being raised in another family.
	Genetic couples' interest in disposition of embryos appears to play a significantly greater role in their decision-making than it does for gamete donors.

$20,000 and $50,000. The financial comparison between HET and infant adoption, however, must include the appreciation that the costs for HET are per cycle. The higher costs of infant adoption reflect the fees paid for a "take home" baby, whereas with HET, a couple might attempt three cycles of HET for a total of approximately $12,000, still not achieving a live birth.

Given this pool of interested couples, it was puzzling to read Arthur Caplan (2003) say: "Almost no one who is going to spend $10,000 per try to use IVF is going to want to try it with another infertile couple's frozen embryo whose chances of properly developing grow less with every year it is frozen." Caplan mis-states two issues here. First, in reference to the cost, contrary to his figure, the literature shows that on average, HET is much less expensive than IVF. All of the medical journal articles on the topic (Robertson, 1995, p. 890;Van Voorhis et al., 1999; Eisenberg and Schenker, 1998; Lee and Yap, 2003) make specific mention of the fact that one of the main reasons that the practice would be so desirable for many couples is that it is much cheaper than IVF-ET, IVF-ICSI, or IVF with either or both donor sperm/egg. The reasons for the lower cost are attributable to the original couple having paid the costs associated with the creation and storage of the embryos. The recipient couple will pay the costs of thawing and transfer along with the costs of any testing and preparation of the potential gestational mother. Indeed, the Snowflake program is currently at about $10,000, but most other programs offering HET are significantly less expensive. For example, the costs are about $3,600-$4,000 at the National Embryo Donation Center (NEDC) in Tennesee and this program includes a model built on the adoption framework (Keenan, 2007, forthcoming; http://www.embryodonation.org/).

Second, regarding the likelihood of success of HET: Admittedly, IVF success rates are on average higher with fresh embryos than with frozen, but a review of the literature shows that the success rate for HET is no different than the success rate for IVF with a frozen embryo that is genetically related to a couple, as stated in the chart above. Currently, the data shows 20% to 40% success rate per cycle for HET (Keenan, 2007, forthcoming). Moreover, HET is also medically less risky for the woman embryo adopting as she will not have to take fertility drugs[7] nor have oocytes harvested (http://www.embryodonation.org/).

If there is such a ready pool of those who would like to do HET, if the procedure works fairly well, and relatively speaking, it is not expensive, why then are not more of these available embryos being transferred to other couples? The current data shows that it is not merely about a lack of knowledge concerning the availability of HET (as would justify the creation of the HHS grants); it seems in fact to be about an unwillingness to donate.[8] Kovas et al. (2003) showed that 89.5% of couples in that study opted to discard rather than to donate their embryos and that this was not necessarily linked to their views on the moral status of the embryo (Lee and Yap, 2003). Burton and Sanders (2004) showed only 15% of their couples were willing to donate.

Some of the most common reasons given by couples reluctant to donate embryos to others include: having unknown children, the possibility of sibling marriage and legal ramifications (Burton and Sanders, 2004). The concerns about having unknown children and the possibility of genetic sibling marriage are shared by individuals who consider becoming or using gamete donors, but I note here that it is equally shared by those women or couples who consider placing their soon to be born infant for traditional adoption. In fact, the number of children born from one person's gamete donation theoretically could be a much higher number than the number of children likely to be born from HET (though some clinics currently limit donations from any one donor). Consequently, the likelihood of genetic sibling marriage, though objectivity low in all these cases, is potentially much greater for the practice of conception using donor gametes that it ever would be for HET or even as it is with traditional infant adoption. The likelihood of future genetic sibling marriage using HET is really more similar to what might be the case in traditional adoption, a practice long in place.

There is yet another reason given by those with "spare" embryos for why they decline donation to other couples. According to the data, the following is the most influential reason for why couples chose against HET: lack of control over the choice of the recipients for their embryos. The norm in reproductive medicine is to require couples to agree to anonymous donation with no knowledge of the outcome of the donation for the genetic couple (Kingsberg, Applegarth and Janata, 2000, p. 217). One study reported that of the total respondents, 72% of clinics offered HET and of those, only 24% allowed donors "some" control over who receives their embryos. Strikingly, Newton, McDermid, Tekpetey and Tummon found in their study that those couples with "spare" embryos willing to consider HET "were more likely to want information about the outcome of any donation and more receptive to the idea of future contact with a child" (2003, p. 883).

While a minority of clinics are now considering what is called "conditional" donation (where the couples may stipulate conditions on who the recipients of their embryos may be or on the degree of future contact they wish to have with any child), it is still a rare policy. The justification of clinician control of donation appears to do with adapting current practices in reproductive medicine to HET. As is the case with donating gametes, embryos—frozen and fresh—are considered to be (philosophically and legally) the property of the individuals from whence they came. This "material" in turn is handled in a manner similar to the dispositions of tissue and organs. Once an individual voluntarily agrees to release his tissue for donation, unless otherwise specified prior to the donation, he does not have a right to control the specifics of the donation because technically it is no longer his property. With HET, once consent is given by the couples to release the embryos for HET, either the physician chooses the embryos for transfer, or she allows the recipient couple to chose from embryos with different genetic parents.

In the face of current practice of physician controlled HET there appears to be support from the lived experience of those with frozen embryos for the view that affording genetic parents greater control over the selection of prospective social parents as well as more access to the outcome of the donation may result in an increase in the amount of embryos available for donation. Why might this be the case and how does the answer challenge the status of HET as yet another ART procedure similar in kind to all the others?

I hold that there is a relevant distinction between the phenomena of individuals donating gametes and that of couples donating embryos and that this difference is precisely why HET is distinct from most other ART procedures: with the case of frozen embryos, the intention of the couple to whom they "belong" was to have children—to parent the children thus born. This intention is never present with gamete donation. As I have said elsewhere, "To equate couples' interest in the disposition of their embryos to those of gamete donors…is to gravely mistake the reality of these families' lives and to inappropriately conflate the two practices" (Brakman, Fall 2005, p. 11).

The embryo is created from the genetic material from both members of the couple for the purpose of having children. The couple always had a sense that these embryos were "ours" (Lacey, 2005)—egg and/or sperm donors do not share that sense. The dispositional decision regarding *their* "spare" embryos is about the fruit of a couple's marriage and about the future of their family—not a decision about a piece of genetic information. Consider Leon Kass' view:

> …Since one's own is not the own of one but of two, the desire to have a child of one's own is a *couple's* desire to embody, out of the conjugal union of their separate bodies, a child who is flesh of their separate flesh made one. This archaic language may sound quaint, but I would argue that this is precisely what is being celebrated by most people who rejoice at the birth of Louise Brown [first IVF baby], whether they would articulate it this way or not. Mr. and Mrs. Brown, by the birth of their daughter, embody themselves in another, and thus fulfill this aspect of their separate sexual natures and of their married life together. They also acquire descendants and a new branch of their joined family tree (Kass, 2002, pp. 96–97).

Not by way of endorsing the practice, but rather by explanation, Kass is recognizing that the genetic/biological embodiment of themselves in a child is satisfying to the couple and important by way of preserving their combined lineage. If this is the reason why many couples engaged in IVF and froze the embryos to begin with, it is hard to see how they can now, easily, "donate" them. This may also explain in part why the donation rate is low among couples for whom IVF was successful. Not only is there the worry of full genetic siblings, but these couples are now acutely, if not in most cases painfully, aware of the fact that the embryos will become children—their genetically related children.

Independent of these couples' views regarding the moral status of embryos prior to using IVF, the data shows that the majority of couples who have used IVF now view their frozen embryos as either children or 'virtual' children (Lacey, 2005, pp. 1665–1667). In one study, couples said that they could not merely donate the embryos as so much genetic material, but rather "embryos were attributed a personhood that lack physical presence but contained biology and spirituality. In this sense they acquired a virtual personhood" (Lacey, 2005, p.1665). Many of those interviewed in this particular study said they felt like this would be placing their child for adoption and that they could not do this.

The legal ramifications concerning HET will be addressed in greater detail in section IV, however, law is currently developing to clarify that donors (whether of gametes or of embryos) are not parents legally and that the woman who gives birth is the legal mother and her husband the legal father (The Uniform Parentage Act, 2002).[9] In terms of the law, then, HET appears to be considered similar to other ART procedures especially artificial insemination by donor (AID) and IVF-ET with donor eggs and not with traditional adoption (more on this comparison with adoption below).

What this review of the data tells us is that although HET occurs in fertility clinics and is currently subject to the same laws that govern children born as a result of gamete donation/IVF, it appears to have significantly different considerations than other ARTs, largely having to do with the fact that the embryo was created often times from the gametes of both members of the couple. The similarities between HET and ART appear to be primarily descriptive (see chart at beginning of section). The final three entries under "differences" in the chart highlight its admittedly unique nature among ART options as a) the social parenting of a child who is not genetically related to either parent and b) who also may have full genetic siblings living in another family. [10] Given the systemic dis-ease of couples with frozen embryos to the way in which HET generally is offered in reproductive medicine and given the current practice's foundation in a paradigm that embraces the primacy of property and decision-making based solely on the rights of the couple involved, let us now turn to a comparison of HET and the current practice of domestic infant adoption.

III. HET AND THE PARADIGM OF OPEN ADOPTION

As seen in chart 2, HET has both similarities and differences with infant adoption. The similarities are those that are intrinsic to the natures of both HET and domestic infant adoption: the lack of genetic relatedness between the rearing parents and the child, the possibility of having full genetic siblings not raised in the same family, the similarity of the reasons given by couples for their reluctance to chose either HET or adoption, and a number of psychological, social and ethical issues that arise for not only the genetic parents and the recipients who become the social parents, but most especially for the individual who is born and raised as a result of these practices. (In the adoption world, these three parties are referred to as the members of the *adoption triad*—birth parents, adoptive parents and child.)

Of note is the second similarity listed in the chart and mentioned earlier in this chapter—the views of the genetic couples regarding why they do not choose to give their embryos to others as similar to the views of those couples or women who chose to terminate pregnancies because they do not feel comfortable choosing traditional adoption. These reasons include the possibility of a future sibling marriage and the unacceptability of having one's child raised in another family (Nachtigall et al., 2005, p. 433). This concern speaks also to the shared lived experience of couples with "spare" embryos and those who are deciding about a crisis pregnancy.

CHART 2. Comparison of HET with Traditional Infant Adoption

The Similarities	The Differences
No genetic connection between social parents and child.	Pregnancy and birth are experienced by the social mother with HET and the pregnant woman becomes biologically related to the child. This means that pre-natal risks to the child are more controlled for in HET than may be the case in infant adoption. HET also affords greater privacy for the adoptive parents versus adopting an infant. Finally, due to pregnancy, HET is 100% more medically risky pre-birth and physically draining pre and post birth for the rearing mother in HET than is the case for the rearing mother in infant adoption.
Reasons given for reluctance to give embryos to other couples: having unknown children, possibility of sibling marriage, legal ramifications (Burton and Sanders, 2004) are similar to reasons given for reluctance of birth parents to place infant for adoption.	HET is significantly less costly than infant adoption ($3,600 to $10,000, compared to $9,000 to $35,000), the "take home baby" rate is 20% to 40% per try for HET, compared to total costs for infant adoptions.
Nurture is considered the basis of parental bond over nature. This "helps acceptance that one's donated embryo evolves into another couple's child. It presupposes differentiating between parents and genitors"(Laruelle and Englert, 1995). Literature in traditional open adoption shows bonds strong between adoptive parents and children.	HET has a more definite time line than infant adoptions (9 months per try vs. waiting time of 1 month to 5 years +).
The possibility of the existence of full siblings in another family exists for HET and infant adoption and raises the same sorts of ethical and psychosocial challenges.	76% of clinics do not allow donors any control over who receives their embryos (Kingsberg et al., 2000). However, all infant adoption professionals allow for choice by birth parents of closed, semi-open or fully open adoptions, with birth parents choosing adoptive parents in all but closed adoptions.
Psychosocial implications for parents of raising a child who has no genetic linkage with either parent.	
Disclosure issues with both child and others about nature of conception and genetic heritage is present for HET and infant adoption.	
Psychosocial implications for genetic parents of the existence of a related child for whom they are no longer considered the parents will be present for those who give embryos to others as it is for birth parents in infant adoption.	
Need to attend to the emotional and psychosocial developmental needs of a child who has genetic links to another man and woman.	

What of the differences between HET and traditional adoption? As listed in the chart, HET affords a couple the opportunity to control the physical maternal environment, perhaps to begin bonding sooner, and to experience the birth of the child whom they will raise but to whom they are unrelated. HET is also more medically risky for the woman than infant adoption; it may be generally less expensive; and the waiting time is on average shorter. All these are matters of fact and do not seem on first blush to have much, if any, normative weight.

What is of interest here, though, is the role of having been pregnant with the child one has adopted. What if any moral significance gestational parenthood provides that makes it different than traditional adoption has been insufficiently analyzed to date. I am unable to go into this particular issue here, yet pregnancy raises the issue that HET may not be the same as infant adoption in a moral sense. This possibility, however, does not undercut the more modest claim I am advancing here: that HET is more analogous to traditional adoption—particularly in regard to the experience of the couples—than it is to any other ART procedure.

There is, however, one difference in practice between the two procedures that may account for the reluctance of couples to donate. This is the issue raised in Section II, that most HET couples do not have control over the specific choice of the couples who will receive their embryos. However, currently in domestic adoption of infants, it is the birth parents who usually choose the adoptive family. Infant adoptions in the United States are characterized today as "open" to distinguish them from the way in which adoptions had been handled prior to fifteen years ago in the United States. These adoptions were what is now referred to as closed and usually entailed secrecy. Adoptive parents and especially adopted individuals lacked information regarding biological heritage. For birth parents, a lack of information about the family chosen to parent the child and a lack of continued knowledge of the well being of the child was also standard practice. (These types of adoptions may still be performed but they are chosen arrangements by the birth parents and are referred to as closed adoptions.) Currently, open adoption, as the practice is known, is a spectrum concept, ranging from limited contact (an exchange of first names and non-identifying information, possibly also including a pre-birth meeting between birth and adoptive parents) to the practice of a complete sharing of all identifying information, as well as an ongoing personal relationship among birth parent(s), child and adoptive parent(s). Open adoption starts with the biological or birth parents choosing the adoptive parents from among a group of prospective candidates and, after placement, receiving updates about the progress of the child (semi-open) or even having a continuing relationship with the adoptive parents and/or child (fully open) (Brakman, 2003, pp. 61–62).

The adoption literature is rich with recent data showing that open procedures make for greater satisfaction and closure for the birth parents, greater empathy for birth parents by the adoptive parents as well as far less fear of birth parents reclaiming children. Most importantly, however, open adoption is being shown to provide greater emotional and psychological stability for the adoptee (Brakman, 2003 p. 62).

With HET, the option of choice of recipient parties on the part of genetic parents is not standard practice. Part of the justification for physician control of the disposition of embryos seems to include not only the view that the embryos are genetic material equivalent to gametes or tissue, but a belief that it is in the best interests of the genetic couple to "let go" fully of the embryos once consent is given and that it is in the best interests of the recipient couple to know that there will be no further intrusion from the genetic couple.

However, the concerns about desiring to know the outcome of HET and one's willingness to have contact with a child in the future indicate that the practices of infant adoption in this regard may be very well suited to HET. In a study of 49 couples in Canada who had frozen embryos available to donate, Newton et al. (2003, p. 883) found:

> [T]hose most likely to participate in ... [HET] hold views more congruent with a model of 'embryo adoption' than with a model of traditional medical donation....Rather than preferring an anonymous, disinterested gift, typical of tissue and organ donation, individuals in our study who were willing to consider ED [embryo donation], wanted to be part of a potential child's life in terms of providing information about themselves that might be important to the child, and were more open to some form of future relationship with the child. This approach has much in common with the growing movement toward open adoption whereby birthparents and adoptive parents exchange identifying or non-identifying information.

Given the similarity in lived experiences and concerns of couples with excess frozen embryos and those considering placing their child through open adoption, it is reasonable to think that the extensive literature on adoption as well as the developed practices of domestic adoption would be useful in the understanding of the practice of HET, the evaluation of the policies surrounding it and the myriad psycho-social and ethical issues that will follow from it.[11]

Though the situation of donor couples and birth parents appear analogous between HET and infant adoption, some have argued that those on the other side of the equation–the adoptive parents and those who wish to use HET—can not be treated analogously. Robertson has said, "Although sometimes termed an 'embryo adoption,' the procedure of embryo donation is not equivalent to postnatal adoption of a born infant and therefore need not entail as rigorous social screening. If the recipient couple is otherwise acceptable for infertility treatment, requiring them to pass parental fitness tests that are not required of other infertility patients would appear to be discriminatory" (Robertson, 1995, p. 887).

To answer Robertson, I will pose a question: Why should adoptive parents have to pass parental fitness tests when any girl/woman can "keep" her biological baby? Is not this discriminatory? The answer is "no" of course for a few reasons. The state has an interest is protecting the welfare of its most vulnerable citizens. Biological parents' rights are recognized as a natural right in our law. When someone is placing her child for adoption, she has a right to expect that our society will have safeguards in place to help her do this. The same could be said of HET. Couples who have embryos that will become their genetic children—whom they consider to be their *virtual* children—have a right to expect safeguards in place if they choose to give the embryos to others.

My response to Robertson then is that while HET recipient couples are going to be treated unequally in comparison to how other fertility patients are treated, this is not treating them unfairly because the situation by definition is different with HET than with all other ARTs; the "gestational" couple would be choosing to bear and rear children from another couple's valued embryos. No such prior interest in genetic material exists for any other ART and certainly does not exist for those who donate gametes under the intention to never parent the children that may come into existence as a result of their actions.

In the preceding two sections, I demonstrated the important differences between HET and most forms of ART which account for the ill-fit of the practice paradigm of ART for HET, as well as the relevance and usefulness of the operative adoption paradigm for HET. Given this, I now turn to the debate surrounding the use of the adoption terminology to refer to HET.

IV. WHAT'S IN A NAME? 'ADOPTION' AND THE MORAL STATUS OF THE EMBRYO

I chose the term *HET* until this point because it presents at first blush as a purely descriptive term. However, such a clinical term is one that in my view neglects to attend to the personal dimension and meaning of this practice, something I hope this chapter has shown is crucial to understanding its nature and how we ought to evaluate it. Interestingly, *HET* is not the term employed by clinicians and in fact is used by default among those who do not wish to embrace other obvious candidates.

Embryo donation is the term employed throughout the fertility industry/reproductive medicine. Embryo donation evolved from the current practice and nomenclature regarding gamete donation—the use by a couple of sperm and/or oocytes from other individuals to create *in vitro* an embryo that will be implanted in the would-be social mother. *Embryo donation* appears at first to be a choice that in fact focuses on the actions of those who contribute their gametes. One might also postulate that *donation* makes the contribution/significance of the gamete donors (read genetic parents) seem more distant and clinical. There is a prevailing understanding in reproductive medicine that this distance is also aesthetically appealing to former fertility patients/donor couples who do not want to see themselves as "giving up" their children. Therefore, employing the term embryo donation makes it seem to the couples with embryos that they are making a gift of their unused genetic material to other infertile couples (http://www.embryoadoption.org/GenPracticalUsage.asp). However, in light of the discussion above, I believe that the term embryo "donation" not merely obscures but rather exacerbates the discomfort of couples by not speaking to their reality and experience. To their mind, they are not merely donating tissue, but rather something much more monumental.

Embryo rescue appears to be used by some moralists who view the practice as responding to the plight of vulnerable human beings who need saving. This term, unlike the others we have considered thus far is rooted in a moral vision of the status of the embryo as a moral person. Some who use this term even argue that embryos ought to be gestated (by women or even artificial wombs), but that these same women need not (some even argue should not) become the social parents of the children thus born.

But does embryo *rescue* work as a descriptor of HET, apart from its implications about the moral status of the embryo? John Berkman explores the adequacy of *rescue* by discussing the concept itself:

> The quintessence of the notion of rescue is that first, it is done in an emergency situation at significant risk to the rescuer, and second, the rescuer has little or no prior or subsequent relationship with the person rescued. While "rescue" at times gets stretched to apply to situations where one of these characteristics is absent, these two features lie at the heart of the heroic and altruistic character ascribed to the action of the paradigmatic rescuer (2003, p. 323).

Berkman goes on to say that HET is neither an emergency case nor a situation where a woman has a "transitory" relationship with the embryo (e.g., the gestational relationship is not transitory, nor is the social parenting relationship.) For these and other reasons (such as the oddity of referring to the "decision to become a parent (either gestational or adoptive mother) as an altruistic one" in the main), he dismisses rescue as a worthy candidate for HET (p. 324). While more could be said here about the aptness of "rescue," for all the reasons above, I take Berkman to be right about the ill fit of the term. I turn now to a consideration of the term embryo *adoption*.

In the first part of this chapter I argued how the framework or paradigm operative in traditional adoption is more apt for HET and even better suited to it than the paradigm currently in use in ART. However, bioethicists along with others have not embraced the adoption terminology. For example, George Annas, has been quoted as saying, "I think you can adopt children. Someone can give you an embryo" (Mulrine, 2004, http://www.usnews.com/usnews/health/articles/040927/27babies.b1.htm). Aside from this one statement, I can find no further development or argument for Annas' position. Another example is from the AMA itself (Arekapudi, 2002):

> The term "embryo adoption" was originally coined by Nightlight Christian Adoptions. The director of Snowflakes, the agency's embryo program, explained that 'we use the adoption language and materials with the hopes of setting a precedent that someday the court will say embryos need to be handled like any other child.' Knowledge of the origin of the term embryo adoption has fueled sharp criticism of the Bush administration [for the creation of the Public Awareness Grants on Embryo Adoption] by abortion rights groups.

There are two possibilities for the rejection of adoption language by mainstream bioethicists: either 1) there is a view that adoption language is not accurate because embryos are not children or 2) a view that we cannot use the term adoption because it will make people think of embryos as children, whether they are or not. I begin with a consideration of 1. Here the argument might go something like this: Unlike the term "embryo rescue," "embryo adoption" does fit well, but since it is predicated on a particular moral vision of the status of embryos (that they are children and therefore have moral standing), it is the wrong terminology. The only argument appears to be this three step one:

1. *Children are adopted.*
2. *Embryos are not children (or "persons" in some versions of the argument).*
3. *Therefore, "embryo adoption" is not an accurate term.*

The focus of the opposition seemingly has been on premise two. Those who view the embryo as an entity equal in worth to any other human dispute the truth value of premise 2, which states that embryos are not children. Those who do not hold that the embryo has moral worth appeal to premise 2 to deny the appropriate applicability of adoption terminology. The fact is we need not engage here and now on the question of the moral status of the embryo in order to accept the suitability of the language of adoption. The more crucial premise is really number one. It says: "children are adopted." It does not say *only* children are adopted. Pets are adopted. Legislation is adopted. Even countries are adopted. Declaring that embryos are adopted or can be adopted does not logically imply anything about embryos being children. Adoption is about a permanence of relationship (Institute for Adoption Information, 2005), about taking something on that was not obviously or initially considered belonging to or part of one, and by a person's will, making a commitment to support, protect and in some cases, yes, love. Given that adoption may be used with entities other than children, we do not have to agree on the moral status of the embryo in order to see that there is nothing inherently mistaken about using adoption language to describe HET.

This point, taken together with the following: a) that the data showing that those couples who have created the embryos do consider the embryos at least as 'virtual' children; b) that these same couples have a vested interest in the fate of the embryos; and c) that indications show families created from this procedure will encounter many of the same psychological, sociological and ethical issues that individuals in the traditional adoption triad meet, means that the term *adoption* is actually the most accurate language to describe this practice currently.

There is, however, a different point being made in regard to adoption terminology concerning the law that bears special attention and to which I now return. Robertson (1995, p. 891) claims, "The most significant point is that there is no 'child' to be adopted because in most legal systems embryos are not legal persons." Legally, adoption applies to a born child. *Wikipedia, the free encyclopedia* (http://en.wikipedia.org/wiki/Embryo_adoption) has the following under the entry Embryo Adoption:

> "Adoption" is only an informal term for this procedure. Since embryos are not considered to be children under the law, they cannot be legally "adopted." Thus, an embryo adoption is legally a transfer of ownership of embryos (ordinarily up to nine embryos are transferred in one adoption because pregnancy may not result with the first

attempt). Use of the term adoption is controversial because embryos are not universally considered to be children, nor are they considered so under the law.

Recently, Mary Anderlik Majumder (2005, pp. 10–11) addressed the difficulties of applying current (infant) adoption law to embryo adoption, showing that subsuming embryo adoption under adoption law would make the procedure more precarious for the recipient couple as the donor couple would retain their legal rights as parents throughout the pregnancy.

The answer to the question of legal standing and procedure is that surely laws may be amended, or new regulations proposed specific to embryo adoption that acknowledge its unique place in the spectrum between the donation of genetic material and the placement of a child after birth. It seems backwards to allow the current law to dictate how we will consider a practice that was not even possible when adoption laws were initially formulated, nor would this be an accurate reflection of the intention of such laws. Finally, the legal issue does not have to address the moral status claim as we could have specific laws governing prenatal adoption and postnatal adoption.

I turn at this point to the second possibility for why bioethicists and others reject adoption: 2) we cannot use the term adoption because it will lead people *to think of embryos as children*, whether they really are or not.

In the essay for *MSNBC* Caplan (2003) said, “This [Embryo Adoption Public Awareness Grants] is a nice way [for the Bush administration] to score points with those who advocate the view that embryos are actual babies and should not be used for research purposes” (http://www.msnbc.com/id/3076556/print/1/displaymode/1098/). He went on to say, “using terms like ‘adoption’ encourages people to believe that frozen embryos are the equivalent of children. But they are not the same. In fact, infertile couples who want children can frequently make embryos but they cannot make embryos that become fetuses or babies” (2003).

Caplan is making at least two claims here: first, that the use of the term “adoption” encourages people to believe that frozen embryos are equivalent to children. I could respond to this argument by saying that there is no proof this will happen, that those who are worried about it could counter act this with their own campaigns, but I respond rather by saying, yes, using embryo adoption might encourage people to think of embryos as children. The more pertinent question here is why is this a problem, especially in light of the fact that an overwhelming number of couples with frozen embryos themselves perceive of the embryos as something like virtual children? Is the concern that people could be just wrong in thinking this? If this is Caplan’s worry, it hardly seems worth his making a fuss.

Is Caplan’s concern rather that such a view might contribute to undercutting laws or public opinion in related areas, such as abortion rights or human embryonic stem cell research? Whether such undercutting would happen as a result of calling this practice “embryo adoption” is again difficult to say, but if as I have shown the terminology is the most accurate and if as I have also shown, the paradigm of adoption is also most accurate for understanding and analyzing the practice, then the principle here seems to be that truth and/or accuracy of language should be subsumed to political interests. The unacceptability of this view is I think obvious to all. As philosopher-bioethicists—seekers of truth regarding ethics in medicine and science—we decry ideology driven medicine and policy. Let us be ever vigilant not to succumb to ideology driven analysis ourselves.

I turn now to Caplan's second point in the quote above, the fact that many embryos die before birth makes them unequivalent to children. Dying before birth works as an arbiter of being a child only if one assumes that there is something about having been born that grants moral status. But to engage in this argument is to engage the debate about whether or not embryos are children, something that I have shown is unnecessary for accepting the terminology of embryo adoption.

Is Caplan alternatively making an empirical point—saying that people who embryo adopt might not get a "take home baby" because there is a high risk of fetal loss in such a practice and therefore those who self-describe as embryo adopting think they are going to get a child (just like those who turn to infant adoption think they are getting a child), but that in fact there is a greater likelihood that the procedure will fail and the couple will have no baby in nine months? On this view it would seem that Caplan's concern is that adoption is not the right framework to use because it promises too much to the couple who are contemplating this practice—namely getting a child. This concern however does not really distinguish embryo adoption from what happens in domestic infant adoption, where couples who have been chosen by birth parents prior to birth are then disappointed when the birth parents change their minds after the baby is born and decide to parent the child themselves.[12]

In conclusion, I argue the concern that *embryo adoption* as a descriptor may somehow indirectly encourage a view of the embryo as a child does not seem reason enough not to use it. The language of adoption does not *necessarily* imply the embryo is already a child and in fact since the paradigm of adoption is a better fit for understanding and carrying out this practice, *embryo adoption* seems apt and completely defensible as not only the rubric for understanding the practice but as the correct terminology as well.

V. TOWARDS A NEW PROPOSAL…

It may be too early to tell, but the Public Awareness Campaigns for Embryo Adoption have yet to raise the profile in this country of the practice. It is still considered more often than it is actually done. A program focusing on the social, ethical and clinical issues embryo adoption raises should be developed to facilitate and encourage more couples to consider this option for disposition (Fuscaldo and Savulescu, 2005; Kovas et al., 2003; Kingsberg et al., 2000). As Lacey argues, "Given the historical potency of practices of relinquishment, it seems unlikely that the numbers of patients who select embryo donation can be increased unless embryo donation can be metaphorically 're-framed'" (2005, p. 1668).

I argue that embryo adoption programs should be developed and standardized, at least nationally, based on the paradigm of the open adoption of infants in the United States. If the biological couples could chose the recipients, if the degree of future contact could be negotiated among the options of closed, semi-open, and open, and if the process of screening the adoptive parents and the literature on the psychological benefits of disclosure were more uniformly and explicitly attended to, then it is likely that the rates of embryo adoption would rise with the satisfaction level of the genetic couples.

Others have argued for guidelines to be established for recipient acceptance into embryo adoption programs and many of these look to be very similar to the procedures of adoption such as: educational interviews addressing fertility loss, reasons couples have chosen this option, plans for disclosure. Also recommended is complete family and social history, including substance abuse, domestic violence, sexual abuse and legal problems (Kingsberg et al., 2000, p. 219). Again, there is a mechanism in place in traditional adoption that covers such concerns—the homestudy procedure, which is the hallmark for adoptive parent suitability.

Part of the "re-framing" that Lacey calls for, to my thinking, should also include what we call these programs and that is why I have suggested attending to the language we use and why I have rejected the suggestion, for example, of Majumder that it is "good enough" to have pragmatic agreements to let clinics/agencies and advocacy groups use different names regardless of what those terms actually specify or connote.

While adoption as a paradigm for understanding the phenomenon is accurate, the word itself, however, admittedly focuses on the recipients or those who take on the embryo or child. Considering that one of the main concerns here is to address the lived experience of the genetic couples who are having a difficult time releasing their embryos, then perhaps what is needed is a term that speaks directly to the actions and perspective of the genetic couple but still is within the adoption paradigm. Along this line, I then propose the term embryo *placement* as the name for the practice of relinquishing embryos to other couple, a name that focuses directly on the actions of the donating couple.[13] Open infant adoption includes, for example, the birth parents "placing" their child for adoption. Given the extent of the birth parents' involvement in choosing the adoptive parents, this term is more accurate than "giving up" or "surrendering" a child, as was the more appropriate term for the practice in adoption in the past.

Reproductive medicine then could offer Embryo Placement/Embryo Adoption Programs. Embryo placement refers to that aspect of the practice where those with "spare" embryos place them with other couples. Embryo adoption refers to the option for other couples who choose to gestate an embryo already in existence.[14]

In conclusion, I have illustrated how giving one's embryos to another family to gestate and raise is better understood conceptually and in practice within the paradigm of adoption than within the paradigm that is currently governing the ART world. Moreover, the criticisms by some well known bioethicists of the use of adoption language appear to result from a failure to analyze the practice in depth and also from a failure to provide a detailed argument about the language of adoption. I have proposed a new term from within the adoption paradigm more accurately to describe and thus encourage genetic parents to consider the practice. Finally, by drawing on the adoption paradigm, I allude to ways in which such a paradigm might be adapted for Embryo Placement/Embryo Adoption Programs in reproductive medicine.

NOTES

[1] An earlier version of this paper was presented October 31, 2004 at the 2004 Annual Meeting of The American Society for Bioethics and Humanities, and again on February 7, 2006 at Villanova University. In addition, sections of this paper originally appeared in my brief essay, "Ethics and Embryo Adoption," *The Lahey Clinic Medical Ethics Journal* Spring 2005: 1–2 and also "Dialogue: The Politics of Embryo Transfer," *The Lahey Clinic Medical Ethics Journal* vol. 12, Issue 3, Fall 2005: 10–11 and reprinted in *Biomedical Ethics: A Multidisciplinary Approach To Moral Issues In Biology and Medicine* (University Press of New England, forthcoming 2007).

[2] HET can also be pursued using an embryo created from a donor egg or/and sperm so that the embryo is genetically unrelated to both the woman who gestates it and her husband who will be the parent of the child when born. Additionally, I recognize that some HETs are not going to result in a gestating woman subsequently becoming the rearing parent of the child she bears. Both of these practices are bracketed in this analysis as I focus here on the "simple" case of couples with embryos created from their gametes who are considering giving those embryos to other couples to gestate and raise as their children.

[3] Press Release for U.S. Senate Committee on Appropriations, November 6, 2001, found at http://www.appropriations.senate.gov.releases/record.cfm?id=179516; A copy of the grant announcement can be found at The Catalog of Federal Domestic Assistance, http://12.46.245.173/cfda/cfda.html. As of 2007, however, the name of these federal grants has been changed to Embryo Donation and/or Adoption Public Awareness Campaign. This was no doubt due to political wrangling over the use of the word adoption.

[4] I was inspired to recall Austin's relevance to this project while reading the work of Stanley C. Brubaker.

[5] I wish to acknowledge the work of Mary B. Mahowald whose forthcoming article on embryo adoption in *The Ethics of Embryo Adoption and the Catholic Tradition* explicitly addresses casuistry in this way and inspired me to see the connections in my work in this chapter.

[6] Those with this view often prefer the use of the term, "embryo rescue". Arguments to this effect currently are found in the literature on the Catholic tradition and HET. See *National Catholic Bioethics Quarterly*, Spring 2005.

[7] The woman attempting pregnancy will use hormonal mediations to prepare her uterus for implantation of the embryo(s).

[8] Caplan was also implying that HET is not successful and that this is the reason why the procedure is so rarely done. However this also appears to have been factually inaccurate. The data that shows 60% of frozen embryos survive the thawing process and there is a 22% implantation rate of success nationwide (http://www.usnews.com/usnews/health/articles/040927/27babies.b1.htm), though Keenan (forthcoming) has a much greater success rate.

[9] National Conference of Commissioners on Uniform State Laws, Uniform Parentage Act (2002) SS 102, 702. Only six states have accepted this policy as of this writing, though some states are creating specific laws concerning embryo donors' lack of parental rights.

[10] There are a number of ART practices that fit a) but not b).

[11] The practical difference between the two situations is that couples with frozen embryos may delay their decision about the disposition of the embryos for years (and thus in effect make the decision to not decide). Pregnant women have no such luxury.

[12] Agencies and lawyers typically prepare prospective adoptive parents that there is a 50–50% chance a woman will change her mind about placing the child for adoption once she actually gives birth and sees the child.

[13] Embryo placement is also preferable to the currently used embryo donation for it can be used to distinguish between donating an embryo to research and placing an embryo with another family. Currently embryo donation can apply to donations to either research or other couples.

[14] I am grateful to James M. Youakim for initially suggesting the term *placement* to me and to Allen Levine and Bradford Wilson who helped me think through different aspects of my argument.

BIBLIOGRAPHY

The American Society for Reproductive Medicine (ASRM) (2004). 'Guidelines for cryopreserved embryo donation,' *Fertility and Sterility*, 82(supplement 1), S16–S17; S20–S21.

Arekapudi, S. (2002). 'Adopting the unborn,' [On-line]. Available: http://www.ama-assn.org/ama/pub/category/9153.html.

Austin, J.L. (1961; 3rd ed. 1979). *Philosophical Papers*. J.O. Urmson and G.J. Warnock (Ed.) Oxford: Oxford University Press, 130.

Bangsboll, S., Pinborg, A., Andersen, C.Y., & Andersen, A.N. (2004). 'Patients' attitudes towards donation of surplus cryopreserved embryos for treatment or research,' *Human Reproduction*, 19(10), 2415–2419.

Bartholet, E. (1999). *Family Bonds: Adoption, Infertility, and the New World of Child Production*. Boston: Beacon Press.

Berkman, J. (2003). 'Gestating the Embryos of Others,' *The National Catholic Bioethics Quarterly*, 3(2), 309–329.

Brakman, S.V. & Scholz, S. (2006). 'Adoption, ART, and a re-conception of the maternal body: toward embodied maternity,' *Hypatia*, 21(1), 54–73.

Brakman, S.V. (Fall 2005). 'Dialogue: The politics of embryo transfer,' *The Lahey Clinic Medical Ethics Journal*, 12(3), 10–11.

Brakman, S.V. (Spring 2005). 'Ethics and embryo adoption,' *The Lahey Clinic Medical Ethics Journal*, 12(2), 1–2.

Brakman, S.V. (2003). 'Open adoption and the ethics of disclosure to children,' *Philosophy in the Contemporary World*, 10(1), 61–67.

Burton, P.J. & Sanders, K. (2004). 'Patient attitudes to donation of embryos for research in western Australia,' *Medical Journal of Australia*, 180(11), 559–561.

Caplan, A. (2003). 'The problem with 'embryo adoption': Why is the government giving money to 'Snowflakes?' June 24, 2003 [On-line]. Available: http://www.msnbc.com/id/3076556/print/1/displaymode/1098/.

Eisenberg, V.H. & Schenker, J.G. (1998). 'Pre-embryo donation: ethical and legal aspects,' *International Journal of Gynaecology and Obstetrics*, 60(1), 51–57.

Embryo Adoption Awareness Campaign. 'Practical usage of the terms "adoption" and "donation"', [On-line] Available: http://www.embryoadoption.org/GenPracticalUsage.asp.

ESHRE Task Force on Ethics and Law. (2002). 'Gamete and embryo donation,' *Human Reproduction*, 17(5), 1407–1408.

Eydoux, P., Thepot, F., Fellmann, F., Francannet, C., Simon-Bouy, B., Jouannet, P., Bresson, J.L., & Siffroi, J.P. (2004). 'How can the genetic risks of embryo donation be minimized? Proposed guidelines of the French Federation of CECOS (Centre d'Etude et de Conservation des Oeufs et du Sperme),' *Human Reproduction*, 19(8), 1685–1688.

Feast, J. (2003). 'Using and not losing the messages from the adoption experience for donor-assisted conception,' *Human Fertility*, 6, 41–45.

Fuscaldo, G. & Savulescu, J. (2005). 'Spare embryos: 3000 reasons to rethink the significance of genetic relatedness,' *Reproductive BioMedicine Online*, 10(2), 164–168.

Hoffman, D.I., Zellman, G.L., Fair, C.C., Mayer, J.F., Zeitz, J.G., Gibbons, W.E., & Turner, T.G., Jr. (2003). 'Cryopreserved embryos in the United States and their availability for research,' *Fertility and Sterility*, 79(5), 1063–9.

Institute for Adoption Information (2005). *A Journalist's Guide to Adoption* [On-line]. Available: http://www.adoptioninformationinstitute.org/JGuide.html.

Kass, L. R. (2002). *Life, Liberty, and the Defense of Dignity: The Challenge for Bioethics*. San Francisco: Encounter Books.

Keenan, J. (forthcoming). 'Development of a national embryo donation center,' in S.V. Brakman and D.F. Weaver (Eds.), *The Ethics of Embryo Adoption and the Catholic Tradition*, Dordrecht: Springer.

Kingsberg, S.A., Applegarth, L.D., & Janata, J.W. (2000). 'Embryo donation programs and policies in North America: survey results and implications for health and mental health professionals,' *Fertility and Sterility*, 73(2), 215–220.

Kovas, G.T. Breheny, S.A., & Dear, M.J. (2003). 'Embryo donation at an Australian university in-vitro fertilization clinic: issues and outcomes,' *Medical Journal of Australia*, 178(3), 127–129.

Lacey, S. de (2005). 'Parent identity and 'virtual' children: why patients discard rather than donate unused embryos,' *Human Reproduction*, 20(6), 1661–1669.

Laruelle, C. & Englert, Y. (1995). 'Psychological study of in vitro fertilization-embryo transfer participants' attitudes toward the destiny of their supernumerary embryos,' *Fertility and Sterility*, 63(5), 1047–1050.

Lee, J. & Yap, C. (2003). 'Embryo donation: a review,' *Acta Obstetricia et Gynecologica Scandinavica*, 82, 991–996.

Majumder, M.A. (Fall 2005). 'Dialogue: The politics of embryo transfer,' *The Lahey Clinic Medical Ethics Journal*, 12(3), 10–11.

McGee, G., Brakman, S.V., & Gurmankin, A. (2001) 'Debate: Disclosure to children conceived with donor gametes,' *Human Reproduction*, 16(10), 2033–6.

Mecker, L. (2002). Kate Michelman, quote in Philadelphia Inquirer article, 'Agency to promote 'embryo adoption,' August 21, 2002 [On-line]. Available: http://www.philly.com/mld/inquirer/news/nation/3906970.htm?1c.

Mulrine, A. (2004). 'A home for frozen embryos,' *U.S. News and World Report*, September 27 [On-line]. Available: http://www.usnews.com/usnews/health/articles/040927/27babies.b1.htm.

Nachtigall, R.D., Becker, G., Friese, C., Butler, A., & MacDougall, K. (2005). 'Parents' conceptualization of their frozen embryos complicates the disposition decision,' *Fertility and Sterility*, 84(2), 431–434.

National Conference of Commissioners on Uniform State Laws, Uniform Parentage Act (2002) SS 102, 702.

National Embryo Donation Center (NEDC). [On-line]. Available: http://www.embryodonation.org/

Newton C.R., McDermid, A., Tekpetey, F., & Tummon, I.S. (2003). 'Embryo donation: attitudes toward donation procedures and factors predicting willingness to donate,' *Human Reproduction*, 18, 4:878–884.

Nightlight Adoptions: Snowflake Program. [On-line]. Available: http://www.nightlight.org/ snowflakeadoption.htm.

Robertson, J.A. (1995). 'Ethical and legal issues in human embryo donation,' *Fertility and Sterility*, 64(5), 885–894.

Press Release for U.S. Senate Committee on Appropriations, November 6, 2001. Available: http://www.appropriations.senate.gov.releases/record.cfm?id=179516; A copy of the grant announcement can be found at The Catalog of Federal Domestic Assistance Available: http://12.46.245.173/pls/portal30/CATALOG.PROGRAM_TEXT_RPT.SHOW?p_arg_

Van Voorhis, B.J., Grinstead, D.M., Sparks, A.E.T., Gerard, J.L., & Weir, R.F. (1999) 'Establishment of a successful donor embryo program: medical, ethical, and policy issues,' *Fertility and Sterility*, 71(4), 604–608.

Wikipedia, the free encyclopedia. [On-line]. Available: http://en.wikipedia.org/wiki/Embryo_adoption

CHAPTER 14

BRODY ON PASSIVE AND ACTIVE EUTHANASIA

F.M. KAMM
Harvard University, Cambridge, MA, U.S.A.

In this article I will consider Baruch Brody's views on passive and active euthanasia.[1] Some of Brody's important claims are that (1) there is a moral difference between killing and letting die, but even if this is true we cannot conclude that killing a patient is not permissible even if letting him die is, and, (2) if we conclude that killing a patient is sometimes permissible it should not be because we assume that there is no moral difference between killing and letting die. I agree with these claims but wish to examine the details of his arguments for them. In section I, I will summarize and critically reflect on his views about the moral distinction between killing and letting die and its possible clinical significance. In section II, I will summarize and critically examine his views about how to draw the killing/letting die distinction.[2]

I.

A.

Brody thinks it is a mistake to conclude on the basis of one set of cases, such as James Rachels presents, where all factors are held constant other than that one is a killing and the other a letting die that there is no moral difference between killing and letting die. (Rachels' cases involve someone who will let a child who has slipped in the bathtub drown in order to inherit his money (letting die), and someone who will push a child down in the water to drown him in order to inherit his money (killing).) Even if we believe in one set of cases that we should condemn the killing and letting die equally, this need not mean that we should condemn them equally in other cases. Furthermore, Brody argues, there are different ways in which killing and letting die could make a moral difference: (1) to degree of condemnation; (2) to whether there are duties not to kill and not to let die; (3) to the strength of the two duties relative to each other and to other duties; (4) to the efforts required to meet each duty.

I think Brody is correct to consider these various dimensions to which the killing/letting die distinction could make a difference.[3] For it might be that the

M.J. Cherry and A.S. Iltis (eds.), Pluralistic Casuistry, 211–218.

efforts required to fulfill the duty to aid and the duty not to kill are equally high considered one at time, and yet the duty not to kill would take precedence over the duty to aid if one had to choose which to perform.

However, Brody is also aware that if, in a given circumstance, we had to make a greater effort to fulfil one duty than another, it still might be true that in the same circumstance we should choose to carry out the duty judged less strenuous by the efforts test rather than the duty judged more strenuous by the efforts test.[4] For example, I may have to make a great effort to keep a business obligation but it would be supererogatory of me to make as great an effort to save someone from death. Yet, if the choice arose as whether to keep the business obligation (at minor effort) or to save someone's life (at minor effort), I should do the latter. This phenomenon raises the problem that the four dimensions he describes, on which killing and letting die might differ, might point in different directions as to which form of conduct is morally more significant. If so, then without further explanation, testing on the dimensions would not settle the question of which form of conduct was weightier. I do not, in fact, believe that killing and letting die yield conflicting results when tested on these various dimensions. (For example, I do not think that we should choose to save a life rather than not kill, at least when the person in both cases is currently independent of us and other factors are held constant.) However, I will not prove that here.

Using the measures of how great an effort we have to make to avoid killing versus to save life, Brody thinks that not killing is shown to be the weightier duty. He says (1996, p. 161) that if someone will kill you unless you kill someone else, you may not kill the other person. However, if someone will kill you unless you let someone die, you may leave the person to die. My concern with Brody's argument here is that in the killing case, the death would be intended as a means, but in the letting die case, it is not clear that the death would be intended rather than merely foreseen. Brody has not clearly equalized all factors besides killing and letting die in his two cases and, therefore, we cannot tell if it is only that distinction that makes the moral difference. To avoid this problem, we might hold all other factors constant, even including intending death as a means. For example, may we let someone die when we aim at his death as a mere means to saving our life, because only if the person is dead will a villain not kill us? Alternatively, we could hold not intending death constant in both cases. For example, we may let someone die rather than run a great risk to our life to save him. But may we not run a great risk to our life to avoid doing an act that we foresee will certainly kill someone? I think not.

B.

What is the clinical relevance of this moral distinction between killing and letting die? Brody argues (p. 162) that one implication is that while we need not fund all sorts of lifesaving treatments, we should not kill people in order to avoid paying for their support. (Since he thinks that withdrawing treatment is letting die, this seems to commit him to the view that it is permissible to terminate lifesaving treatment in patients because we do not want to pay for their support. Arguably, some may resist the permissibility of this bedside decision even if they agree that we need not invest at a social level so that lifesaving treatments are available.)

Is Brody entitled to reach this conclusion? For recall again, in order to be sure that it is the killing/letting die distinction that is accounting for our different judgments, we must hold all other factors constant. When we refuse to fund lifesaving, we foresee the deaths, we do not intend them. By contrast, if we were to kill people to avoid paying for their treatment, we would be intending their deaths. Some might, therefore, conclude that it is not the killing/letting die distinction on its own that accounts for his conclusion. To deal with this objection, Brody might consider a case of letting die that involved intending death to compare with the killing case that involved intending death. So, would it be permissible to let someone die as a means of getting their organs, if the organs could be used to save other people instead of our having to use funds for more expensive life saving treatments? If not, then the clinical implication Brody wishes to derive from the difference between killing and letting die would not hold. Now consider a case where the killing involves foreseen death rather than intended death, to match the merely foreseen death in a letting die case. Would it be permissible not to spend money on improving a medical procedure that we foresee will otherwise cause unintended deaths? If spending this money to prevent doctors unintentionally killing people were more important than spending money on life-saving equipment, this would be a clinical implication to support Brody's view. But if spending the money on avoiding unintentional killing were no more important than spending money on life-saving aid, we would still not have derived a clinical implication from the fact that we should make greater efforts to avoid killing than to save life. Indeed, it would show that the effort (or cost more generally) test does not always work to distinguish killing and letting die.

My point in considering these cases is to raise a methodological concern that Brody does not pay enough attention to holding all factors aside from killing and letting die constant. It is only by doing this that one can conclude that it is this distinction and not some other that is making a moral difference.[5]

C.

If there is a moral distinction between killing and letting die on some dimensions, what significance does it have for the issue of voluntary euthanasia, where this is assumed to involve intending death on the grounds that it is in the interest of the patient. (I am here concerned with Brody's discussion of euthanasia and so will not discuss any views he might have on physician assisted-suicide, which is different from active euthanasia and may differ from passive euthanasia.) Brody first argues that any duty to give life-saving aid disappears when a competent adult makes clear that she does not want the aid and waives her right to it. Importantly, Brody wants to argue that this will be true even if she and the doctor, who does not give the aid, both intend the patient's death because they believe it is in her interest. That is, he agrees that the patient need not reject the aid merely on the grounds that it itself is too unpleasant or intrusive (which would then not involve seeking death as in euthanasia), but on the grounds that she seeks passive euthanasia because life itself is too burdensome (p. 165).

Then he makes the important point that even if killing is morally distinct from letting die in some ways, this does not show that the right not to be killed may not be waived, just as the right to be aided can be waived. This might leave the way open for the permissibility of voluntary active euthanasia, because it may imply that the duty not to kill is no longer in force, if the duty is merely the correlative of a waivable right not to be killed. Similarly, the duty to aid was no longer in force if someone waived his right to be aided, thereby allowing for passive euthanasia. Hence, he claims, we do not have to argue for voluntary active euthanasia on the grounds that passive euthanasia is permissible *and* there is no moral difference between killing and letting die. Nor do we have to conclude that voluntary active euthanasia is impermissible merely on the grounds that there is a moral difference between killing and letting die.

I agree with this argument, but I would like to point out two ways in which it does not go far enough. First, Brody's argument claims that it is permissible not to provide aid when a patient waives his right to it. The stronger additional conclusion is that it is *impermissible* to provide the aid when the competent adult patient waives his right and also makes clear he does not want the aid. Waiving a right to aid need not always imply absence of a desire for it. For example, someone may waive his right in order that you will be free to decide whether to aid him or another person, but may also be very happy if you chose to aid him. The person who rejects aid because he intends to die, can make it impermissible for a doctor to aid because aiding would involve interfering with the person against his will. It would also involve acting against the interests of the person, if death is indeed in his interest. Hence there is often a duty to let die, rather than just a permission to do so. By contrast, suppose the right not to be killed were waived and the only grounds for a duty not to kill were such a right, and the patient desired to be killed, and it was in his interest to die. It would still not be true that there is a duty to kill comparable to the duty to let die. This is, in part, because if we do not kill someone, we are not interfering with him as an autonomous being, though we may not be promoting his autonomy in the sense of helping him carry out his wishes.

The second point is that when a competent adult waives a right and also desires to die, it still may not be in his interests to die. Nevertheless, I think we still have a duty not to interfere by providing life-saving aid. (This will not be euthanasia.) However, if he waives a right not to be killed and desires to be killed, I do not think it is necessarily even permissible to kill him if this would be against his interests. As Philippa Foot noted (1977), we have a duty not to violate someone's rights but we can also have a duty not to act against her interests even when this would not violate her rights. This, however, will not be an argument against voluntary active euthanasia, as euthanasia only involves killing that is in the interest of the person killed.

Despite his argument that the great strenuousness of the duty not to kill does not show that the right correlative to it cannot be waived, Brody is not sure that voluntary active euthanasia is permissible. He suggests that we investigate whether there are grounds for the wrongness of killing that go beyond any right of the person not to be killed. (So though we would not be wronging the person in killing, we would still be doing wrong.) It may also be, he suggests, that the right not to be killed is not waivable, even if the mere difference in strenuousness of the duty not to kill and the duty not to let die does not show this. I think there is reason to doubt the last claim. For the denial of waivability suggests that it could never be a necessary condition (even if not a sufficient condition) for killing someone that he has agreed to be killed. In other words, it suggests the claim that whenever other factors weigh in favor of killing someone, it could never be true that the fact that he has not agreed to be killed could stand in the way of killing him. I doubt this is true.

II.

I have discussed Brody's views on the question of whether there is a moral difference between killing and letting die and the possible clinical significance of this distinction. Now let us consider his views on how to draw the killing/letting die distinction. Brody believes that withholding and withdrawing life support involve letting die, regardless of the intention with which one does it, and giving a lethal drug involves killing. He thinks this is our intuitive judgment and the role of ethical theory is to provide a more precise characterization of the killing/letting die distinction that does not undermine these initial judgments. He considers two views: (1)that an act or omission that involves intending someone's death earlier than it would otherwise have occurred is a killing, and (2) that killing but not letting die involves causing someone's death earlier than it would have occurred. (His focus on bringing about an earlier death seems incorrectly to exclude the possibility that one kills someone at exactly the time that he would have died anyway. I shall ignore this issue.)

Consider his first claim. Brody correctly, I believe, argues against the view that intention makes for a killing. His grounds for doing so, however, are merely conservative. That is, he assumes (p. 169) that we need an account that makes DNR, not giving antibiotics, and other withdrawals of treatment that may result in death, be lettings die, even when there is an intention to have the patient die for their own good. (Possibly, he will also want it to come out true that these are permissible lettings die rather than impermissible ones.) A nonconservative reason to reject the intending account of killing is that there are clear cases of killing that do not involve intending death; for example, when one runs over someone because one is determined to get somewhere fast. In the clinical context, giving morphine for pain relief when one foresees with certainty that it will also stop the heart is a killing that is often considered permissible. Consider also a case where the omission to aid A is justified by the duty to save B, C and D instead; and yet, the agent who omits to aid A fulfils his duty to the greater number instead of helping A only because he has recognized A as his enemy, and intends his death. Shall we say in this case that the agent kills A in virtue of his intention, even though another agent who would save B, C and D instead of A, not intending A's death, would not kill A? I think not.

Consider his second claim. Brody considers the view that says killing involves *causing* earlier death, whereas letting die is only a necessary condition of an earlier death whose cause is the underlying disease condition of the patient. Some of his concerns with this account are that: (a) It is not clear how to draw the distinction between causes and necessary conditions. (b) Withdrawals of food result in death from starvation, not from the underlying disease condition. Is this then a killing? (c) Most withdrawals of aid are now considered lettings die but it is not clear that they will come out as such on the cause versus necessary condition account. (d) How is one to deal with the fact that when a doctor deliberately withdraws life-saving treatment intending death we are thought to have a letting die, but when a greedy nephew does the same, we are said to have a killing?[6] Here, Brody provides the following (nonconservative) answer: The greedy nephew does not kill. "He brought about conditions in which the patient's underlying medical problem caused the death and from a reprehensible motive" (p. 169). This, Brody thinks, is as morally bad as a killing.

I think that Brody's analysis of the killing as causing earlier death view is not correct. Let us start with his last point. Suppose the doctor who deliberately withdraws treatment, intending death, does so from as reprehensible a motive as the greedy nephew. Is what she does as morally bad as what the nephew does? I do not think so, even if what she does is impermissible because she has a duty to provide life support. Furthermore, I think that what the nephew does is kill someone, while the doctor lets die (perhaps impermissibly). I believe that what accounts for the difference between the nephew and the doctor is that the doctor stops assistance that she herself (or some entity whose agent she is) is providing.[7] By contrast, the nephew is interfering with assistance that someone else is providing. This means that such things as proprietary rights over the aid (i.e., am I stopping my aid or someone else's for whom I am not an agent) can be crucial to determining if someone is killing or letting die. (Note, that on this account when a person himself pulls out the doctor's machine and dies of an underlying disease, he is not killing himself but letting himself die because he is removing *himself*—something he has rights over that is required in order that assistance be given—from the process.)

Consider the cases where withdrawal of treatment is a killing even when the patient dies from the underlying disease condition (such as the nephew's interference with aid). Does the agent cause the patient's death? If not, the idea that killing necessarily involves causing death will be wrong. I suggest that a distinction might be drawn between causing death and introducing the cause of death. Hence, Brody's presentation of the causing death/providing a necessary condition for death may be too simple. When the nephew or the doctor withdraw treatment, I think they both cause death by removing protection against it, but that need not mean that they introduce the cause of death. If they removed food, I think they would also remove protection from the cause of death (starvation), even though the patient does not die of an underlying disease. When the doctor removes the things he is providing, though he causes death, he does not kill because he neither introduces the cause of death nor does he interfere with life sustaining procedures someone else is providing. When the nephew causes death by removing what interferes with it, he also kills (as argued for above) though he does not introduce the cause of death.

Sometimes withdrawing treatment can also introduce the cause of death which can be referred to as inducing death. For example, suppose that in a hospital there is faulty wiring and when the doctor unplugs life support an electric shock is produced that causes the patient's death before anything else can (Faulty Wiring Case).[8] In this case, the doctor kills the patient because he introduces the cause of death. I believe that it may be no less permissible for him to do this than to withdraw the treatment when there is no faulty wiring, even if voluntary active euthanasia is not, in general, permissible. This can be because, if the patient has requested to be disconnected, he should not be required to remain connected to treatment he does not want merely because the shock will cause his death before anything else will.

But suppose the patient did not request the disconnection. In the Faulty Wiring Case, as in the case where the doctor withdraws the treatment and the wiring is not faulty, the patient will only lose out on life he would have had by way of the doctor's life support system. This is a crucial part of what makes the killing in the Faulty Wiring Case have the same moral status as letting die, whether permissible or impermissible.[9]

Hence, I think that Brody may be wrong not to distinguish between causing death and introducing the cause of death, he is wrong not to distinguish whether something is a killing or a letting die on the basis of a proprietary relation to the life support, and he is wrong to focus on contrasting motives of an agent (e.g., a nephew versus the doctor) as the basis for determining how morally bad the termination of aid is.

NOTES

[1] My discussion is an examination of the arguments in his "Withdrawal of Treatment versus Killing of Patients" (1996). All references to Brody, unless otherwise noted, are to that article.

[2] This follows the order in which he himself discusses these issues in the article I am examining.

[3] I discuss similar dimensions on which to test the killing/letting die distinction in my "Killing and Letting Die: Methodological and Substantive Issues" (1983) and *Morality, Mortality,* Vol. 2 (1996).

[4] He cites my discussion in "Supererogation and Obligation" (1985), where I tried to prove this.

[5] I emphasize this in my work cited in note 3. In drawing attention to the intention/foresight distinction, I do not mean to suggest that I think that an agent's intention determines the permissibility of an act. It may be the role of death as a causal means or as the only possible effect of an act, whether it is intended or not, that has a role in determining permissibility. I am only concerned with suggesting that Brody has not answered questions that need to be answered.

[6] Brody cites Shelly Kagan in regard to this question.

[7] I have argued for this view in, for example, "Killing and Letting Die: Methodological and Substantive Issues" and *Morality, Mortality,* Vol. 2.

[8] I first presented this case in "Ronald Dworkin on Abortion and Assisted Suicide" (2001).

[9] Above I noted that Brody accepts that a duty which is more strenuous by the measure of how much effort we must make to perform it can be weaker by the measure of which duty we ought to perform when we must choose. He says this may make trouble for his argument against abortion, which is based on the idea that the duty not to kill is stronger than the duty to aid (as measured by the efforts test). But I think that what really makes trouble for his argument against the permissibility of abortion (based on the idea that the duty not to kill is stronger than the duty to aid) is the fact that in being killed the fetus only loses life it would have received by way of the woman's life support system. This will make killing it in some ways analogous to what happens when the doctor kills the patient who does not want to die in the Faulty Wiring Case. However, what the doctor does may be as impermissible as his not fulfilling a duty to aid a patient who wants aid. By contrast, the woman (herself or through an agent) who would kill in an abortion, may have no duty to provide the aid whose termination involves killing. For more on this analysis of abortion see my *Creation and Abortion* (1992).

BIBLIOGRAPHY

Brody, B. (1996). 'Withdrawal of treatment versus killing of patients,' in T.L. Beauchamp (Ed.), *Intending Death: The Ethics of Assisted Suicide and Euthanasia.* Upper Saddle River, NJ: Prentice Hall.

Foot, P. (1977). 'Euthanasia,' *Philosophy & Public Affairs*, 6(2), 85–112.

Kamm, F.M. (1983). 'Killing and letting die: methodological and substantive issues,' *Pacific Philosophical Quarterly*, 64, 297–312.

Kamm, F.M. (1985). 'Supererogation and obligation,' *The Journal of Philosophy*, 82, 18–38.

Kamm, F.M. (1996). *Morality, Mortality*, vol. 2. Oxford: Oxford University Press.

Kamm, F.M. (2001). 'Ronald Dworkin on abortion and assisted suicide,' *The Journal of Ethics*, 5(3), 221–240.

Kamm, F.M. (1992). *Creation and Abortion.* New York: Oxford University Press.

SECTION V

RESPONSE TO FRIENDS AND CRITICISMS

CHAPTER 15

COMMENTS ON THE ESSAYS

BARUCH A. BRODY

Center for Medical Ethics and Health Policy, Baylor College of Medicine, Houston, Texas and Department of Philosophy, Rice University, Houston, Texas

I. OPENING COMMENTS

When Ana and Mark first contacted me to raise the idea of this volume, I was very honored but also quite concerned. Being honored is easy to understand; it is hard to imagine a greater honor than the desire of your students and your colleagues/friends to produce such a volume. The concern is also not that hard to understand; we have all seen far too many volumes like this consisting of many essays of modest value unrelated to each other and to the work of the honoree. After some reflection, I agreed to the project with the request that the essays relate to my work, raising critical suggestions and offering extensions and/or alternatives. As I reviewed the essays in this volume, it was clear that this request has been fully honored, so I want to thank the contributors both for their work on this volume and for their respecting my request. Naturally, special thanks are owed to Ana and Mark for their extensive efforts in bringing this project to completion. Perhaps it would help in thanking them, as well as the many contributors, if I add that the appearance of such a volume after the last few very hard years is the finest tribute I can imagine.

In my office, I have a plaque given to me by Maureen when she finished her studies quoting Rilke on the teacher's task to transform his students into "many different human beings." Maureen went on to thank me for encouraging her "independence of mind." That plaque expresses an approach to teaching that has been my approach for the 40+ years that I have been a full time professor. I learned it from Peter Hempel in my first semester at Princeton. He assigned us an essay written by a group of students in a previous semester which raised a serious technical problem with his theory of explanation, and he seemed delighted with their critical independence of mind. The contrast between his attitude and the attitude of the head of the rabbinical seminary I had just left, who seemed to me to demand reverent acceptance of his views, was great, and I decided to follow Peter. The essays in this volume convince me that this happened, and that delights me.

In this spirit, I turn to the essays written both by these students and by colleagues/friends. I will agree with some of their points, disagree with others, and

M.J. Cherry and A.S. Iltis (eds.), Pluralistic Casuistry, 221–231.

propose alternatives that hopefully incorporate the strengths of both viewpoints. This is how progress is slowly made in philosophy. Respectful but critical dialogue leads to a better understanding of the issues and of the alternative possibilities for resolving them.

II. PLURALISTIC MORAL THEORY

My fundamental commitment in moral theory is a commitment to moral pluralism, to the view that moral conclusions about the rightness or wrongness of actions are justified by arguments whose premises are a variety of different and independent moral considerations (moral appeals). Or to put it another way, there are a variety of different and independent right-making characteristics of actions. In earlier years, I stressed the appeal to consequences, to a variety of procedural or substantive rights, to considerations of justice, and to the virtuous character traits reflected by the action. I have more recently added to that list appeals to special obligations and appeals to deontological side constraints. But whatever the full list, this approach stands in sharp contrast to monistic moral theories (e.g., utilitarianism, natural rights theories, Kantianism) that stress the legitimacy of only one moral appeal.

Moral pluralism is attractive for two very different reasons. To begin with, it offers a very plausible account of the history of moral philosophy in the last three centuries. A striking fact is that during that period a wide variety of monistic theories were advocated, each of which had considerable plausibility but each of which was found lacking. On the pluralistic account, this is all very understandable. Each of the theories stressed a legitimate moral appeal, so each had considerable plausibility. But each of these theories maintained that the stressed moral appeal was the only legitimate moral appeal, and that is why they were found lacking. Secondly, moral pluralism offers a very plausible account of deep intra and inter personal moral ambiguity (moral ambiguity which remains even after there is total agreement about the morally relevant facts). This type of moral ambiguity arises from the uncertainty about which of the relevant moral appeals has greater significance in a given case.

Kevin offers an interesting metaphor for the metaphysical claims of moral pluralism. He suggests that the moral world of the pluralist is like a room with different types of furniture. That metaphor strikes me as very helpful, and I would like to explain why, drawing upon what I have learned about these matters from my wife Dena, who is an interior designer. One can design rooms with many different types and styles of furniture whose total effect is still very pleasing and harmonious. It requires, however, that the pieces be suited to each other. But if they sharply conflict with each other, the results can be jarring and disturbing. On my account, which type of room is the moral world? Given the centrality of deep moral ambiguity in my theory, I cannot help but suspect that it is the latter type of room. At least, it is for those, like myself, who would prefer a less ambiguous moral world. And that is a shame, given that we began with perfectly pleasing bits of furniture.

The change in the list of moral appeals accepted in my pluralistic theory offers an approach to addressing the important points raised by Larry and Haavi. In particular, the addition of special obligations enables me to incorporate Larry's concept of professional obligations, while the addition of deontological side constraints enables me to incorporate Haavi's suggestion about research ethics.

Larry has been in the forefront of the movement to restore professional medical ethics to a central place in discussions of medical ethics. His efforts can be divided into two interrelated components: a ground-breaking study of the introduction of that notion in the writings of Gregory and Percival and a series of writings in which he applies that concept to resolving various controversies in medical ethics. His suggestion is that a pluralistic moral theory must be supplemented by a pluralistic professional medical ethics, one which contains both principles and virtues.

I have always been concerned about these claims. Larry has quite correctly pointed out the fallacy of attempting to derive them from some essentialist account of the nature of medicine. But what then is their basis? Is it some professional consensus (does that exist, and if it does, why should any individual physician feel bound by it)? Is it some societal imposed rules (and if so, doesn't their validity depend upon the justification for the rules)? And what then is their content? In his essay in this book, Larry opines that physicians are required in appropriate cases to at least discuss the option of termination of a pregnancy and that physicians are required to refuse to implement rationing decisions that require foregoing evidence-based standards of care. How does he know that these are consequences of a proper pluralistic professional medical ethics?

But let us put these concerns aside for now. Let us suppose that there is such a professional medical ethics, generating a clearly understood set of obligations and virtues. It is likely that there are other professions (e.g., lawyers, accountants) that have a (perhaps partially different) set of obligations and virtues. And there are other non-professional relations (e.g., familial) that generate still another set. All of these should be covered in a general moral theory under the rubric of special moral obligations, and that is how I now treat them. If Larry is right about the existence of these professional obligations, then they can be incorporated into an appropriate general moral theory. But, of course, this way of thinking still must face the same issues raised above.

Many of the same points need to be made about Haavi's suggestions about the obligations of researchers to subjects. These are strong obligations, as Haavi points out, but they are not the obligations of a fiduciary. Haavi suggests that these obligations can be placed under the rubric of side-constraints, and that is a suggestion quite compatible with my revised version of pluralism, which incorporates such constraints.

There is, however, an important theoretical question raised by this suggestion, a question that has important practical implications. These obligations can be viewed as special obligations of researchers to their subjects or they can be viewed as deontological side-constraints on the behavior of researchers. There is an important difference between these two ways of categorizing the obligations of researchers. If they are conceived of as special obligations to the subjects, the subjects can release the researchers from those obligations. If they are conceived of as side constraints, then they may constrain the researchers no matter how the subjects feel. Saying that requires adopting a different approach to side-constraints than Nozick's, but it is the approach I am adopting. On this approach, deontological side-constraints are free-floating constraints in that they constrain certain types of behavior because they are impersonally wrong. (The behavior may also be wrong because it violates an obligation to an effected individual, but that is a separate matter.) Consider, for example, the proposed side-constraint of not diminishing biological diversity by destroying an entire species, even if no person is wronged.

There are, it seems to me, reasons for treating the obligations to research subjects as special obligations from which subjects can release the researchers. Consider Haavi's briefly-mentioned obligation to halt a research study in certain cases. Perhaps the clearest example of this is when the treatment group is suffering substantially more short term mortalities than the control group. It would be normal practice to stop the trial, supposing that the researchers have an obligation to the subjects to do so, an obligation usually built into consent forms. But suppose the researchers think that this trend may be reversed in the long run and they want to continue the research to test that hypothesis. Suppose moreover they re-consent present subjects and consent future subjects after telling them about this imbalance in the short term mortality rate. Suppose finally that the subjects agree to participate. In such cases, I would suppose that the trial may continue because the subjects have released the researchers from their obligation to halt the trial. If this is so, then it would be better to categorize these obligations as special obligations of researchers to their subjects, rather than as side-constraints on researcher behavior. In fairness to Haavi, however, it should be noted that she may not be building into deontological side-constraints the feature of being free floating.

A standard concern about pluralistic theories is how they deal with the problem posed by different appeals supporting doing conflicting actions in a given case. The question of whether to continue the just-described research protocol is a good example. The beneficial consequences for society of continuing the research combined with the consent of the subjects support continuing the research but free floating side-constraints on the behavior of researchers support stopping the trial. I have argued that there is no lexical priority of one appeal over the others and that talk of balancing the appeals rests upon a misleading metaphor; all that we can do is to make a judgment as to which appeals take precedence over the others in a given case. Sometimes, I have argued that this is a virtue of my approach, for it offers an explanation of both intra- and interpersonal deep moral ambiguity about particular cases. At other times, wanting to make my theory as action-guiding as possible, I have attempted to provide guidance for these judgments in the form of criteria as to when a given appeal has greater or lesser significance in a given case. There is nothing contradictory about adopting both of these attitudes, but it does reflect my own mixed feeling about this issue of conflicting appeals.

Janet offers an interesting suggestion to mediate between these mixed feelings. The suggestion builds upon two thoughts: (1) there are natural cognitive constraints on our judgments that lead ideal judgers to make the right judgment, and in making our judgments, we need to reflect upon what an ideal judger would decide; (2) the actual moral judgments we make (including judgments about what the ideal judger would decide) are distorted by various factors that make us less than ideal judgers and that is why we are left with intra and interpersonal deep moral ambiguity. This suggestion is at the end rejected by her, in part because of her doubts about the existence of these constraints and in part because of skeptical doubts as to whether judging under these constraints is judging in ways that are truth discovering.

I realize now that my introduction of the idea of natural constraints was very incomplete, and probably misleading, so let me try to do a little better. Sometimes, people say that this judgment approach can support any conclusion, because there are no constraints on what judgments will emerge. It is against this concern that I introduced the idea of natural constraints that limit the judgments that can honestly be made. It was meant to play only this very limited role of excluding some judgments, and nothing more. Even with these constraints, there are many remaining judgments that could be made, and that is why deep moral ambiguity exists. This does not mean, as Janet suggests as one of her possibilities, that there may be more than one morally correct judgment; it only means that we cannot be sure whether our judgment is the correct one. But it also does not mean that our judgments are not action guiding; it just means that we should remain open to revising our judgments, and therefore our actions, upon further reflection. None of this, of course, is a response to Janet's two legitimate skeptical concerns about natural constraints, and she may be right in thinking that the introduction of natural constraints was not helpful.

Maureen's concerns about these crucial intuitive judgments go beyond the purely abstract question of why we should trust them (the concerns she nicely explains in her section on the normative critique). In the following two sections, on the psychological and sociological critique, she succinctly summarizes a wide variety of reasons for mistrusting them. In doing so, she has deepened my worries about the judgment approach. But she also offers a direction for a way out. It involves empirical research to better define the psychological barriers faced by judgers and to better identify the institutions for decision making that are responsive to these limitations. It also involves incorporating the results of this research into the process of making judgments, attempting to mitigate the barriers to trustworthy judgments. I understand the first step better than I understand the second step, but this does seem to be a promising research project if we don't want to give up any sense of trust in the judgments we need to make.

III. MORAL EPISTEMOLOGY

I have always considered myself to be an intuitionist. For me, this means that the fundamental data for moral inquiry are the intuitions we have about the rightness and wrongness of particular acts in particular circumstances. But these intuitions are only the first epistemological step in developing a moral theory. The next step is hypotheses generation, in which we formulate generalizations that are supposed to systematize and explain the intuitions. In my moral theory, these are generalizations both about the validity of various moral appeals and about the factors that give the legitimate appeals greater or lesser significance in a given case. It is these generalizations that are then used to analyze troubling cases and to help generate the final judgments we have to make about those cases. We can make mistakes at any stage in this process, and because this is so, we should treat the products of each stage as tentative and open to revision.

This approach to moral epistemology is case-based (casuistric) in two very different ways. It begins all moral reflection with intuitions about particular cases that seem clear cut, and it sees as the end product of all of this moral reflection a judgment about a particular troublesome case with all of its troubling complexity. But it is important to remember that there is an in-between theorizing process. Laurie tells us about her colleague who wanted to create a data base of judgments the committee had made, from which a framework could be developed which could serve as the basis for judgments in future cases. Laurie argues that this is wrong because "The particularities of the case and the partialities of the family…all will vary." How different is that colleague's proposal from the theorizing process I advocate? It is hard to say without knowing more about the structure of her proposed data base. But I would not be as quick to dismiss her suggestion entirely; there is something there to be elaborated upon and developed into a plausible proposal. Laurie is one of the best reporters of cases (storytellers) in bioethics, in part because she rightly emphasizes the particularities of the case; my only point is that developing a theoretical framework to help deal with troubling cases may require abstracting away from some of those particularities.

It is important to note that this whole epistemological process is a "product of human reason," accessible to all those who want to reason about morality. It is also important to note that the results of this process are moral judgments "binding upon all," even if not all would accept them because they make different judgments. In these two ways, my approach to moral reasoning is similar to the approach to moral reasoning found in natural law theorists. But there are, of course, major differences: these include my emphasis on starting with intuitions about particular cases and my emphasis upon the tentativeness of the results of this process of moral reasoning. And there is my acceptance of moral obligations of God's people that are dictated by the commandments of God directed towards His people, and this is in tension with the rationality and universality aspect of the natural law tradition.

Andy is sympathetic to Novak's natural law approach to the universal components of Jewish ethics (the Noahide commandments), and I think now that he is right. Those seven commandments are just the sort of ethical constraints that might emerge from a pluralistic casuistry of the type I am advocating, without the invoking of any divine commandments. They would then be commandments binding on all, with their universally binding character knowable by natural reasoning processes of the type I have described. The differences that Andy notes, especially about tentativeness, would still remain, but that would not negate the similarities. They would be different from other obligations, ritual or moral, knowable only through divine commandment.

Let me take this last point one step further. It is a common (but by no means universal) Talmudic practice to begin a discussion by citing an accepted ruling in a given case. Attempts are then made to understand the basis of that ruling. This understanding is used to decide a different troubling case. Most crucially, in the cases I am thinking of, there is little if any explicit reference to divine commandment, neither in the initial ruling nor in the development of the general understanding. This is, of course, not all that different from the type of reasoning that you find in traditional common law opinions. And, most crucially, this type of reasoning is very similar to the casuistry I have been describing. It might well be that my advocacy of this way of doing moral philosophy grew out of my Talmudic training.

All of this sheds much light on the crucial issues raised by Tris, who is committed to emphasizing the lack of a purely rational morality in the Jewish tradition. This is certainly plausible in the case of those many commandments shaping the ritual life and of those moral commandments binding only upon God's people. But in light of my newer understanding, I am not so sure about this in the case of the basic moral prohibitions, those common to the Noahide law and to the full halachik tradition. To begin with, the Talmudic discussions often refer to extending the requirements imposed by those commandments beyond what is strictly required in light of the need to do what a righteous man would do. This latter point requires an appeal to some standard of behavior independent of the commandments, and this may be a purely rational component. Moreover, and more importantly, the Talmudic argument often appeals to purely rational arguments in the discussion of these basic moral prohibitions (e.g., outside of the defense context, you cannot kill an innocent party to save your life because "who says that your blood is redder than his.") So there may well be important components of a rational morality in Jewish tradition, even if other components are based on divine commandment.

IV. SUBSTANTIVE END OF LIFE ISSUES

One of the issues on which I have written extensively is end of life decision making. My earliest work in bioethics discussed some of these issues in the context of analyzing the ethics of abortion, but these views were developed more fully in my later work on the ethical issues I encountered in my clinical practice in consulting about dying patients.

Let me say a few words about my views on abortion and then relate them to Sarah's essay. I continue to hold the symmetry view, the view that the beginning of life as a human being with moral status and the end of life as a human being with moral status needs to be treated symmetrically. In my earlier work, misled by a simplistic understanding both of brain death and of developmental neurophysiology, I concluded that fetuses acquired that status at around six weeks (the traditional 40 days) after conception. I no longer believe that, in part because of some work that Amir Halevy and I did about the complexities of brain death, and I am really quite unsure on the point at which fetuses become human beings with moral status. But I am sure that we need to find an appropriate conceptual scheme for thinking about fetuses before then. Sarah's suggestion that thinking about ex utero spare embryos as potential adoptees is an interesting possibility for the context she is addressing, providing that we can do so without thereby treating them as human beings with moral status or without thereby prohibiting their use for research.

Over the years, my views on the broader end of life issues have changed somewhat, so perhaps it would be best to begin by summarizing my current views. I have always been committed to the view that there is a fundamental moral distinction between the obligation not to kill and the obligation not to let someone die. The former is a more demanding obligation, requiring much greater sacrifices. Nevertheless, even the obligation not to kill may be overridden, but not merely in order to save one's own life ("Who says that your blood is redder than his."). What more is required is a matter requiring much further analysis. Moreover, the obligation not to kill does not disappear just because the party to be killed requests that he be killed (so that request does not by itself justify the killing, and does not serve as a justification for voluntary active euthanasia). This is because the obligation not to kill is based not only on the right of each person not to be killed but also on the deontological side constraint of not killing; the request overrides the right but not the free-floating side-constraint. The obligation not to let someone die is much less demanding. I am not required to give one of my kidneys to a stranger in order to prevent him from dying, because I am not required to undergo the burdens of the surgery and the increased risk of hypertension over time in order to save a stranger's life. (An important question is whether there are people for whom I am required to undergo these burdens?) Moreover, in at least some cases, the obligation not to let someone die does disappear when the party in question requests that he be allowed to die. The comparative role of that request and of the person's objective situation in justifying letting them die is another matter requiring much further analysis. Finally, even if dying persons or those who speak for them request that measures be taken to stop them from dying, the obligation not to let them die may not exist. The role of their objective condition is relevant to the determination that the obligation does not exist, but even more relevant are the personal values of the potential savers of the life.

It is this last observation which represents the moral foundation of the Houston Futility Policy which I helped develop. The ethics committees of the major hospitals in Houston created a task force, chaired by Amir Halevy and myself, which developed a policy for an institution and its staff to refuse to provide requested life prolonging therapy. This policy has been adopted with modifications into Texas law. Notice, as many have not, that this is not a policy enabling an institution and its staff to prevent the patient from receiving the requested therapy; it only enables the institution and its staff to refuse to be the providers. Transfers to other institutions whose staff are willing to provide the care is explicitly allowed by our policy, and has occurred in a number of cases since the policy was adopted. (I am aware of several and have been personally involved in at least one.) Clint correctly points out that this is a policy designed to enable the institution and its providers to be faithful to their personal values (e.g., not producing uncompensated suffering, not providing unseemly care, responsibly stewarding limited resources). If they were required to provide the requested care, they would have to violate those values. This faithfulness is how I understand the virtue of integrity. But preserving integrity need not be compromised when other providers provide the requested care, and that is why transfers are allowed for under this policy. An interesting question, with many practical implications, is whether preserving integrity is compatible with facilitating a transfer. Clint correctly points out that integrity is not a virtue if the relevant personal value is objectively immoral, but that is not a problem for our policy since the above-cited values seem to be at least objectively acceptable.

This understanding enables me to respond to Bob's theoretical concerns. Bob's first theoretical objection is that we are appealing to some defensible notion of professional integrity, and he is skeptical that it exists. I agree with this skepticism, but that is not the notion of integrity to which our policy appeals. Our policy appeals to the virtue of providers being faithful to their personal values, even while recognizing that others in the same profession need not share those values. Institutions cannot have personal values, but those who create them and/or govern them may build their personal values into the mission of the hospital, and the institutions should then be faithful to those values. No notion of professional integrity is involved in our policy, although I suspect that we were not always as careful as we might have been to make that point clear. This also explains why transfers are appropriate; they are transfers of the patient to providers whose personal values are not challenged by providing the requested care. Finally, this also answers Bob's question as to why provider integrity wins over patient autonomy. It does not, because all that the policy does is to allow individuals not to provide care they find personally objectionable; the patient can still receive the care from other providers, and that has happened in some of our cases. If our policy had been in place in the Schiavo case, and if it were determined that the parents were the appropriate decision makers, then she would have been kept alive by care provided by other care givers.

Bob also has a number of practical concerns, and they are worth addressing in general terms (others have the rights to the numeric data summarizing Houston futility practice). There definitely needs to be extensive efforts at mediation before the policy is invoked; in fact, that is called for under the Houston policy (although not by the Texas law). It has been my experience that most cases are resolved through such a process; it is remarkable to me how few cases actually get to a full futility review. The process must be structured so as not to turn it into an automatic collegial endorsement of the views of the doctor. We have had a reasonable percentage of cases in which the physician's views have been rejected. And, as noted above, the transfer option has been a real one. No doubt, further improvements in process are possible, and that is on the agenda of the Texas legislature in 2007. Bob is right that it is not worth doing any of this to save costs; it is, however, worth doing it in a limited number of cases to preserve the value of personal integrity.

One of the issues over which providers and surrogates often disagree is the issue of artificial nutrition and hydration (ANH) for persistent vegetative (PVS) patients. Loretta quite correctly notes that I have suggested on a number of occasions that such care should be treated as a special case. Unfortunately, I often used language implying that it was special because it was "basic care" rather than a medical intervention. She offers a variety of cogent reasons for rejecting that classification of ANH, and I think that she is right. On other occasions, I argued that a failure to provide nutrition and hydration to a party dependent on you to supply it would be a form of killing (think of the parents who fail to feed an infant). That claim depends both upon a causal analysis of killing (to kill someone is to cause their death and not merely to not prevent their death) and upon a particular application of that analysis to feeding cases, and I am not satisfied with my attempts to articulate either, although I think that they are still viable possibilities. I would like to suggest another possibility. Loretta concedes that even PVS patients should be kept warm and dry, whatever the wishes of their surrogates, even though a failurc to do so cannot make them uncomfortable. Why? If, as I suspect, the answer to that question is that we have an obligation to human beings dependent upon us to meet certain of their *basic physiological needs,* then perhaps that justifies the claim that we must also meet their basic physiological need for nutrition and hydration, even by *non basic modalities of care*. Crucially, in response to Loretta's concern about an inconsistency in my views, allocation of resource considerations could take precedence over these obligations, and that may explain why rationing ANH may be justified; it might well be different if failing to provide ANH were killing. Also, since I agree with Loretta that the case of the patient being kept warm and dry shows us that surrogate wishes about the patient having these needs met is irrelevant, then the obligation to meet those needs must be grounded in a deontological side constraint against failing to meet them.

Frances raises a large number of important questions, and I find it helpful to begin by listing our points of agreement and then reflecting upon two major points of disagreement. Frances and I are in agreement that (a) there is a moral difference between killing and letting die, (b) this difference is reflected in a variety of ways (e.g., effort required to meet the obligation, degree of justifiable condemnation for failing to meet the obligation) (c) it could still be permissible in some cases to kill, and (d) this may occur in cases where the person has waived the right not to be killed.

The first of our major disagreements is about the role of intentions, where Frances ascribes to them greater significance that I would. All of this comes out in her discussion of equalizing all factors besides killing versus letting die. She asks: "May we let someone die when we aim at his death as a mere means to saving our life, because only if the person is dead will a villain not kill us?" Unlike her, I think that the answer to that question is that we may. It is still a letting die, and that is the crucial point. This may well be one of those cases of deep interpersonal moral ambiguity about a case leading to different theoretical conclusions. Similarly, my intuition is that letting anencephalics die as a means of getting hearts for babies with hypoplastic left hearts may well be permissible, even if killing them to get those hearts would not be, despite the fact that both the killing and letting die are intended as a means to get the hearts for transplants.

Disagreements both about particular intuitions and about the methodology of incorporating them into our moral theories lie behind our disagreements on how to draw the distinction between killing and letting die. To begin with, I am perfectly happy with her claim that I am insisting that any account of that distinction keeps DNR orders as orders to let die. But while she describes this as "merely conservative," I feel that it is a matter of trying to preserve widely held intuitions that underlie contemporary end of life practice, and that is what my moral epistemology calls for. But my moral epistemology also allows for rejecting some intuitions in order to preserve a good systematization, and that is why I develop the account I have of the doctor versus the nephew. I do explain the intuition that what the nephew does is very wrong, while insisting that it is still only a letting die by withdrawing life support. Many, myself included, also intuit that he is killing, and that is the intuition that I have to reject to maintain my systematization, But, unlike her, my intuition is that the doctor who withdraws the life support for motives as reprehensible as the nephew's is acting as badly as the nephew. That is why I reject her view that the proprietary relation to the life support is relevant. Again, my methodology explains the occurrence of deep interpersonal moral ambiguity both about particular cases and about systematizations.

Let me conclude by returning to the theme with which I began. As is evident from my responses, my students and my colleagues/friends have honored me by seriously commenting upon and challenging my work. For that, I am very thankful to them. And I am especially thankful to Ana and Mark for the work they did in producing a volume of this type. Finally, I think this volume illustrates the rabbinic claim that "much have I learned from my teachers, more from my friends, but the most from my students."

APPENDIX: SELECTED BIBLIOGRAPHY OF THE WORKS OF BARUCH A. BRODY

BOOKS

(1975) *Abortion and the Sanctity of Human Life*, M.I.T. Press, Cambridge.
(1980) *Identity and Essence*, Princeton University Press, Princeton.
(1988) *Life and Death Decision-Making*, Oxford University Press, New York.
(1995) *Ethical Issues in Drug Testing, Approval, and Pricing*, Oxford University Press, New York.
(1998) *The Ethics of Biomedical Research: An International Perspective*, Oxford University Press, New York.
(2004) *Taking Issue*, Georgetown University Press, Washington, D.C.

EDITED VOLUMES

(1968) *Science: Men, Methods and Goals*, W.A. Benjamin, New York.
(1970) *Reid's Philosophical Works*, M.I.T. Press, Cambridge.
(1970) *Moral Rules and Particular Circumstances*, Prentice-Hall, Englewood Cliffs.
(1970) *Readings in the Philosophy of Science*, Prentice-Hall, Englewood Cliffs.
(1973) *Logic: Theoretical and Applied*, Prentice-Hall, Englewood Cliffs.
(1974) *Philosophy of Religion: The Analytic Approach*, Prentice-Hall, Englewood Cliffs.
(1976) *Beginning Philosophy*, Prentice-Hall, Englewood Cliffs.
(1979) *Mental Illness*, D. Reidel Publishing Company, Dordrecht.
(1983) *Ethics and its Applications*, Harcourt Brace Jovanovich, New York.
(1986) *Bioethics: Readings and Cases*, with H. T. Engelhardt, Jr. (co-author and co-editor), Prentice-Hall, Englewood Cliffs.
(1988) *Moral Theory and Moral Judgments*, D. Reidel Publishing Company, Dordrecht.
(1989) *Readings in the Philosophy of Science*, second edition, with R. Grandy (co-editor), Prentice-Hall, Englewood Cliffs.
(1989) *Suicide and Euthanasia: Historical and Contemporary Perspectives*, D. Reidel Publishing Company, Dordrecht.
(1991) *Bioethics Yearbook - Vol. I*, with B. Andrew Lustig, L. McCullough, and H.T. Engelhardt, Jr. (co-editors), D. Reidel Publishing Company, Dordrecht.
(1992) *Bioethics Yearbook - Vol. II*, with B. Andrew Lustig, L. McCullough, and H.T. Engelhardt, Jr. (co-editors), D. Reidel Publishing Company, Dordrecht.

(1992) *Philosophy of Religion: The Analytic Approach*, second edition, Prentice-Hall, Englewood Cliffs.
(1993) *Bioethics Yearbook - Vol. III*, with B. Andrew Lustig, L. McCullough, and H.T. Engelhardt, Jr. (co-editors), D. Reidel Publishing Company, Dordrecht.
(1998) *Clinical Ethics in Surgery,* with L. McCullough and J. Jones (co-editors), Oxford University Press, New York.
(1998) *Readings in Social and Political Philosophy*, with G. Sher (co-editor), Harcourt Brace Jovanovich, New York.
(2000) *Medical Ethics: Codes, Opinions, and Statements*, with M. Bobinski, L. McCullough and M. Rothstein (co-editors), BNA Books, Washington, D.C.
(2001) *Medical Ethics: Introductory Essays*, with M. Bobinski, L. McCullough, and M. Rothstein (co-authors), BNA Books, Washington, D.C.

ARTICLES AND BOOK CHAPTERS

(1967) 'Natural Kinds and Real Essences,' *Journal of Philosophy* (64)14, 431–446.
(1967) 'The Equivalence of Act and Rule Utilitarianism,' *Philosophical Studies* 18 (6), 81–87.
(1968) 'Confirmation and Explanation,' *Journal of Philosophy* 65(10), 282–289.
(1969) 'Choosing and Doing,' *Philosophical Studies* 20(6), 92–95.
(1971) 'On the Ontological Priority of Physical Objects,' *Nous* 5(2), 139–155.
(1971) 'Is There a Philosophical Problem about the Identity of Substances?' *Philosophia* 1(1-2), 43–59.
(1971) 'Abortion and the Law,' *Journal of Philosophy* 68(12), 357–369.
(1971) 'Reid and Hamilton on Perception,' *Monist* 55, 422–441.
(1972) 'De Re and De Dicto Interpretations of Modal Logic,' *Philosophia* 2(1-2), 117–136.
(1972) 'Towards an Aristotelian Theory of Scientific Explanation,' *Philosophy of Science* 39, 20–31.
(1972) 'Sommers on Predictability,' *Philosophical Studies* 23(1–2), 138–140.
(1972) 'Thomson on Abortion,' *Philosophy and Public Affairs* 1(3), 335–340.
(1972) 'Locke on the Identity of Substances,' *American Philosophical Quarterly* 9, 327–334.
(1973) 'Why Settle for Anything Less than Good Old-Fashioned Aristotelian Essentialism,' *Nous* 7, 351–365.
(1973) 'Abortion and the Sanctity of Human Life,' *American Philosophical Quarterly* 10, 133–140.
(1974) 'An Impersonal Theory of Personal Identity,' *Philosophical Studies* 26(5-6), 313–329.
(1974) 'More Confirmation and Explanation,' *Philosophical Studies* 26(1), 73–75.
(1974) 'Morality and Religion Reconsidered,' in B. Brody (ed.), *Philosophy of Religion: The Analytic Approach*, second edition, Prentice-Hall, Englewood Cliffs, pp. 592–603.

(1975) 'Voluntary Euthanasia and the Law,' in M. Kohl (ed.), *Beneficent Euthanasia*, Prometheus Books, Buffalo, pp. 218–232.
(1975) 'On the Humanity of the Fetus,' in R. Perkins (ed.), *Abortion: Pro and Con*, Schenkman, New York, pp. 35–46.
(1975) 'The Reduction of Teleological Sciences,' *American Philosophical Quarterly* 12, 69–76.
(1975) 'Fetal Humanity and the Theory of Essentialism,' in F. Ellison (ed.), *Sex and Philosophy*, Prometheus Books, Buffalo, pp. 338–355.
(1976) 'Reid, Hume, and Kant,' in S. F. Barker and T. L. Beauchamp (eds.), *Thomas Reid: Critical Interpretations*, Temple University Press, Philadelphia, pp. 8–13.
(1977) 'Kripke on Proper Names,' *Midwest Studies in Philosophy* 2, 64–69.
(1977) 'Szasz on Mental Illness,' in H.T. Engelhardt, Jr. and S.F. Spicker (eds.), *Mental Health: Philosophical Perspectives*, D. Reidel Publishing Company, Dordrecht, pp. 251–257.
(1977) 'Leibniz's Metaphysical Logic,' *Rice University Studies* 63(4), 43–55.
(1978) 'Religious, Moral, & Sociological Issues: Some Basic Distinctions,' *Hastings Center Report* 8 (4), pp. 13.
(1978) 'The Problem of Exceptions in Medical Ethics,' in R. McCormick and P. Ramsey (eds.), *Doing Evil to Achieve Good*, Loyola University Press, Chicago, pp. 54–68.
(1979) 'Intuitions as a Source of Objective Moral Knowledge,' *Monist* 62, 445–456.
(1980) 'Morality and Rational Self-Interest,' in J. Arthur (ed.), *Morality and Moral Controversies: Readings in Moral, Social, and Political Philosophy*, Prentice-Hall, Englewood Cliffs, pp. 8–15.
(1981) 'Health Care for the Haves and Have-Nots,' in E. Shelp (ed.), *Justice and Health Care*, D. Reidel Publishing Company, Dordrecht, pp. 151–159.
(1981) 'Work Requirements and Welfare Rights,' in *Moral Foundations of Income Support Policy*, Littlefield & Adams, Totowa.
(1981) 'Faith Healing for Childhood Leukemia,' *Hastings Center Report* 11 (1), 10–11.
(1981) 'Marriage, Morality, & Sex Change Surgery,' *Hastings Center Report* 11 (4), 8–9.
(1983) 'Legal Models of the Patient-Physician Relation,' in E. Shelp (ed.), *Clinical Encounter: The Moral Fiber of the Patient-Physician Relation*, Kluwer Academic Publisher, Dordrecht.
(1983) 'Towards a Theory of Respect for Persons,' *Tulane University Studies* 31, 61–76.
(1983) 'The Use of Halachik Material in Discussions of Medical Ethics,' *Journal of Medicine and Philosophy* 8(3), 317–328.
(1983) 'Redistribution Without Egalitarianism,' *Social Philosophy and Policy* 1, 71–87.

(1985) 'The Interaction Between Ethics and Economics in Planning Health Care for the Aged,' in *Aging*, Springer-Verlag Publishing, Heidelberg.
(1985) 'The DNR Order in Teaching Hospitals,' with A. Evans (co-author), *Journal of the American Medical Association* 253, 2236–2239.
(1986) 'The Vitamin E Study: A Case Study in the Ethics of Experimentation,' in A. McPherson (ed.), *Retinopathy of Prematurity*, Decker Publishing Co., Hamilton, Ontario.
(1986) 'HCFA Data Release—Government Abuse or Patient Right,' *Hospital Physician* 22(6), 36–7, 40.
(1985) 'The International Defense of Liberty,' *Social Philosophy and Policy* 3(1), 27–42.
(1986) 'Practicing CPR Procedures,' in K. Iserson, A. Sanders, D. Mathieu, and A. Buchanan (eds.), *Ethics in Emergency Medicine*, Lippincott Williams and Wilkins, Philadelphia.
(1986) 'Resuscitating a Patient with no Vital Signs,' in K. Iserson, A. Sanders, D. Mathieu, and A. Buchanan (eds.), *Ethics in Emergency Medicine*, Lippincott Williams and Wilkins, Philadelphia.
(1986) 'Ethical Problems in the Placement of Patients,' *Current Concepts in Rehabilitative Medicine* 2, 17–81.
(1987) 'Medico-legal Considerations and Compliance Problems in the Critically Ill Substance Abuser,' with E. Boisaubin (co-author), *Problems of Critical Care* 1(1) 1139–1149.
(1987) 'Justice and Competitive Markets,' *Journal of Medicine and Philosophy* 12 (1), 37–50.
(1987) 'The Role of Private Philanthropy in a Free and Democratic Society,' *Social Philosophy and Policy* 4(2), 79–92.
(1987) 'Quasi-Libertarianism and the Laetrile Controversy,' in H. T. Engelhardt, Jr. and A. Caplan (eds.), *Scientific Controversies*, Cambridge University Press, Cambridge.
(1988) 'Ethical Questions Raised by the Persistent Vegetative Patient,' *Hastings Center Report* 18(1), 33–37.
(1988) 'Wholehearted and Halfhearted Care: National Policies vs. Individual Choices,' in S. Spicker, S. Ingman and I. Lawson (eds.), *Ethical Dimensions of Geriatric Care*, D. Reidel Publishing Co., Dordrecht, pp. 79–93.
(1988) 'The Macro Allocation of Health Care Resources,' in H.M. Sass and R.U. Massey, *Health Care System*, D. Reidel Publishing Co., Dordrecht.
(1988) 'Justice in the Allocation of Resources to Disabled Citizens,' *Archives of Physical Rehabilitation and Medicine* 69, 333–336.
(1988) 'Withholding Medical Treatment from the Severely Demented Patient: Decisional Processes and Cost Implications,' N. Wray and T. Bayer et al. (co-authors), *Archives of Internal Medicine* 148(9), 1980–1984.
(1989) 'Growing into Rights,' *Second Opinion* 10, 66–71.

(1989) 'Jewish Views on Suicide,' in B. A. Brody (ed.), *Suicide and Euthanasia: Historical and Contemporary Perspectives*, D. Reidel Publishing Company, Dordrecht.

(1989) 'The President's Commission: The Need to be More Philosophical,' *Journal of Medicine and Philosophy* 14 (4), 369–383.

(1989) 'Can Scoring Systems be Ethically Interfaced with ICU Patient Management,' *Problems in Critical Care* 3, 662–670.

(1989) 'An Evaluation of the Ethical Arguments Commonly Raised Against the Patenting of Transgenic Animals,' in W.H. Lester (ed.), *Animal Patents: The Legal, Social, and Ethical Issue*s, Macmillan Publishers, Ltd., New York, pp. 141–153.

(1989) 'The Baylor Experience in Teaching Medical Ethics,' *Academic Medicine* 64(12), 715–718.

(1990) 'Religion and the New Reproductive Technology,' D. Bartels, R. Priester, D. Vawter and A. Caplan (eds.), *Beyond Baby M: Ethical Issues in New Reproductive Techniques*, Humana Press, Totowa.

(1990) 'Quality of Scholarship in Bioethics,' *Journal of Medicine and Philosophy* 15 (2), 161–178.

(1990) 'Prediction of Dangerousness in Different Contexts,' in R. Rosner and R. Weinstock *Ethical Practice in Psychiatry and the Law*, Plenum Press, New York.

(1990) 'Ethics in Family Practice,' with W. Holleman (co-author), in R. Rakel (ed.), *Family Practice*, Saunders, Philadelphia, pp. 127–136.

(1990) 'The Economics of the Law of Rodef,' *Svara*, 1(1), 67–71.

(1991) 'Physicians and Rationing,' *Texas Medicine* 87(2), 86–90.

(1991) 'Ethical and Legal Issues in Clinical Pediatric Oncology,' in D. Fernbach and T. Vietti (eds.), *Clinical Pediatric Oncology, 4th Edition*, C.V. Mosby Company, St. Louis.

(1991) 'Why the Right to Health Care is Not a Useful Concept for Policy Debates,' in T. Bole and W. Bondeson (eds.), *Rights to Health Care,* Kluwer Academic Publishers, Dordrecht.

(1991) 'The Use of the Natural Death Act in Pediatric Patients,' *Critical Care Medicine*, with L.S. Jefferson, B.C. White, P.T. Louis, D.D. King, and C.E. Roberts (co-authors), *Critical Care Medicine* 19 (7), 901–905.

(1991) 'The Impact of Economic Considerations on Clinical Decision Making,' with N. Wray, S. Bame, C. Ashton, N. Petersen and M. Harward (co-authors), *Medical Care* 29(9), 899–910.

(1992) 'Scientific Responses to Moral, Cultural, and Religious Concerns,' in D.J. Roy (ed.), *Bioscience: Society*, John Wiley & Sons, Hoboken.

(1992) 'Special Ethical Issues in the Management of PVS Patients,' *Law, Medicine and Health Care* 20 (1-2), 104–115.

(1992) 'Ethical Issues Raised by the Clinical Use of Prognostic Information,' in Evans et al. (eds.), *Prognosis of Neurological Disorders*, Oxford University Press, New York.

(1992) 'Important Legal Concepts in Critical Care,' in J. Civetta, R. Taylor, and R. Kirby (eds.), *Critical Care, 2nd Edition*, JP Lippincott Company, Philadelphia.

(1992) 'Ethical Reflections on International Health Care Expenditures,' in E. Matthews and M. Menlown (eds.), *Philosophy and Health Care*, Avebury Publishing Company, Brookfield.

(1992) 'The Challenge of Follow-up for Clinical Trials of Somatic Cell Gene Therapy,' with F.D. Ledley, C.A. Kozinetz and S.G. Mize (co-authors), *Human Gene Therapy* 3(6), 657–663.

(1992) 'Case Managers and Eccentric Clients,' in A. Caplan and R. Kane (eds.), *Ethics and Conflict in Management of Home Care*, Springer, New York.

(1993) 'Hardwig on Proxy Decision-Making,' *Journal of Clinical Ethics* 4(1), 66–67.

(1993) 'The Physician as Professional and the Physician as Honest Businessman,' *Archives of Otolaryngology* 119 (5), 495–497.

(1993) 'Liberalism, Communitarianism, and Medical Ethics,' *Law and Social Inquiry* 18(2), 393–407.

(1993) 'Addressing Empirical Research in Ethics,' *Theoretical Medicine and Bioethics* 14(3), 211–219.

(1993) 'Brain Death: Reconciling Definitions, Criteria, and Tests,' with A. Halevy (co-author), *Annals of Internal Medicine* 119(6), 519–525.

(1993) 'Methodological and Conceptual Issues in Health Care System Comparisons,' with R. Lie (co-author), *Journal of Medicine and Philosophy* 18 (5), 437–463.

(1994) 'Acquired Immunodeficiency Syndrome and the Americans with Disability Act,' with A. Halevy (co-author), *American Journal of Medicine* 96(3), 282–288.

(1994) 'The Destruction of the Temple and the Rise of Judaism,' in V. Kealman (ed.), *Critical Moments in Religious History*, Mevcer University Press.

(1994) 'Is Cost a Barrier to Screening Mammography,' with C.I. Kiefe, S.V. McKay, and A. Halevy (co-authors), *Archives of Internal Medicine* 154 (11), 1217–1224.

(1995) 'Is Futility a Futile Concept,' with A. Halevy (co-author), *Journal of Medicine and Philosophy* 20(2), 123–44.

(1995) 'Limiting Life Prolonging Medical Treatment: A Comparative Analysis of the President's Commission and the New York State Task Force,' in R. Bulger, E. Bobby, and H. Fineberg (eds.), *Society's Choices: Social and Ethical Decision Making in Biomedicine*, Institute of Medicine, National Academy Press, Washington, D.C., pp. 307–334.

(1995) 'The Role of Futility in Health Care Reform,' with A. Halevy (co-author), in R. Mishbin et al. (eds.), *Health Care Crisis*, University Publishing Group, Frederick.

(1995) 'Pluralistic Moral Theory,' *Revue Internationale de Philosophie* 49(193), 323–340.

(1995) 'Ethical Issues Raised by Intensive Care,' in K. Wildes (ed.), *Critical Choices and Critical Care,* Kluwer Academic Publishing, Dordrecht.
(1996) 'Medical Ethics Education: Past, Present, and Future,' with E. Fox and R.M. Arnold (co-authors), *Academic Medicine* 70(9), 761–769.
(1996) 'The Low Frequency of Futility in an Adult ICU Setting,' with A. Halevy and R.C. Neal (co-authors), *Archives of Internal Medicine* 156(1), 100–104.
(1996) 'Conflicts of Interests and the Validity of Clinical Trials,' in R. Spence, D. Shim, and A. Buchanan (eds.), *Conflict of Interest in Clinical Practice and Research,* Oxford University Press, Oxford, pp. 407–417.
(1996) 'Diffusion of Innovative Approaches to Managing Hypoplastic Left Heart Syndrome,' with W. Caplan, T. Cooper and J. Garcia-Prats (co-authors), *Archives of Pediatrics and Adolescent Medicine* 150, 487–490.
(1996) 'Public Goods and Fair Prices; Balancing Technological Innovation with Social Well-Being,' *Hastings Center Report* 26(2), 5–11.
(1996) 'Resource Consumption and the Extent of Futile Care Among Patients in a Pediatric ICU Setting,' with R. Sachdeva and L. Jefferson (co-authors), *Journal of Pediatrics* 128(6), 742–747.
(1996) 'A Multi-institutional Collaborative Policy on Medical Futility,' with A. Halevy (co-author), *Journal of the American Medical Association* 276(7), 571–574.
(1996) 'Withdrawal of Treatment versus Killing and Letting Die,' in T. Beauchamp (ed.), *Intending Death*, Prentice Hall, Upper Saddle River.
(1996) 'Medical Futility: Philosophical Reflections on Death,' in K. Hoshino (ed.), *Japanese and Western Bioethics*, Kluwer Academic Publishers, Dordrecht.
(1996) 'The Interrelationship of Ethical Issues in the Transition from Old Paradigms to New Technologies,' with T. Cooper and J. Garcia-Prats (co-authors), *Journal of Clinical Ethics* 7(3), 243–250.
(1997) 'Ethical Issues Raised by Managed Care,' *Texas Medicine*, 93(2), 43–45.
(1997) 'In Case of Emergency No Need for Consent,' *Hastings Center Report* 27(1), 7–12.
(1997) 'Whatever Happened to Research Ethics,' in R. Carson and C. Burns (eds.), *Philosophy of Medicine and Bioethics*, Kluwer Academic Publishers, Dordrecht.
(1997) 'Research Ethics: International Perspectives,' *Cambridge Quarterly of Healthcare Ethics* 6(4), 376–384.
(1997) 'Dimensions of Quality of Life Expressed by Men Treated for Metastatic Prostate Cancer,' with J. Clark, N. Wray, C. Ashton, B. Giesler and H. Watkins (co-authors), *Social Science and Medicine* 45 (8), 1299–1309.
(1997) 'When are Placebo-Controlled Trials No Longer Appropriate?' *Controlled Clinical Trials* 18(6), 602–612.
(1998) 'Medical Futility and Scoring Systems in the Intensive Care Unit,' *Current Opinions in Critical Care* 4, 139–141.
(1998) 'Religion and Bioethics,' in H. Kuhse and P. Singer (eds.), *Companion to Bioethics*, Blackwell Publishers, Oxford, pp. 45–46.

(1998) 'Responding to Requests for Initiation of Dialysis in Dying Patients,' *Seminars in Dialysis* 11, 305–307.

(1998) 'The Houston Process-Based Approach to Medical Futility,' with A. Halevy, *Bioethics Forum* 14(2), 10–17.

(1998) 'Research on the Vulnerable Sick,' in J. Kahn, A. Mastroianni and J. Sugarman (eds.), *Beyond Consent*, Oxford University Press, New York, pp. 32–46.

(1999) 'How Much of the Brain Death Must Be Dead?' in S.Younger, R. Arnold and R. Schapiro (eds.), *The Definition of Death: Contemporary Controversies*, Johns Hopkins University Press, Baltimore, pp. 71–82.

(1999) 'Protecting Human Dignity and the Patenting of Human Lives,' in A. Chapman, *Perspectives on Genetic Patenting,* AAAS Press, Washington, D.C.

(1999) 'Assessing the Performance of Utility Elicitation Techniques in the Absence of a Gold Standard,' with B. Giesler et al. (co-authors), *Medical Care* 37 (6), 580–588.

(1999) 'Managing Disagreement in the Management of Short Bowel and Hypoplastic Left heart Syndrome,' with T. Cooper and J. Garcia-Prats (co-authors), *Pediatrics* 10(4), e48.

(2000) 'Jewish Reflections on Life and Death Decision Making,' in E. Pellegrino and R. Faden (eds.), *Jewish and Catholic Bioethics*, Georgetown University Press, Washington, D.C., pp. 17–24.

(2000) 'A National Survey of Policies on Disclosure of Conflicts of Interest in Biomedical Research,' with C.B. Anderson, S.V. McCrary, L.B. McCullough, R. Morgan & N. Wray (co-authors), *New England Journal of Medicine* 343(22), 1621–1626.

(2000) 'A Historical Introduction to the Requirement of Obtaining Informed Consent from Research Subjects,' in L. Doyal and J. Tobas (eds.), *Informed Consent in Medical Research,* BMJ Books, London, pp. 7–14.

(2000) 'A Trial for Comparing Methods for Eliciting Treatment Preferences from Men with Prostate Cancer,' with J. Souchok et al. (co-authors), *Medical Care* 38(10),1040–50.

(2001) 'Defending Animal Research: An International Perspective,' in E. F. Paul and J. Paul (eds.), *Why Animal Experimentation Matters,* Transaction Publishers, Somerset.

(2001) 'Making Informed Consent Meaningful,' *IRB: Ethics and Human Research* 23(5), 1–5.

(2001) 'A Psychometric Study of the Measurement Level of the Rating Scale,' with K. Cook et al. (co-authors), *Social Science and Medicine* 53(10), 1275–85.

(2002) 'The Ethical Recruitment of Subjects,' *Medical Care* 40 (4), 269–70.

(2002) 'Philosophical Reflections on Clinical Trials in Developing Countries,' in M. Battin, R. Rhodes, and A. Silver (eds.), *Justice in Research*, Oxford University Press, Oxford, pp. 197–211.

(2002) 'Freedom and Responsibility in Genetic Testing,' *Social Philosophy and Policy* 19, 343–59.

(2002) 'Pharmacogenetics: Ethical Issues and Policy Options,' with A. Buchanan, A. Califano, J. Kahn, E. McPherson and J. Robertson (co-authors), *Kennedy Institute for Ethics Journal*, 12(1), 1–15.

(2002) 'Pharmacogenetic Challenges for the Health Care System,' with J.A. Robertson, A. Buchanan, J. Kahn, and E. McPherson (co-authors), *Health Affairs* 21(4), 155–167.

(2002) 'Bioethics Consultation in the Private Sector,' with N. Dubler, J. Blustein, A. Caplan, J.P. Kahn, N. Kass, B. Lo, J. Moreno, J. Sugarman, and L. Zoloth (co-authors), *Hastings Center Report* 32(3), 14–20.

(2002) 'A Randomized Placebo Controlled Trial of Arthroscopic Surgery for Osteoarthritis of the Knee', with B. Moseley, K. O'Malley, N.J. Petersen, T.J. Menke, D.H. Kuykendall, J.C. Hollingsworth, C. M. Ashton, and N.P. Wray (co-authors), *New England Journal of Medicine* 347 (2), 81–88.

(2002) 'Ethical Dilemmas in Hyperbaric Medicine,' with E. Chan (co-author), *Underwater and Hyperbaric Medicine* 28(3), 123–130.

(2002) 'Ethical Issues in Clinical Trials in Developing Countries,' *Statistics in Medicine* 21, 2853–58.

(2002) 'Noncompliance with Medical Follow-up after Pediatric Intensive Care,' with M. McPherson et al. (co-authors), *Pediatrics* 109 (6), e94.

(2003) 'Is the Use of Placebo Controls Ethically Permissible in Clinical Trials of Agents Intended to Reduce Fractures in Osteoporosis,' with N. Dickey and S. Ellenberg et al. (co-authors), in *Journal of Bone and Mineral Research* 18(6), 1105–1109.

(2003) 'The AMA's Position on the Ethics of Managed Care,' in W.B. Bondeson and J.W. Jones (eds.), *The Right to Health Care*, Kluwer Academic Publishers, Dordrecht.

(2003) 'Expanding the Scope of Disclosures of Conflicts of Interest: The View of the Stakeholders,' with C. Anderson, S.V. McCrary, L. McCullough, R. Morgan and N. Wray (co-authors), *IRB* 25(1), 1–8.

(2004) 'The Ethics of Controlled Clinical Trials,' in G. Khushf (ed.), *Handbook of Bioethics*, Kluwer Academic Publishers, Dordrecht, pp. 337–352.

(2005) 'Consensus and Controversy in Clinical Research,' with L.B. McCullough, and R.R. Sharp (co-authors), *Journal of the American Medical Association* 294(11), 1411–1414.

(2006) 'Community Consultation in Emergency Research,' with C. Contant et al. (co-authors), *Critical Care Medicine* 34(8), 2049–2052.

(2006) 'Intellectual Property and Biotechnology: the U.S. Internal Experience – Part I,' *Kennedy Institute of Ethics Journal* 16(1), 1–37.

(2006) 'Intellectual Property and Biotechnology: the U.S. Internal Experience – Part 2,' *Kennedy Institute of Ethics Journal* 16(2), 105–128.

INDEX

Abbott, W. M., 125, 129
abortion, 2, 12, 30, 40, 110, 118–119, 127, 142, 193, 203, 205, 218, 227
Abraham, 114, 123
accountability, 33–34, 36
Adam, 124
adoption, 7, 139, 192–210
adultery, 91, 94, 110, 124
AIDS, 73
allocation of resources, 8–9, 39
American Academy of Neurology, 171, 177
American Academy of Pediatrics, 81, 176–177
American Medical Association, 18, 135, 169–170, 177
American Society for Bioethics and Humanities, 23, 143, 208
Andersen, A. N., 209
Andersen, C. Y., 209
Anderson, E., 189
Andrews, K., 138, 147
anencephalic infants, 9, 18
Angell, M., 134, 147, 150, 163
Anscombe, G. E. M., 117, 126, 129
Anselm of Canterbury, 126
Appelbaum, P. S., 157, 160, 163
Applegarth, L. D., 196, 210
Aquinas, T., 88–90, 94, 120, 126–127
Archimandrite Sophrony, 127
Aristotle, 24, 32, 44, 67, 78, 89–90, 128, 190
Arnold, R., 2, 14, 139, 144, 146–148, 229
artificial nutrition and hydration, 8–9, 14, 17, 43, 101, 168–177, 230
asceticism, 126
Ashton, M., 20
atheism, 117, 126
Athens, 116, 122
Atkinson, S., 137, 147
Aulisio, M., 144, 147
Austin, 1
Austin, J. L., 193, 208–209
authority, 2, 16, 24, 27–29, 31, 33, 42, 106, 111–112, 114–115, 118–119, 122, 128, 154, 170, 172, 176
autonomy, 11, 18, 25, 31, 35, 86, 134–136, 139, 143, 182, 214, 229
axiology, 2, 89–90

Baby Doe, 101, 177
Back, A. L., 139, 147
Baker, R., 27, 35
Bangsboll, S., 209
Barrett, J. F., 125, 129
Bartholet, E., 209
Bartholome, W. G., 70, 81
Bartholomew I, Ecumenical Patriarch, 114, 125, 129
Basil the Great, Saint, 127
Battin, M., 18, 178
Baylor College of Medicine, 15, 23, 35, 65, 109, 133, 221
Beauchamp, T., 23, 26, 35, 40–42, 44, 48, 70, 73, 81, 86, 97, 178, 218
Becker, G., 210
ben Noah, 119
Benson, P., 163
Berg, J., 160
Berkman, J., 203, 209
Berkowitz, L., 82
Bernasconi, R., 108
bestiality, 91, 94
best-interests standard, 176
Bihari, D., 147
bioethics, 1–3, 11–12, 15, 23–27, 29, 33–35, 37–47, 65–66, 69–74, 78–81, 85, 102,

104, 106–107, 109–110, 118–120, 122–124, 127, 129, 134–135, 146, 149, 167, 181, 193, 226–227
clinical, 65, 78
Jewish, 13, 85, 92, 96, 107, 226
secular, 13, 109
Birdzell, L. 16
Biros, M. H., 4, 18
Bishop, J. P., 25, 35
blasphemy, 91, 94, 110
bnai Noah, 12–13, 119–120, 122–123
Boedder, B., 125, 129
Bopp, J., 175, 177
Bornstein, H., 130
brain dead, 7, 17
brain stem functioning, 8
Brakman, S. -V., 2, 14, 16, 197, 200, 209–210, 228
Brandt, R. B., 62, 64
Braunack-Mayer, A., 80–81
Breheny, S. A., 192, 210
Bresson, J. L., 209
Brett, A. S., 135, 148
Britain, 28
Brock, D. W., 176, 178
Brody, B. A., 1–60, 62–66, 75–76, 81, 85–88, 90–97, 99, 101, 104–111, 118–120, 123–124, 129, 133–150, 157–158, 161, 163, 167–175, 177, 181–190, 193–194, 211–218
Brody, H., 148, 152, 158, 160, 164
Brubaker, S. C., 208
Buchanan, A. E., 19, 75, 78, 80–81, 176, 178
Buckley, M., 126, 129
Burant, C., 18, 20
Buridan's Ass, 79
Burton, P. J., 195–196, 199, 209
Butler, A., 210
Byock, I. R., 176, 178

Cahill, M. T., 154, 163
California, 48, 97, 101
Callahan, D., 135, 147
Campbell, C. S., 18–19
cancer, 100, 137, 147, 162, 164, 181
Caplan, A., 178, 195, 205–206, 208–209
Carnahan, S. J., 4, 19
Carson, T. L., 61, 62, 64–65
casuistry, 43–45, 47, 65–67, 69–72, 74–77, 80, 85–86, 88, 90, 96, 104, 208, 227
Jewish, 12
Cefalu, W. T., 160, 163
Center for Medical Ethics and Health Policy, 23, 35, 133, 221
Chambers, T., 108
Cherry, M. J., 1, 221, 231
Chervenak, F. A., 30–31, 36
children, 12, 42, 53–54, 70, 78, 91, 100–101, 103, 105, 108, 123, 128, 177, 187–188, 192–193, 195–206, 208–210
Childress, J., 40–42, 44, 48, 70, 73, 81, 86, 97
Chomsky, N., 53, 64
Christ, 112, 125
Christianity, 13, 110, 113–114, 116, 119, 123, 125–126
Orthodox, 13, 110–112, 114, 116–119, 122–125, 127, 129
Roman Catholic, 111–114, 123, 125–127, 129, 171
Western, 113–114, 116–117, 119, 123, 126
Chrysostom, Saint John, 125, 127
Churchill, L. R., 146–147
clinical equipoise, 149, 151, 161
clinical ethics, 133
Clouser, K. D., 26, 34, 35, 170, 174, 178
coercion, 3
cognition, 8
cognitive constraints, 12, 53–64
cognitive dissonance, 71–72, 77, 79
cognitive psychology, 67–68, 71, 77–78
cognitive science, 75
Coleman, C., 160
community consultation, 4, 73–74
compassion, 11, 31
competence, 11, 27–28, 31, 112
confidentiality, 139, 156
conflicts of interest, 3, 5–6, 34, 154, 158, 160, 162
conscience, 29–30, 34, 135
consciousness, 7–8, 128
consensus, 7–8, 65–66, 70–74, 76, 134, 141–142, 171, 173, 176, 223

consent, 3–4, 10, 14, 16, 19, 30, 70, 73, 102, 158–159, 169–170, 173, 176–177, 196, 201, 224
waivers of, 4–5
consequences, 2, 10, 17, 24–25, 32, 43–46, 48, 65–66, 70, 88, 126–127, 134, 139, 157, 162, 168, 176, 183–184, 186, 193, 222–224
Cooter, R., 162–163
cosmology, 125
Coss-Bu, J., 148
cost-effectiveness, 25, 33–34, 88
courage, 11, 79, 105, 114
Critchley, S., 108
Crone, R. K., 178
Cruzan, N., 70–71, 101
cultural diversity, 39–40
Culver, C., 170, 174, 178
Curran, C., 128–129
Curtis, J. R., 135, 139, 147

Darley, J. M., 82
Davies, M., 126, 129
Davis, K. B., Jr., 153–154, 162–163
De Lacey, S., 192, 197, 206–207, 210
De Meo, D. L., 147
De Ville, K. A., 168–169, 174, 178
dead donor rule, 10, 17, 19
Dear, M. J., 192, 210
death, 2, 6–10, 12, 17–19, 31–32, 102–105, 123, 127, 133–137, 141, 145, 168–170, 212–217, 227–228, 230
brain-oriented, 7–8, 100
higher-order brain, 7–9
whole body death, 7, 10
Declaration of Helsinki, 150
DeMott, D. A., 153–154, 162–163
Denmark, 141
Dennett, D., 68, 80–81
deontology, 39, 42
Desbiens, N., 148
dialysis, 181–182
Dickey, N., 18
dignity, 9, 17–18, 70, 108
Dionysios the Areopagite, 125
disease, 10, 17, 27–28, 31–32, 66, 160, 216–217
disease, iatrogenic, 32
dissonance stress, 71
distress, 31–32, 71
Djulbegovic, B., 161, 164
do-not-resusitate order, 145, 215, 231
Doerflinger, R., 175, 178
Doherty, C., 18
Douglas Butler, J., 178
Doyal, L., 19
drug approval, 3

Edwards, P., 190
Eisenberg, V. H., 194–195, 209
Elder Joseph the Hesychast, 126
Elder Paisios of Mt. Athos, 126
Elder Paisios of Romania, 126
Elder Porphyrios, 127
electroencephalograms, 8
Eliezer, Rabbi, 115–116, 127
elitism, 78
Ellenberg, S. S., 18
Elliott, C., 71, 80–81
Emanuel, E. J., 158, 163
embryo, 2, 14, 119, 191–210, 228
adoption of, 14, 191–208
donation of, 191, 202
frozen, 191
rescue, 191
transfer, 2, 191
emotional sensitivity, 11, 79
end-of-life decision making, 2, 38, 107, 227
Engel, L. W., 164
Engelhardt, H. Tristram Jr., 2, 12–13, 40–42, 48, 70, 81, 117, 119, 124, 126–127, 129, 227
England, 20, 27, 36, 138
Englert, Y., 199, 210
Enkin, M., 150, 165
epistemology, 2, 39, 44–47, 67, 71, 74–75, 87, 89–90, 96, 109, 112, 120, 128, 226, 231
moral, 67, 75, 78, 80, 85, 87, 93, 226, 231
Epstein, I., 130
essentialism, 32
ethical controversies, 5
ethics committee, 143–144, 228
ethics of care, 73–74
ethics, clinical, 14, 39, 47, 105–106, 149
Europe, 13, 110, 124

euthanasia, 2, 8, 12–14, 18, 167–170, 174–175, 177, 211, 213–214, 217, 228
  active, 168–169, 213–215
  passive, 168, 213–214
Evagrios the Solitary, 124
expert testimony, 29
exploitation, 3–4, 158–159
Eydoux, P., 191, 209

Faden, R., 176, 178
Fair, C. C., 209
Feast, J., 209
Fellmann, F., 209
feminism, 73
Ferranti, J., 178
Festinger, L., 71, 81
fiduciary relationship, 149–155, 157–160, 162–163
Fine, A., 182, 190
Fine, R., 143–144, 147
Finn, P. D., 154, 162–163
Finnis, J., 88–90, 97
Firth, R., 61–62, 64
Fisher, E. S., 148
Fiske, A., 53, 64
Flanagan, O., 67–68, 80–81
Florida, 174
Food and Drug Administration, 3–4, 16, 19
Foot, P., 214, 218
fornication, 123
Fost, N., 4, 18–19
Fox, M., 95–96
Frader, J., 135, 148
Francannet, C., 209
Frankel, T., 154, 162–163
Frankena, W. K., 186, 190
fraud, 6
Freedman, B., 150, 162–163
Friese, C., 210
Fuchs, J., 114, 129
Fuscaldo, G., 206, 209
futility, 2, 8, 14, 17, 133–148, 181–182, 185, 229
  imminent demise, 136
  lethal condition, 136
  physiological, 136
  qualitative, 136

Gabriel, G. S., 130
gametes, 192–193, 196–198, 201–202, 208
Gampel, E., 143, 147
Gelsinger, J., 16
gender, 12, 31, 72
gender reassignment surgery, 12
Genovo, 16–17
Gentiles, 12, 93, 109–110, 118–120, 124, 127
Gerard, J. L., 210
Gert, B., 168–170, 174, 178
Gibbard, A., 80
Gibbons, W. E., 209
Glare, P., 138, 147
Glass, K. C., 150–151, 157, 163, 165
God, 12–13, 32, 91–93, 97, 99, 101, 104, 110–118, 120–129, 226–227
Goldman, A., 80
Goldner, J. A., 16–17, 19
Gornall, T., 125, 129
Gospels, 106
Grady, C., 163
Greece, 125
Greenberg, J., 72, 82
Gregory Palamas, Saint, 125
Gregory, J., 27–29, 32, 35, 223
Grinnell, F., 25, 35
Grinstead, D. M., 210
Grisso, T., 157, 160, 163
Grunberg, S. M., 160, 163
guardians, 3, 14, 168–177
Guess, H. A., 164
Gurmankin, A., 210

Haakonssen, L., 27, 35
halakhic material, 12–13, 91, 93–95, 109–113, 115, 118–120, 124–125, 127–128
halakhic requirements, 13
Halevy, A., 6–7, 9–10, 14, 17, 19, 134–138, 140–143, 147, 190, 228
Hall, J. B., 4, 20, 36
Harris, J., 19
Hausman, D. M., 18–19
health care, 1–2, 8–9, 14, 17, 29, 33–34, 36–38, 40, 47, 100, 124, 133, 136–137, 139, 141–144, 146, 162
Heaney, R. P., 18

Hegel, G. W. F., 128–129
Heidegger, M., 117, 126, 129
Hempel, P., 221
Hippen, B., 65
HIV, 3, 73, 151, 156
Ho, D., 65, 81
Hobbes, T., 69, 190
Hoffman, D. I., 191, 209
Hoffmaster, B., 80–81
Holder, A. R., 150, 157, 163
Hollingswroth, J. C., 20
Holloway, M. R., 125, 129
holocaust, 66
Holy Spirit, 125
homosexuality, 91, 94, 123
Houston, 14, 23, 109, 140, 142–143, 147, 182, 185, 221, 228–229
Houston Bioethics Network, 140
human rights, 78
Hume, D., 67–68, 78–79, 89–90, 97
Hurka, T., 190
hypertension, 33, 228

ideal intuitor, 61–64
ideal observer, 61–62, 64
idolatry, 91, 94–95, 110
Iltis, A. S., 1, 221, 231
in vitro fertilization, 191, 194–195, 197–198
incest, 91, 94, 124
informed consent, 3–5, 18, 30, 74, 104, 139, 146, 162
injury, 28, 31–32, 106, 138, 157
innovation, 1–2, 16, 152
institutional review boards, 4
integrity, 5, 9, 11, 29–31, 33, 120, 124, 135, 139, 141–143, 145, 175, 182–190, 229–230
intensive care unit, 7, 13, 17, 101–104, 134, 138, 144–146
intuitionism, 11, 25, 44, 48, 66–68, 71, 75, 77, 85–88, 90–91, 93, 96, 194
intuitions, 11, 24–25, 44, 46, 48, 50, 54–56, 66, 71–72, 77, 81, 86–91, 96, 108, 111, 114, 117–118, 121–123, 164, 183, 193–194, 215, 225–226, 231
Isaac the Syrian, Saint, 123, 125, 129
ius civile, 94
ius gentium, 94

Jacobson, P. D., 154, 163
Jakobovits, I., 129
Janata, J. W., 196, 210
Jecker, N., 135, 148
Jefferson, L. S., 148
Jehovah's Witness, 79, 104
Jerusalem, 116
Jesus, 128
Job, 127
John of San Francisco, Saint, 126
John Paul II, Pope, 17, 114, 129, 168, 170–172, 175, 178
Johnson, M., 36, 80
Jolson, L., 100
Jonas, H., 126, 129
Jones, J. W., 126, 135
Jones, K. C., 129
Jones, M., 147
Jonsen, A., 42, 48, 65, 81, 90, 97, 135, 148
Jouannet, P., 209
Joyce, G. H., 125, 129
Judaism, 18, 94–95, 97
  Orthodox, 13, 110–118, 123, 127–129
justice, 2, 12, 25, 33, 65, 70, 74, 78, 86, 88, 105, 107, 124, 183, 192, 222

Kahn, J., 19
Kamm, F., 2, 14, 16, 218, 230
Kant, I., 68, 117, 120, 126, 190
Karlawish, J. H., 4, 20
Kass, L., 197, 210
Kaveny, M. C., 125, 129
Keenan, J., 195, 208, 210
Kelley, M., 2, 11, 16, 81, 221, 225
Kelly, G., 127, 129
Kelly, S., 72, 74, 82
Keown, J., 19
Key, C., 100
Kierkegaard, S., 123
killing, 14, 43, 45, 167–168, 170, 177, 211–218, 228, 230–231
  killing vs. letting die, 14, 168, 211–218, 230–231
King, N., 4, 20, 178
Kingsberg, S. A., 196, 199, 206–207, 210
Kipnis, K., 4, 20, 160

Kirshner, H. S., 146–147
Koop, E. C., 175, 178
Kopelman, L., 2, 14, 80–82, 168–169, 174, 176–178, 182, 189–190, 230
Koran, 106
Kovach, K., 150, 164
Kovas, G. T., 192, 195, 206, 210
Krone, M. R., 147
Kuhn, T., 128–129
Kusek, J. W., 164
Kuykendall, D. H., 20

Lakatos, I., 129
Lantos, J., 135, 137, 147–176, 178
Laruelle, C., 199, 210
Lee, J., 194–195, 210
Leibniz, G. W., 126
Leo XIII, Pope, 113, 129
Lerner, B. H., 163–164
Levinas, E., 104–105, 108
Levine, A., 209
Levine, R. J., 18, 159, 164
Lewis, R. J., 18
libertarian, 124
Lidz, C. W., 157, 160, 163
Lilly, C. M., 100, 146–147
Lkeinman, A., 164
Locke, J., 190
Lustig, B. A., 2, 12, 226
lying, 123
Lynn, J., 148

MacDougall, K., 210
Machiavelli, N., 69
Mahowald, M. B., 208
Maimonides, M., 12, 92–93, 124, 127–129
Majumder, M. A., 205, 207, 210
Makarios, Saint, 124, 129
Malek, J., 2, 11, 16, 225
managed care, 104
Mann, H., 150, 161, 164
Margaret, Sister Mary, 100
Marxism-Leninism, 126
Masterman, M., 129
Mastroianni, A., 19
Matt, D., 108
Matz, D. C., 71, 82
Mayer, J. F., 209
Mayo, T., 143–144, 146–147
McCormick, R., 32, 36
McCullough, L., 2, 10–11, 25, 27–28, 30–32, 35–36, 135, 147, 223
McDermid, A., 196, 210
McGee, G., 210
McWilliams, J. A., 125, 129
Mecker, L., 193, 210
medical ethics, 11, 14, 18, 23, 26–30, 32–35, 66, 73, 81, 85, 91, 109, 124, 133, 223
  de-professionalization of, 34
  Jewish, 2, 149
  Judeo-Christian, 32
  professional, 23, 31–35
medicine, 1, 4, 17, 27–35, 37–38, 65–66, 73, 81, 102, 105–106, 133, 142, 146, 167, 182, 192, 196, 198, 202, 206–208, 223
Melchizedek, 127
memes, 68
Mendelssohn, M., 129
Menke, T. J., 20
mental flexibility, 11, 79
metaphysics, 2, 85, 109, 111, 116–117, 120, 124, 126
Michels, K. B., 150, 164
Middle Ages, 112
Midrash, 13, 106–107
Miles, S. H., 134, 147
Mill, J. S., 67–68, 74
Miller, F. G., 150, 152, 157–158, 160–164
Miller, P. B., 165
model of conflicting appeals, 25, 49–51, 53–55, 57–58, 61–64
modernity, 104, 122, 181
moral,
  agents, 45, 49–52, 56–57, 59, 61, 63, 188
  ambiguity, 11, 37, 39–45, 47, 51, 76, 88, 105–106, 149, 222, 224–225, 231
  appeals, 11, 14, 24–26, 39, 41–46, 65–66, 70, 75–77, 79, 88, 101, 105–106, 182–185, 190, 222–223, 226
  casuistry, 65, 69, 74
  cognitive faculties, 12
  community, 41
  confusion, 54–56, 61

disagreement, 54, 56–62, 76, 79
friends, 41
judgment, 11, 42, 44–45, 49–50, 53, 55–58, 60–64, 66–67, 71–81, 92–93, 183, 225–226
philosophy, 12–13, 24, 46, 67, 74, 78, 105, 110–112, 115–116, 118, 128, 222, 227
pluralism, 4, 11, 38–47, 66–67, 75–76, 85, 87–88, 149, 222–223
pluralistic casuistry, 2, 10–11, 14–15, 51, 76, 81, 85, 90, 194, 226
pluralistic moral theory, 10, 11, 23, 25–26, 31, 33–35, 88, 183, 223
psychology, 11, 75–77, 81
rationality, 115–117, 122–123, 125
side constraints, 158
strangers, 41
theory, 2, 10, 11, 24–26, 30, 42, 46–47, 50–52, 60, 68–69, 78, 86–87, 90, 93, 118, 183, 190, 222–223, 226
vision, 41, 203–204
morality, secular, 13, 41, 110–111, 115, 117, 119, 120–124, 127, 129
Morin, K., 157, 164
Morreim, E. H., 2, 14, 150–151, 156, 161, 164, 223
Moses, 109, 112, 120, 122, 128
Moses, Law of, 112
Mosley, J. B., 3, 16, 20
Mount Sinai, 116
Mulrine, A., 203, 210
murder, 45, 95, 119, 124
Murphy, D., 148
Musgrave, A., 129

Nachtigall, R. D., 198, 210
National Abortion and Reproductive Rights Action League, 193
National Embryo Donation Center, 195, 210
National Placebo Working Committee, 150, 161, 164
National Rural Bioethics Project, 74
natural cognitive constraints, 225
natural law, 12, 42, 85–97, 118, 120, 123, 226
naturalism, 67–69, 74, 80
Neal, R. C., 17, 19, 137, 147
Nelson, R. M., 4, 20
nephrology, 181
neuroscience, 68, 75, 77
New Orleans, 37
Newman, Cardinal Henry, 125
Newport, F., 178
Newton, C. R., 196, 201, 210
nihilism, 81, 125–126
Nikodimos, Saint, 124, 129
Ninevites, 127
Noah, 91, 109–110, 119–120, 124, 127–128
Noahide commandments, 12, 86, 91–95, 97, 226
Noonan, J., 125
North Atlantic Treaty Organization, 133
North Carolina, 164, 170, 189
Norton, M. I., 71–72, 82
Novak, D., 86, 91–97, 226
Nozick, R., 66, 82, 158, 164, 224
Nuremberg Code, 16

O'Brien, R. L., 18
O'Malley, K., 20
Office of Technology Assessment, 133
Ohio, 17, 164
ontology, 39, 41, 43, 45–47
oocytes, 194–195, 202
Oregon, 141
organ donation, 9, 201
organ harvesting, 8–9

pain, 1, 16, 18, 31–32, 101, 103–104, 106, 168, 187, 215
Palmer, G. E. H., 129
papal infallibility, 125
paradigm, 44, 65–66, 70–73, 122, 128–129, 192–193, 198, 202–203, 205–208
paradigm shift, 128
Parham, T., 16
Paris, J. J., 176, 178
Park, D. R., 147
Parker, J. C., 2, 14, 16, 229
Patel, C., 189
Paul, Saint, 18, 36, 127
Payne, K., 169, 178
Pearlman, R. A., 147
pediatrics, 2, 70

Pellegrino, E., 32, 36
Peperzak, A., 108
Percival, T., 27–29, 33–36, 223
permanently vegetative state, 8–9, 13–14, 17, 19, 32, 105, 136, 167–177, 230
permission, 3, 41, 104, 214
Perry, J. E., 146–147
personhood, 9, 168, 175, 197
persons, respect for, 11, 25, 31, 41, 86, 88, 183
Petersen, N. J., 20
philosophy of law, 2, 85
philosophy of medicine, 1–2
philosophy of religion, 2, 109
physician-patient relationship, 25, 29, 31, 34, 151, 157, 161
Pinborg, A., 209
Plato, 78, 92–93, 111, 121–123
political theory, 2, 68, 75
pornography, 140
Porter, D., 28, 35–36
Porter, R., 36
post-modernism, 68–69, 72, 125
practical knowledge, 44–45
practical reason, 6, 12, 88–92, 126
pregnancy, 30, 161, 191, 194, 198–200, 204–205, 208, 223
principlism, 82, 86
professional integrity, 2, 29–30, 32–34, 139, 141–143, 146, 229
property, intellectual, 3
Purtilo, R. B., 18
Pyszcynski, T., 72, 82

Quill, T. E., 169, 178
Quine, W. V. O., 67, 82

Rachels, J., 178, 211
Rasmussen, L., 65, 81
rationing, 33, 223, 230
Rawls, J., 65, 82, 190
Reagan, R., 178
Reardon, F., 178
reflective equilibrium, 65, 67, 73
regulation
  of biomedical research, 3
  governmental, 3–4, 29, 101, 172, 174, 205
relativism, 40, 89, 128
religion, 30–31, 38, 97, 107, 111, 113, 119, 175
renal failure, 181
Repenshek, M., 171, 178
research ethics, 1–3, 14, 37, 39, 47, 69, 73, 149, 160, 223
  double-blinding, 6
  emergency research, 3–4
  financial relationships, 5
  recruitment of research subjects, 3–4, 16, 149, 224
  pediatric assent to research, 70
  placebo-controlled trials, 5, 16, 161, 164
Resnick, D., 160
rheumatology, 160
Rhodes, R., 18, 178
Rice University, 109, 133, 221
Richardson, H. S., 81–82
rights, 10, 15, 17, 24–25, 65–66, 70, 73, 78, 88, 135, 158, 172–174, 183–184, 186, 190, 192, 198, 201, 203, 205, 208, 214, 216, 222, 229
  as side-constraints, 66, 158, 223, 224
Rilke, R. M., 221
Rix, B. A., 141, 148
robbery, 91, 93, 95, 110
Robertson, J., 191–192, 194–195, 201–202, 204, 210
Rodwin, M. A., 153–154, 162, 164
Romanides, J., 124, 130
Rorty, R., 125–126, 130
Rosenberg, A., 16, 80–82
Rosenstein, D. L., 152, 157–158, 160, 164
Ross, W. D., 48, 65, 75–76, 82, 86
Roth, L. H., 160, 163
Rothman, K. J., 150, 164
Rubin, S., 135, 148
Runge, J. W., 18
Ruse, M., 53, 64

Sachdeva, R. C., 137, 148
Sachs, G. A., 178
sanctity of human life, 18
Sanders, K., 195–196, 199, 209
Sanhedrin, 91, 110, 115, 119
Saver, R. S., 150, 164

Savulescu, J., 206, 209
Schafer-Landau, R., 64
Schaff, P., 127, 130
Schapiro, R., 147–148
Schenker, J. G., 194–195, 209
Schiavo, T., 70, 144, 146–147, 174–175, 177, 229
Schmidt, T. C., 18, 20
Schneiderman, L., 135, 137, 148
Scholasticism, 94
Scholz, S., 209
Schwarzschild, S., 127, 130
Scotland, 27
Scott, A. W., 154, 162, 164
Scripture, 94, 111, 114–115
self-effacement, 11, 31
self-sacrifice, 11, 31
Seltzer, D., 146
Sextus Empiricus, 128
sexual orientation, 31
Shafer-Landau, R., 62
shame, 68, 222
Shannon, T. A., 178
Shapiro, S. H., 150, 165
Sharpe, V. A., 34, 36
Shepherd, J. C., 153–154, 162, 164
Sherrard, P., 129
Shewmon, D. A., 18, 20
Shimm, D. S., 19
Siffroi, J. P., 209
Silvers, A., 18, 178
Siminoff, L. A., 18, 20
Simon-Bouy, B., 209
Sinnott-Armstrong, W., 55, 64
Slosar, J. P., 171, 178
Slote, M., 80
Smart, J. J. C., 184, 190
Snyder, L., 178
social psychology, 71, 77
Society for Health and Human Values, 23
Society of Critical Care Medicine, 17, 20
sociology, 66, 72, 74, 77, 120
Socrates, 121
Sommers, T., 80–82
Sonna, L. A., 147
Sparks, A. E. T., 210
Spece, Roy G. Jr., 19
sperm, 194–195, 197, 202, 208
Spinoza, Benedict, 127–128, 130
Silouan the Athonite, Saint, 127
Starbucks, 38
Steinberg, D., 157, 161, 164
stem cell research, 38, 205
Stinson, P., 176, 178
Stinson, R., 176, 178
Stocking, C., 178
Strong, C., 65, 82
Stuikel, T. A., 148
suffering, 1–2, 17, 31–32, 101, 106, 181–182, 185, 189, 224, 229
Sugarman, J., 19
suicide, 12, 18, 40, 45, 55, 107, 123, 167, 174, 178, 218
  physician-assisted, 32, 54, 167, 178, 213
supervenience, 46, 87
surrogate decision maker, 8, 10, 13–14, 134, 139–140, 145–146, 182, 230
Symeon the New Theologian, Saint, 125

Talmud, 13, 106, 110, 116, 119
Taylor, R. M., 178
technology transfer, 3
Tekpetey, F., 196, 210
Tennessee, 149, 161, 195
Teno, J. M., 145,148
Texas, 1, 14, 16, 23, 109, 143–144, 182, 189, 221, 228–229
Texas Advance Directive Act, 14, 143
Texas Medical Center, 16, 182
theft, 124
theology, 13, 109–118, 121, 124–126, 128
  medical moral, 124
  natural, 125
  Orthodox Christian, 111, 125
  Western Christian, 126, 128
Thepot, F., 209
Thomasma, D. C., 32, 36
thrombolytic agents, 5
Tobas, J., 19
Tomlinson, T., 135, 148
Torah, 93–95, 116, 120, 124, 127–129
Tosteson, A., 148
Touger, E., 127, 129
Toulmin, S., 42, 48, 65, 81, 91, 97

transplantation, 6–7, 9
triage, 9, 168
Troug, R., 134–135, 148
Tummon, I. S., 196, 210
Turner, T. G., Jr., 209
Tuskegee syphilis experiments, 16

United States, 3, 34, 38, 70, 73, 191, 200, 207
United States Supreme Court, 154
University of Montana-Missoula, 74
University of Pennsylvania, 16
Urmson, J. O., 209
usury, 123
utilitarianism, 42, 184, 190, 222

Van Voorhis, B. J., 195, 210
Vandello, J. A., 82
Vatican Council, II, 125–126
Vattimo, G., 125–126, 130
Veatch, R., 17, 20, 27, 29, 36
Virik, K., 147
virtue theory, 39
virtues, 10–11, 24–25, 29, 31, 44–46, 65, 67, 78–79, 88, 183, 223
  professional, 29, 31
Vitoria, F. de, 123, 128
Vlachos, H., 124, 130

Walbert, D. F., 178
Walter, J. J., 171, 178
Ware, K., 129
Warnock, G. J., 209
Wathen, J. F., 126, 130
Weaver, D. F., 210
Weijer, C., 18, 150, 157, 161, 163–165
Weinberg, J., 145, 148
Weir, R. F., 210
Wendler, D., 163
Whitehead, A. 126
Wicclair, M. R., 143, 148
Wildes, K. Wm., S. J., 2, 11, 16, 38, 42, 48, 222
Williams, B., 66, 68, 82, 184, 190
Williams, E., 130
Wilson, B., 209
Wilson, E., 53, 64
Wilson, J. M., 16
Winslade, W., 163
Wittgenstein, L., 32
Wood, W., 71, 77, 82
World War II, 16
worship, 115, 118, 124, 127–128
Wray, N. P., 20

Yap, C., 194–195, 210
Youakim, J. M., 209
Youngner, S., 18, 20, 135, 144–148

Zabala, S., 126, 130
Zeitz, J. G., 209
Zellman, G. L., 209
Zientek, D., 19
Zoloth, L., 2, 12–13, 102, 106, 108, 226

NOTES ON CONTRIBUTORS

Robert M. Arnold, M.D., Director of Clinical Training, Acting Co-Director of Research, Associate Professor of Medicine and Psychiatry, Montefiore University Hospital, University of Pittsburgh, Pittsburgh, Pennsylvania

Sarah-Vaughan Brakman, Ph.D., Associate Professor, Department of Philosophy, Villanova University, Villanova, Pennsylvania

Baruch A. Brody, Ph.D., the Leon Jaworski Professor of Biomedical Ethics and Director of the Center for Medical Ethics and Health Policy at Baylor College of Medicine, Houston, Texas, and the Andrew Mellow Professor of Humanities, Department of Philosophy, Rice University, Houston, Texas

Mark J. Cherry, Ph.D., the Dr. Patricia A. Hayes Professor in Applied Ethics and Associate Professor, Department of Philosophy, St. Edward's University, Austin, Texas

H. Tristram Engelhardt, Jr., Ph.D., M.D., Professor, Department of Philosophy, Rice University, Houston, Texas and Professor Emeritus, Baylor College of Medicine, Houston, Texas

Ana Smith Iltis, Ph.D., Associate Professor, Center for Health Care Ethics, Saint Louis University, St. Louis, Missouri

Frances M. Kamm, Ph.D., Littauer Professor of Philosophy and Public Policy, Kennedy School of Government, Harvard University, Cambridge, Massachusetts

Maureen Kelley, Ph.D., Assistant Professor, Department of Pediatrics, Division of Bioethics, Treuman Katz Center for Pediatric Bioethics, University of Washington School of Medicine, Seattle, Washington

Loretta M. Kopelman, Ph.D., Professor, Department of Medical Humanities, East Carolina University, Greenville, North Carolina

B. Andrew Lustig, Ph.D., the Holmes Rolston III Professor of Religion, Department of Religion, Davidson College, Davidson, North Carolina

Janet Malek, Ph.D., Assistant Professor, Department of Medical Humanities, East Carolina University, Greenville, North Carolina

Laurence B. McCullough, Ph.D., Professor of Medicine and Medical Ethics, Center for Medical Ethics and Health Policy, Baylor College of Medicine, Houston, Texas

E. Haavi Morreim, Ph.D., Professor, Human Values & Ethics, The University of Tennessee, Health Science Center, Memphis, Tennessee

J. Clint Parker, M.D., Ph.D., Brody School of Medicine, East Carolina University, Greenville, North Carolina

Kevin Wm. Wildes, S.J., Ph.D., President, Loyola University New Orleans, New Orleans, Louisiana

Laurie Zoloth, Ph.D., Professor of Medical Humanities & Bioethics and Religion, Director of the Center for Bioethics, Science and Society, Feinberg School of Medicine, Northwestern University, Chicago, Illinois

Philosophy and Medicine

1. H. Tristram Engelhardt, Jr. and S.F. Spicker (eds.): *Evaluation and Explanation in the Biomedical Sciences.* 1975 ISBN 90-277-0553-4
2. S.F. Spicker and H. Tristram Engelhardt, Jr. (eds.): *Philosophical Dimensions of the Neuro-Medical Sciences.* 1976 ISBN 90-277-0672-7
3. S.F. Spicker and H. Tristram Engelhardt, Jr. (eds.): *Philosophical Medical Ethics.* Its Nature and Significance. 1977 ISBN 90-277-0772-3
4. H. Tristram Engelhardt, Jr. and S.F. Spicker (eds.): *Mental Health.* Philosophical Perspectives. 1978 ISBN 90-277-0828-2
5. B.A. Brody and H. Tristram Engelhardt, Jr. (eds.): *Mental Illness.* Law and Public Policy. 1980 ISBN 90-277-1057-0
6. H. Tristram Engelhardt, Jr., S.F. Spicker and B. Towers (eds.): *Clinical Judgment.* A Critical Appraisal. 1979 ISBN 90-277-0952-1
7. S.F. Spicker (ed.): *Organism, Medicine, and Metaphysics.* Essays in Honor of Hans Jonas on His 75th Birthday. 1978 ISBN 90-277-0823-1
8. E.E. Shelp (ed.): *Justice and Health Care.* 1981 ISBN 90-277-1207-7; Pb 90-277-1251-4
9. S.F. Spicker, J.M. Healey, Jr. and H. Tristram Engelhardt, Jr. (eds.): *The Law-Medicine Relation.* A Philosophical Exploration. 1981 ISBN 90-277-1217-4
10. W.B. Bondeson, H. Tristram Engelhardt, Jr., S.F. Spicker and J.M. White, Jr. (eds.): *New Knowledge in the Biomedical Sciences.* Some Moral Implications of Its Acquisition, Possession, and Use. 1982 ISBN 90-277-1319-7
11. E.E. Shelp (ed.): *Beneficence and Health Care.* 1982 ISBN 90-277-1377-4
12. G.J. Agich (ed.): *Responsibility in Health Care.* 1982 ISBN 90-277-1417-7
13. W.B. Bondeson, H. Tristram Engelhardt, Jr., S.F. Spicker and D.H. Winship: *Abortion and the Status of the Fetus.* 2nd printing, 1984 ISBN 90-277-1493-2
14. E.E. Shelp (ed.): *The Clinical Encounter.* The Moral Fabric of the Patient-Physician Relationship. 1983 ISBN 90-277-1593-9
15. L. Kopelman and J.C. Moskop (eds.): *Ethics and Mental Retardation.* 1984 ISBN 90-277-1630-7
16. L. Nordenfelt and B.I.B. Lindahl (eds.): *Health, Disease, and Causal Explanations in Medicine.* 1984 ISBN 90-277-1660-9
17. E.E. Shelp (ed.): *Virtue and Medicine.* Explorations in the Character of Medicine. 1985 ISBN 90-277-1808-3
18. P. Carrick: *Medical Ethics in Antiquity.* Philosophical Perspectives on Abortion and Euthanasia. 1985 ISBN 90-277-1825-3; Pb 90-277-1915-2
19. J.C. Moskop and L. Kopelman (eds.): *Ethics and Critical Care Medicine.* 1985 ISBN 90-277-1820-2
20. E.E. Shelp (ed.): *Theology and Bioethics.* Exploring the Foundations and Frontiers. 1985 ISBN 90-277-1857-1
21. G.J. Agich and C.E. Begley (eds.): *The Price of Health.* 1986 ISBN 90-277-2285-4
22. E.E. Shelp (ed.): *Sexuality and Medicine.* Vol. I: Conceptual Roots. 1987 ISBN 90-277-2290-0; Pb 90-277-2386-9
23. E.E. Shelp (ed.): *Sexuality and Medicine.* Vol. II: Ethical Viewpoints in Transition. 1987 ISBN 1-55608-013-1; Pb 1-55608-016-6

24. R.C. McMillan, H. Tristram Engelhardt, Jr., and S.F. Spicker (eds.): *Euthanasia and the Newborn.* Conflicts Regarding Saving Lives. 1987 ISBN 90-277-2299-4; Pb 1-55608-039-5
25. S.F. Spicker, S.R. Ingman and I.R. Lawson (eds.): *Ethical Dimensions of Geriatric Care.* Value Conflicts for the 21st Century. 1987 ISBN 1-55608-027-1
26. L. Nordenfelt: *On the Nature of Health.* An Action-Theoretic Approach. 2nd, rev. ed. 1995 ISBN 0-7923-3369-1; Pb 0-7923-3470-1
27. S.F. Spicker, W.B. Bondeson and H. Tristram Engelhardt, Jr. (eds.): *The Contraceptive Ethos.* Reproductive Rights and Responsibilities. 1987 ISBN 1-55608-035-2
28. S.F. Spicker, I. Alon, A. de Vries and H. Tristram Engelhardt, Jr. (eds.): *The Use of Human Beings in Research.* With Special Reference to Clinical Trials. 1988 ISBN 1-55608-043-3
29. N.M.P. King, L.R. Churchill and A.W. Cross (eds.): *The Physician as Captain of the Ship.* A Critical Reappraisal. 1988 ISBN 1-55608-044-1
30. H.-M. Sass and R.U. Massey (eds.): *Health Care Systems.* Moral Conflicts in European and American Public Policy. 1988 ISBN 1-55608-045-X
31. R.M. Zaner (ed.): *Death: Beyond Whole-Brain Criteria.* 1988 ISBN 1-55608-053-0
32. B.A. Brody (ed.): *Moral Theory and Moral Judgments in Medical Ethics.* 1988 ISBN 1-55608-060-3
33. L.M. Kopelman and J.C. Moskop (eds.): *Children and Health Care.* Moral and Social Issues. 1989 ISBN 1-55608-078-6
34. E.D. Pellegrino, J.P. Langan and J. Collins Harvey (eds.): *Catholic Perspectives on Medical Morals.* Foundational Issues. 1989 ISBN 1-55608-083-2
35. B.A. Brody (ed.): *Suicide and Euthanasia.* Historical and Contemporary Themes. 1989 ISBN 0-7923-0106-4
36. H.A.M.J. ten Have, G.K. Kimsma and S.F. Spicker (eds.): *The Growth of Medical Knowledge.* 1990 ISBN 0-7923-0736-4
37. I. Löwy (ed.): *The Polish School of Philosophy of Medicine.* From Tytus Chałubiński (1820–1889) to Ludwik Fleck (1896–1961). 1990 ISBN 0-7923-0958-8
38. T.J. Bole III and W.B. Bondeson: *Rights to Health Care.* 1991 ISBN 0-7923-1137-X
39. M.A.G. Cutter and E.E. Shelp (eds.): *Competency.* A Study of Informal Competency Determinations in Primary Care. 1991 ISBN 0-7923-1304-6
40. J.L. Peset and D. Gracia (eds.): *The Ethics of Diagnosis.* 1992 ISBN 0-7923-1544-8
41. K.W. Wildes, S.J., F. Abel, S.J. and J.C. Harvey (eds.): *Birth, Suffering, and Death.* Catholic Perspectives at the Edges of Life. 1992 [CSiB-1] ISBN 0-7923-1547-2; Pb 0-7923-2545-1
42. S.K. Toombs: *The Meaning of Illness.* A Phenomenological Account of the Different Perspectives of Physician and Patient. 1992 ISBN 0-7923-1570-7; Pb 0-7923-2443-9
43. D. Leder (ed.): *The Body in Medical Thought and Practice.* 1992 ISBN 0-7923-1657-6
44. C. Delkeskamp-Hayes and M.A.G. Cutter (eds.): *Science, Technology, and the Art of Medicine.* European-American Dialogues. 1993 ISBN 0-7923-1869-2

Philosophy and Medicine

45. R. Baker, D. Porter and R. Porter (eds.): *The Codification of Medical Morality.* Historical and Philosophical Studies of the Formalization of Western Medical Morality in the 18th and 19th Centuries, Volume One: Medical Ethics and Etiquette in the 18th Century. 1993 ISBN 0-7923-1921-4
46. K. Bayertz (ed.): *The Concept of Moral Consensus.* The Case of Technological Interventions in Human Reproduction. 1994 ISBN 0-7923-2615-6
47. L. Nordenfelt (ed.): *Concepts and Measurement of Quality of Life in Health Care.* 1994 [ESiP-1] ISBN 0-7923-2824-8
48. R. Baker and M.A. Strosberg (eds.) with the assistance of J. Bynum: *Legislating Medical Ethics.* A Study of the New York State Do-Not-Resuscitate Law. 1995 ISBN 0-7923-2995-3
49. R. Baker (ed.): *The Codification of Medical Morality.* Historical and Philosophical Studies of the Formalization of Western Morality in the 18th and 19th Centuries, Volume Two: Anglo-American Medical Ethics and Medical Jurisprudence in the 19th Century. 1995 ISBN 0-7923-3528-7; Pb 0-7923-3529-5
50. R.A. Carson and C.R. Burns (eds.): *Philosophy of Medicine and Bioethics.* A Twenty-Year Retrospective and Critical Appraisal. 1997 ISBN 0-7923-3545-7
51. K.W. Wildes, S.J. (ed.): *Critical Choices and Critical Care.* Catholic Perspectives on Allocating Resources in Intensive Care Medicine. 1995 [CSiB-2] ISBN 0-7923-3382-9
52. K. Bayertz (ed.): *Sanctity of Life and Human Dignity.* 1996 ISBN 0-7923-3739-5
53. Kevin Wm. Wildes, S.J. (ed.): *Infertility: A Crossroad of Faith, Medicine, and Technology.* 1996 ISBN 0-7923-4061-2
54. Kazumasa Hoshino (ed.): *Japanese and Western Bioethics.* Studies in Moral Diversity. 1996 ISBN 0-7923-4112-0
55. E. Agius and S. Busuttil (eds.): *Germ-Line Intervention and our Responsibilities to Future Generations.* 1998 ISBN 0-7923-4828-1
56. L.B. McCullough: *John Gregory and the Invention of Professional Medical Ethics and the Professional Medical Ethics and the Profession of Medicine.* 1998 ISBN 0-7923-4917-2
57. L.B. McCullough: *John Gregory's Writing on Medical Ethics and Philosophy of Medicine.* 1998 [CoME-1] ISBN 0-7923-5000-6
58. H.A.M.J. ten Have and H.-M. Sass (eds.): *Consensus Formation in Healthcare Ethics.* 1998 [ESiP-2] ISBN 0-7923-4944-X
59. H.A.M.J. ten Have and J.V.M. Welie (eds.): *Ownership of the Human Body.* Philosophical Considerations on the Use of the Human Body and its Parts in Healthcare. 1998 [ESiP-3] ISBN 0-7923-5150-9
60. M.J. Cherry (ed.): *Persons and Their Bodies.* Rights, Responsibilities, Relationships. 1999 ISBN 0-7923-5701-9
61. R. Fan (ed.): *Confucian Bioethics.* 1999 [ASiB-1] ISBN 0-7923-5723-X
62. L.M. Kopelman (ed.): *Building Bioethics.* Conversations with Clouser and Friends on Medical Ethics. 1999 ISBN 0-7923-5853-8
63. W.E. Stempsey: *Disease and Diagnosis.* 2000 PB ISBN 0-7923-6322-1
64. H.T. Engelhardt (ed.): *The Philosophy of Medicine.* Framing the Field. 2000 ISBN 0-7923-6223-3

Philosophy and Medicine

65. S. Wear, J.J. Bono, G. Logue and A. McEvoy (eds.): *Ethical Issues in Health Care on the Frontiers of the Twenty-First Century.* 2000 ISBN 0-7923-6277-2
66. M. Potts, P.A. Byrne and R.G. Nilges (eds.): *Beyond Brain Death.* The Case Against Brain Based Criteria for Human Death. 2000 ISBN 0-7923-6578-X
67. L.M. Kopelman and K.A. De Ville (eds.): *Physician-Assisted Suicide.* What are the Issues? 2001 ISBN 0-7923-7142-9
68. S.K. Toombs (ed.): *Handbook of Phenomenology and Medicine.* 2001 ISBN 1-4020-0151-7; Pb 1-4020-0200-9
69. R. ter Meulen, W. Arts and R. Muffels (eds.): *Solidarity in Health and Social Care in Europe.* 2001 ISBN 1-4020-0164-9
70. A. Nordgren: *Responsible Genetics.* The Moral Responsibility of Geneticists for the Consequences of Human Genetics Research. 2001 ISBN 1-4020-0201-7
71. J. Tao Lai Po-wah (ed.): *Cross-Cultural Perspectives on the (Im) Possibility of Global Bioethics.* 2002 [ASiB-2] ISBN 1-4020-0498-2
72. P. Taboada, K. Fedoryka Cuddeback and P. Donohue-White (eds.): *Person, Society and Value.* Towards a Personalist Concept of Health. 2002 ISBN 1-4020-0503-2
73. J. Li: *Can Death Be a Harm to the Person Who Dies?* 2002 ISBN 1-4020-0505-9
74. H.T. Engelhardt, Jr. and L.M. Rasmussen (eds.): *Bioethics and Moral Content: National Traditions of Health Care Morality.* Papers dedicated in tribute to Kazumasa Hoshino. 2002 ISBN 1-4020-6828-2
75. L.S. Parker and R.A. Ankeny (eds.): *Mutating Concepts, Evolving Disciplines: Genetics, Medicine, and Society.* 2002 ISBN 1-4020-1040-0
76. W.B. Bondeson and J.W. Jones (eds.): *The Ethics of Managed Care: Professional Integrity and Patient Rights.* 2002 ISBN 1-4020-1045-1
77. K.L. Vaux, S. Vaux and M. Sternberg (eds.): *Convenants of Life. Contemporary Medical Ethics in Light of the Thought of Paul Ramsey.* 2002 ISBN 1-4020-1053-2
78. G. Khushf (ed.): *Handbook of Bioethics: Taking Stock of the Field from a Philosophical Perspective.* 2003 ISBN 1-4020-1870-3; Pb 1-4020-1893-2
79. A. Smith Iltis (ed.): *Institutional Integrity in Health Care.* 2003 ISBN 1-4020-1782-0
80. R.Z. Qiu (ed.): *Bioethics: Asian Perspectives A Quest for Moral Diversity.* 2003 [ASiB-3] ISBN 1-4020-1795-2
81. M.A.G. Cutter: *Reframing Disease Contextually.* 2003 ISBN 1-4020-1796-0
82. J. Seifert: *The Philosophical Diseases of Medicine and Their Cure.* Philosophy and Ethics of Medicine, Vol. 1: Foundations. 2004 ISBN 1-4020-2870-9
83. W.E. Stempsey (ed.): *Elisha Bartlett's Philosophy of Medicine.* 2004 [CoME-2] ISBN 1-4020-3041-X
84. C. Tollefsen (ed.): *John Paul II's Contribution to Catholic Bioethics.* 2005 [CSiB-3] ISBN 1-4020-3129-7
85. C. Kaczor: *The Edge of Life.* Human Dignity and Contemporary Bioethics. 2005 [CSiB-4] ISBN 1-4020-3155-6
86. R. Cooper: *Classifying Madness.* A Philosophical Examination of the Diagnostic and Statistical Manual of Mental Disorders. 2005 ISBN 1-4020-3344-3
87. L. Rasmussen (ed.): *Ethics Expertise.* History, Contemporary Perspectives, and Applications. 2005 ISBN 1-4020-3819-4

Philosophy and Medicine

88. M.C. Rawlinson and S. Lundeen (eds.): *The Voice of Breast Cancer in Medicine and Bioethics.* 2006 ISBN 1-4020-4508-5
89. M. Bormuth (ed.): *Life Conduct in Modern Times: Karl Jaspers and Psychoanalysis.* 2006 ISBN 1-4020-4764-9
90. H. Kincaid and J. McKitrick (eds.): *Establishing Medical Reality: Essays in the Metaphysics and Epistemology of Biomedical Science.* 2006 ISBN 1-4020-5215-4
91. S.C. Lee (ed.): *The Family, Medical Decision-Making, and Biotechnology: Critical Reflections on Asian Moral Perspectives.* 2007 ISBN 978-1-4020-5219-4
92. H.G. Wright: *Means, Ends and Medical Care.* 2007 ISBN 978-1-4020-5291-0
93. C. Tollefsen (ed.): *Artificial Nutrition and Hydration:* The New Catholic Debate. 2007 ISBN 978-1-4020-6206-3
94. M.J. Cherry and A.S. Iltis (eds.): *Pluralistic Casuistry:* Moral Arguments, Economic Realities, and Political Theory. 2007 ISBN 978-1-4020-6259-9

Printed in the United States
90172LV00002B/277-318/A

9 781402 062599